Atmospheric Boundary Layer Processes, Characteristics and Parameterization

Atmospheric Boundary Layer Processes, Characteristics and Parameterization

Editors

Yubin Li

Jie Tang

MDPI • Basel • Beijing • Wuhan • Barcelona • Belgrade • Manchester • Tokyo • Cluj • Tianjin

Editors
Yubin Li
Nanjing University of Information Science & Technology
Nanjing
China

Jie Tang
Shanghai Typhoon Institute of China Meteorological Administration
Shanghai
China

Editorial Office
MDPI
St. Alban-Anlage 66
4052 Basel, Switzerland

This is a reprint of articles from the Special Issue published online in the open access journal *Atmosphere* (ISSN 2073-4433) (available at: https://www.mdpi.com/journal/atmosphere/special_issues/ABL_Processes).

For citation purposes, cite each article independently as indicated on the article page online and as indicated below:

LastName, A.A.; LastName, B.B.; LastName, C.C. Article Title. *Journal Name* **Year**, *Volume Number*, Page Range.

ISBN 978-3-0365-7430-1 (Hbk)
ISBN 978-3-0365-7431-8 (PDF)

Contents

atmosphere

MDPI

Editorial

Atmospheric Boundary Layer Processes, Characteristics and Parameterization

Yubin Li [1,2,3] and Jie Tang [4,5,*]

1 Collaborative Innovation Center on Forecast and Evaluation of Meteorological Disasters, Key Laboratory for Aerosol-Cloud-Precipitation of China Meteorological Administration, School of Atmospheric Physics, Nanjing University of Information Science and Technology, Nanjing 210044, China; liyubin@nuist.edu.cn
2 Southern Marine Science and Engineering Guangdong Laboratory (Zhuhai), Zhuhai 519000, China
3 China Meteorological Administration Xiong'an Atmospheric Boundary Layer Key Laboratory, Xiong'an New Area, Baoding 071800, China
4 Shanghai Typhoon Institute, China Meteorological Administration, Shanghai 200030, China
5 Asia-Pacific Typhoon Collaborative Research Center, Shanghai 201306, China
* Correspondence: tangj@typhoon.org.cn

The atmospheric boundary layer is distinguished from the rest of the atmosphere due to its unique characteristics, i.e., its direct interaction with the Earth's surface and active turbulence. Understanding the dynamic and chemical processes in the boundary layer is of great importance in weather and air quality forecasting. Recently, with the improvements in observation and simulation techniques, our understanding of atmospheric boundary layer processes and characteristics has significantly improved. For example, ultrasonic anemometers and large-aperture scintillometers can provide information about turbulent exchanges, while the large eddy simulation technique simulates the detailed structure of turbulent eddies. This Special Issue is dedicated to reporting new findings with regard to atmospheric boundary layer processes, characteristics, and parametrization methods, including, but not limited to, turbulent exchange, transportation, and parametrization; boundary layer jets; local atmospheric circulation; surface energy partitioning; atmospheric stability conditions; pollutant distribution and transportation; etc.

This Special Issue has published 12 papers reporting new findings on various aspects of the atmospheric boundary layer.

Four papers were dedicated to understanding the characteristics and parametrization methods of the tropical cyclone atmospheric boundary layer. One of the four papers looked into the vertical eddy diffusivity in the atmospheric boundary layer during landfall of a tropical cyclone, which was observed by three-dimensional ultrasonic anemometers. An exceptional finding in this paper was the variation in the turbulent parameters with regard to the distance to the tropical cyclone center: outside three times of the radius of maximum wind (RMW) from the tropical cyclone center, the turbulent kinetic energy and eddy diffusivity values increased with increasing wind speed; however, in the area that was within one to three times the RMW from the tropical cyclone center, these values decreased slowly with increasing wind speed [1]. The other three papers investigated the turbulent fluxes exchanged over the ocean under tropical cyclone conditions through aircraft eddy-covariance measurements [2], dropsondes observations [3], and numerical simulations [4]. Gao et al. proposed new equations to parameterize the surface drag coefficient over the ocean surface through aircraft eddy-covariance measurements [2]; Ye et al. showed from the dropsondes observations that the relationship between the surface drag coefficient and wind speed varied with the distance from the tropical cyclone center [3]; and Ye et al. also showed from simulations that the surface flux scheme option, which overestimated the enthalpy exchange coefficient, leads to excessive inflow within the boundary layer and larger eyewall updrafts [4].

Citation: Li, Y.; Tang, J. Atmospheric Boundary Layer Processes, Characteristics and Parameterization. *Atmosphere* **2023**, *14*, 691. https://doi.org/10.3390/atmos14040691

Received: 30 March 2023
Revised: 1 April 2023
Accepted: 4 April 2023
Published: 7 April 2023

Three papers reported improvements in models or parameterizations related to the atmospheric boundary layer. The paper by Zhang et al. proposed using the random forest model to correct the simulation results of the Simple Biosphere Model 2, which improved the coefficient of determination between the calculations and measurements by 13–68% [5]. Nofal et al. introduced a more efficient parameterization to obtain the appropriate concentration boundary layer height and internal integral calculation intervals, and the new parameterizations showed an ability to save about 78% of the computation time compared to the original algorithm [6]. Reilly et al. put forward their understanding of the performance of different turbulence length-scale parameterization methods in the numerical weather prediction models using turbulence kinetic energy schemes, and they recommended using the turbulence length-scale formulations which considered the boundary layer height, turbulent kinetic energy and stratification, as these formulations had a satisfactory performance in different flow regimes [7].

Five papers investigated the air–land–sea interface phenomenon and the associated mechanisms through numerical simulations and/or observations. The paper by Wang et al. analyzed the impact erosion of aeolian sand saltation in the Gobi Desert and found that the maximum value of the saltation erosion rate increased due to a power law relationship with the friction velocity [8]. Zhang et al. presented how the atmospheric convective boundary layer varied with different wind shear settings in large eddy simulations, and they found that the increasing wind shear not only enlarged the variances of horizontal winds and temperature, but also enhanced large-scale coherent structures [9]. Moreira et al. demonstrated the spatial characteristics of the winter atmospheric boundary height and the aerosol layer aloft in São Paulo with observations from two simultaneous Lidar methods, and they found that the boundary layer height differences were affected by the cloud and sea breeze mostly, together with other influential factors including the development stages, topographic effects, and the presence of aerosol layers associated with biomass burning events [10]. Han et al. clarified how the periodic tidal elevations modified the sea surface fluxes by making observations from a near-coast platform in the East China Sea, and they found about a 1.5–3.5% mean difference of surface fluxes caused by the tide-dominated sea surface elevation [11]. Zhu et al. revealed that the Madden–Julian oscillation, partially through modifying the local-scale land–sea circulations, affected the diurnal variation and offshore propagation of the Sumatra western coast rainfall [12].

To summarize, the present SI provides not only new findings, but also new methods and new insights regarding the processes, characteristics and parameterization of the atmospheric boundary layer.

Author Contributions: Y.L. conceptualized the theme of this SI and prepared the original draft of this editorial. J.T. reviewed and edited this SI. All authors have read and agreed to the published version of the manuscript.

Funding: This research received no external funding.

Conflicts of Interest: The authors declare no conflict of interest.

References

1. Chen, C. Vertical Eddy Diffusivity in the Tropical Cyclone Boundary Layer during Landfall. *Atmosphere* **2022**, *13*, 982. [CrossRef]
2. Gao, Z.; Zhou, S.; Zhang, J.; Zeng, Z.; Bi, X. Parameterization of Sea Surface Drag Coefficient for All Wind Regimes Using 11 Aircraft Eddy-Covariance Measurement Databases. *Atmosphere* **2021**, *12*, 1485. [CrossRef]
3. Ye, L.; Li, Y.; Gao, Z. Surface Layer Drag Coefficient at Different Radius Ranges in Tropical Cyclones. *Atmosphere* **2022**, *13*, 280. [CrossRef]
4. Ye, L.; Li, Y.; Gao, Z. Evaluation of Air–Sea Flux Parameterization for Typhoon Mangkhut Simulation during Intensification Period. *Atmosphere* **2022**, *13*, 2133. [CrossRef]
5. Zhang, S.; Duan, Z.; Zhou, S.; Gao, Z. Correction to a Simple Biosphere Model 2 (SiB2) Simulation of Energy and Carbon Dioxide Fluxes over a Wheat Cropland in East China Using the Random Forest Model. *Atmosphere* **2022**, *13*, 2080. [CrossRef]
6. Nofal, O.M.M.; Al-Jaghbeer, O.; Bakri, Z.; Hussein, T. A Simple Parameterization to Enhance the Computational Time in the Three Layer Dry Deposition Model for Smooth Surfaces. *Atmosphere* **2022**, *13*, 1190. [CrossRef]

7. Reilly, S.; Bašták Ďurán, I.; Theethai Jacob, A.; Schmidli, J. An Evaluation of Algebraic Turbulence Length Scale Formulations. *Atmosphere* **2022**, *13*, 605. [CrossRef]
8. Wang, Y.; Zhang, J.; Dun, H.; Huang, N. Numerical Investigation on Impact Erosion of Aeolian Sand Saltation in Gobi. *Atmosphere* **2023**, *14*, 349. [CrossRef]
9. Zhang, H.; Yin, J.; He, Q.; Wang, M. The Impacts of Wind Shear on Spatial Variation of the Meteorological Element Field in the Atmospheric Convective Boundary Layer Based on Large Eddy Simulation. *Atmosphere* **2022**, *13*, 1567. [CrossRef]
10. Moreira, G.d.A.; Oliveira, A.P.d.; Codato, G.; Sánchez, M.P.; Tito, J.V.; Silva, L.A.H.e.; Silveira, L.C.d.; Silva, J.J.d.; Lopes, F.J.d.S.; Landulfo, E. Assessing Spatial Variation of PBL Height and Aerosol Layer Aloft in São Paulo Megacity Using Simultaneously Two Lidar during Winter 2019. *Atmosphere* **2022**, *13*, 611. [CrossRef]
11. Han, Y.; Liu, Y.; Jiang, X.; Lin, M.; Li, Y.; Yang, B.; Xu, C.; Yuan, L.; Luo, J.; Liu, K.; et al. Effects of Periodic Tidal Elevations on the Air-Sea Momentum and Turbulent Heat Fluxes in the East China Sea. *Atmosphere* **2022**, *13*, 90. [CrossRef]
12. Zhu, B.; Du, Y.; Gao, Z. Influences of MJO on the Diurnal Variation and Associated Offshore Propagation of Rainfall near Western Coast of Sumatra. *Atmosphere* **2022**, *13*, 330. [CrossRef]

atmosphere

Article

Vertical Eddy Diffusivity in the Tropical Cyclone Boundary Layer during Landfall

Chen Chen

Shanghai Typhoon Institute, China Meteorological Administration, Shanghai 200030, China; chenc@typhoon.org.cn

Abstract: This study investigated surface layer turbulence characteristics and parameters using 20 Hz eddy covariance data collected from five heights with winds up to 42.27 m s^{-1} when Super Typhoon Maria (2018) made landfall. The dependence of these parameters including eddy diffusivities for momentum (K_m) and heat (K_t), vertical mixing length (L_m), and strain rate (S) on wind speed (u_n), height, and radii was examined. The results show that momentum fluxes (τ), turbulent kinetic energy (TKE), and K_m had a parabolic dependence on u_n at all five heights outside three times the RMW, the maximum of K_m and S increased from the surface to a maximum value at a height of 50 m, and then decreased with greater heights. However, K_m and S were nearly constant with wind and height within two to three times the RMW from the TC center before landfall. Our results also found the $|\tau|$, TKE, and K_m were larger than over oceanic areas at any given wind, and K_m was about one to two orders of magnitude bigger than K_t. The turbulence characteristic and parameters' change with height and radii from the TC center should be accounted for in sub-grid scale physical processes of momentum fluxes in numerical TC models.

Keywords: turbulent momentum flux; tropical cyclones boundary layer; eddy diffusivity of momentum; eddy diffusivity of heat; vertical mixing length

Citation: Chen, C. Vertical Eddy Diffusivity in the Tropical Cyclone Boundary Layer during Landfall. *Atmosphere* **2022**, *13*, 982. https://doi.org/10.3390/atmos13060982

Academic Editor: Branko Grisogono

Received: 6 May 2022
Accepted: 14 June 2022
Published: 17 June 2022

1. Introduction

Turbulent mixing within the planetary boundary layer (PBL) of tropical cyclones (TCs) controls the radial and vertical distribution of momentum and enthalpy, consequently, changes the structure and intensity of TCs [1–4]. A better understanding of turbulent mixing would improve the accuracy of TC intensity forecasts [5]. Key parameters in PBL parameterizations are the eddy diffusivities of momentum (K_m) and heat (K_t). These describe turbulent mixing and related fluxes in terms of the vertical gradient of the mean quantities, which control the feedbacks involving momentum, moisture, and heat between the surface and atmosphere. Numerical and observational studies have sought to constrain the turbulent mixing parameterizations in a numerical weather prediction (NWP) model for forecasting both TC track and intensification events over the ocean [6–16]. For example, Nolan et al. [17], Kepert [18], and Zhang et al. [19] found that improved estimations of the eddy diffusivity coefficient in the hurricane boundary layer are beneficial for hurricane prediction over the ocean. Gopalakrishnan et al. [20] reported that the value of K_m controls the intensity of inflow in the boundary layer of TCs, and its parameterization leads to diverse solutions, based on analysis of dropwindsonde observations. More observations are required to improve simulations of hurricane or TC structure, path, and intensity.

Previous observation studies investigating the relationship between the vertical eddy diffusivities (VED) and wind speed (u_n) are limited to improving the turbulent mixing parameterization for TC forecasting. Based on aircraft observations during the Coupled Boundary Layer Air–Sea Transfer experiment [21], Zhang and Drennan [5] and Zhang et al. [18] estimated K_m as a function of u_n at an altitude of 450–500 m over the ocean. Zhao et al. [22] identified a decreasing trend in K_m values in high wind conditions (>40 m s^{-1}) at heights of

500–670 m, using aircraft observations in five hurricanes. Katz and Zhu [23] used observational data to evaluate surface layer flux parameterizations. Based on flux tower observations, Tang et al. [24] concluded that K_m values in a hurricane PBL are greater when the wind is blowing from inland than the ocean at u_n of up to ~30 m s^{-1} during typhoon landfall. Zhao et al. [25] analyzed mean wind profiles by a 350-m height flux tower during a typhoon landfall with u_n up to ~33 m s^{-1}.

However, due to the difficulties in obtaining turbulence data during extreme typhoon conditions [24], previous studies have been mostly limited to low and moderate wind conditions, or in the PBL at >100 m height over the ocean [21–23]. Moreover, changes in surface conditions induce the boundary layer of TCs to differ significantly moving from the oceans to land [11,26]. The abrupt change in K_m values at high wind conditions (>30 m s^{-1}) has rarely been documented during hurricane landfall. In addition, the variations in K_t are unclear, especially at heights of <400 m during high-wind conditions [5]. In theoretical and numerical models, it is typically assumed that the value of K_t is equal to that of K_m, or it is calculated using Prandtl and Schmidt numbers. However, Zhang and Drennan [5] found the K_t is much smaller than K_m, based on ocean observational data obtained from the coupled boundary layer air–sea transfer hurricane experiment [26].

In order to fill the knowledge gap of the lower atmospheric boundary layer characteristics and the vertical profile of the VED at high wind conditions (>30 m s^{-1}) in the inner-core region, this study focused on the turbulent mixing process with u_n up to severe typhoon category (42.27 m s^{-1}) at heights of <200 m in the boundary layer during Super Typhoon Maria's landfall, using high-frequency data (20 Hz) collected from multi-level flux observations on two flux towers. Although there remain difficulties in applying the method over tall, heterogeneous canopies, sloping surfaces, and as extreme typhoon conditions change rapidly making it difficult for an instrument and data system to survive [22,27], these data, after strict quality screening, are useful to constrain the key physical processes [28,29]. This paper is organized as follows: The data and processing methods are introduced in Section 2. Section 3 provides an analysis of the results and a comparison with previous studies. Finally, the conclusions and discussion for future research are presented in Section 4.

2. Material and Methods

2.1. Review of Typhoon Maria

Super Typhoon Maria (2018) was the eighth typhoon to form in the Northwest Pacific Ocean in 2018. Figure 1 shows the track of the typhoon relative to the location of the two eddy covariance (EC) flux towers, and also the temporal changes in its track and intensity using data from the Shanghai Typhoon Institute of China Meteorological Administration (http://www.typhoon.org.cn/, accessed on 1 January 2022) [30]. After a short period of gradual intensification on 6 July, Typhoon Maria made landfall over Lianjiang County, Fujian Province, China, at 0850 local standard time (LST), 11 July 2018, when the maximum wind speed was ~38 m s^{-1}. The location of landfall was ~68 km from the two observation towers. The towers at this time were located in the inner core region of the circulation, with the radius of maximum wind (RMW) at this time being ~60 km [31]. A maximum 10 min average u_n of 42.27 m s^{-1} was observed at 50 m height on the higher observation tower (110 m above the ground, or 170 m above sea level altitude (a.s.l)). After landfall, Typhoon Maria continued its track northward and weakened rapidly.

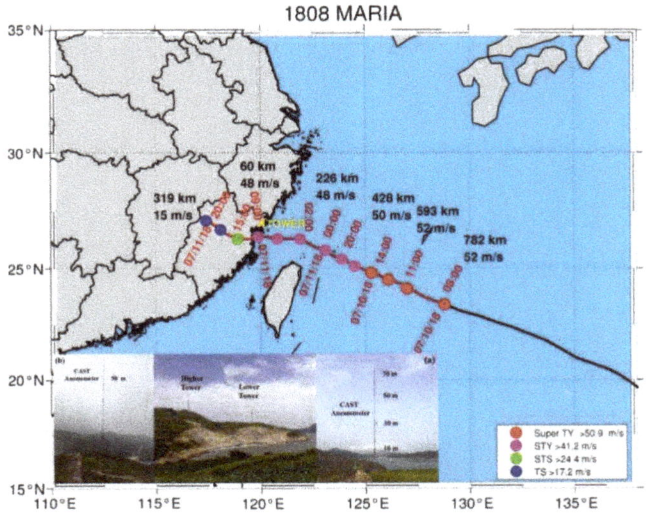

Figure 1. Track of Typhoon Maria and the location, exposure, and instruments on the two eddy covariance flux towers. The coastal observation towers are marked by the yellow arrow. (**a**) Photograph of the coastal observation tower approximately 10 m from the coastline (Figure 1a), which is referred to as the lower tower in this study. The CAST anemometers on the lower tower are 10, 30, 50, and 70 m above the ground (60, 80, 100, and 120 m above sea level). (**b**) Photograph of the coastal observation tower at the top of the hill, which is referred to as the higher tower in this study. The CAST anemometers on the higher tower are 50 m above the ground (170 m above sea level).

2.2. Observational Data and Analysis Method

The observation site was located on the coastline at Sansha, Fujian Province, China, and comprised two EC flux towers. The one meteorology tower (26°55′30″ N, 120°13′46″ E, 60 m above a.s.l.) was approximately 10 m from the coastline (Figure 1a) and is referred to as the low tower. The other meteorology tower (26°55′25″ N, 120°13′54″ E,120 m a.s.l.) was deployed at the top of the island (Figure 1b) and is referred to as the high tower. Each tower was equipped with four three-dimensional (3-D) ultrasonic anemometers (Campbell Scientific, Inc., Logan, UT, USA), with 1.5 m cantilever brackets at 10, 30, 50, and 70 m above the ground. The underlying surface of the two towers is grass that is <0.1 m in height. In this study, the data analyzed were collected from these two EC flux towers during the passage of Typhoon Maria (1808).

2.2.1. Eddy Covariance Method

The turbulent momentum flux τ ($N\ \mathrm{m}^{-2}$) was calculated using the direct eddy covariance (EC) method as follows:

$$\tau = \rho\left(-\overline{w'u'}\mathrm{i} - \overline{w'v'}\mathrm{j}\right) \tag{1}$$

where ρ (kg m^{-3}) is the air density, and u'(m s^{-1}), v' (m s^{-1}), and w'(m s^{-1}) are the along-wind (i), cross-wind (j), and vertical components of the wind fluctuations. The overbar indicates Reynolds averaging.

The turbulent kinetic energy (TKE) can be expressed as follows:

$$\mathrm{TKE} = \frac{1}{2}\left(\overline{u'^2} + \overline{v'^2} + \overline{w'^2}\right) \tag{2}$$

In the surface layer, the surface wind stress can be related to the surface frictional velocity u_* by the standard K theory [32], and K_m is usually parameterized as

$$K_m = ku_* z\Phi_m \tag{3}$$

where Φ is the stability function that is calculated using the Monin–Obukhov length L and the following equation:

$$\Phi_m = \ln\left[\left(\frac{1+x}{2}\right)^2\left(\frac{1+x}{2}\right)\right] - 2\arctan(x) + \frac{\pi}{2} \tag{4}$$

$$x = \left(1 - 16\frac{z}{L}\right)^{1/4} \tag{5}$$

For the EC method, the Monin–Obukhov length scale is calculated as follows:

$$L = -\frac{u_*^3 T_v}{kg\overline{T_v' w'}} \tag{6}$$

where $\overline{T_0}$ is the mean temperature at the reference height z, g is gravitational acceleration, k is the von Karman constant, and $\overline{T_v' w'}$ is the flux of the virtual potential temperature at the observational height z.

The non-dimensional standard deviations of the three-dimensional wind velocity can be written as:

$$\varphi_\alpha(\zeta) = \frac{\sigma_\alpha}{|\theta_*|} \tag{7}$$

where α (= u, v, and w) denotes the longitudinal, lateral, and vertical velocity components, respectively, σ_u , σ_v and σ_w are the standard deviations (σ) of u, v, and w, respectively.

Based on the standard K theory, the strain rate (S) and vertical mixing length (L_m) were estimated as follows:

$$S = \sqrt{\left(\frac{\partial\overline{u}}{\partial z}\right)^2 + \left(\frac{\partial\overline{v}}{\partial z}\right)^2} \tag{8}$$

$$l = \sqrt{\frac{K_m}{S}} \tag{9}$$

where the overbar expresses the average over the length [33].

2.2.2. Data Processing and Quality Control

The data pre-processing mainly involved outlier removal, tilt correction, coordinate rotation corrections, and linear detrending [28]. Schmid et al. [33] proposed a method to examine the power spectra of turbulent fluctuations and this method was used in the present study, as follows.

(1) Spikes in the datasets were removed using the criterion X(h) < (X − 4σ) or X(h) > (X + 4σ), where X(h) denotes the original data, X is the mean over the averaging interval, and σ is the standard deviation [34].

(2) The calculation of VED was omitted when the corresponding u_* was < 0.01 m s^{-1}. No gap filling was used.

(3) Based on the coastline features near the measurement site, the onshore wind direction varied from 52.5° to 227.5°, and the offshore wind direction varied from 272.5° to 5°. The wind data from the back of the three-dimensional sonic anemometer measurements of the lower tower (249–269°) and higher tower (68–88°) were removed due to the turbulent eddies generated by the towers.

(4) Averaging period: A cumulative frequency curve (ogive) can be used to understand the turbulent stationarity and its spatial scale, which is then typically used to determine the appropriate time period required to calculate the turbulent flux [34].

$$Og_{w,x}(f_0) = \int_{\infty}^{f_0} Co_{w,x}(f)df \quad (10)$$

where x (= u, v, and T) denotes the along-wind (u) and cross-wind (v) components, and ultrasonic virtual temperature (T), respectively, $Co_{aw}(f)$ is the cospectrum of $w'u'$, $w'v'$, and $w'T'$ at frequency f. As we can see, for the ogive curves of $w'u'$, $w'v'$, and $w'T'$ shown in Figure 2a, the ogive curves of the $w'u'$, $w'v'$, and $w'T'$ approached a constant value at an averaging period of 5 min when it converged towards 0.005 Hz. Therefore, an averaging period more than 5 min is suitable for calculating the turbulent flux and can represent the total flux. Hence, we chose 10 min for the averaging period to calculate the moment and heat flux.

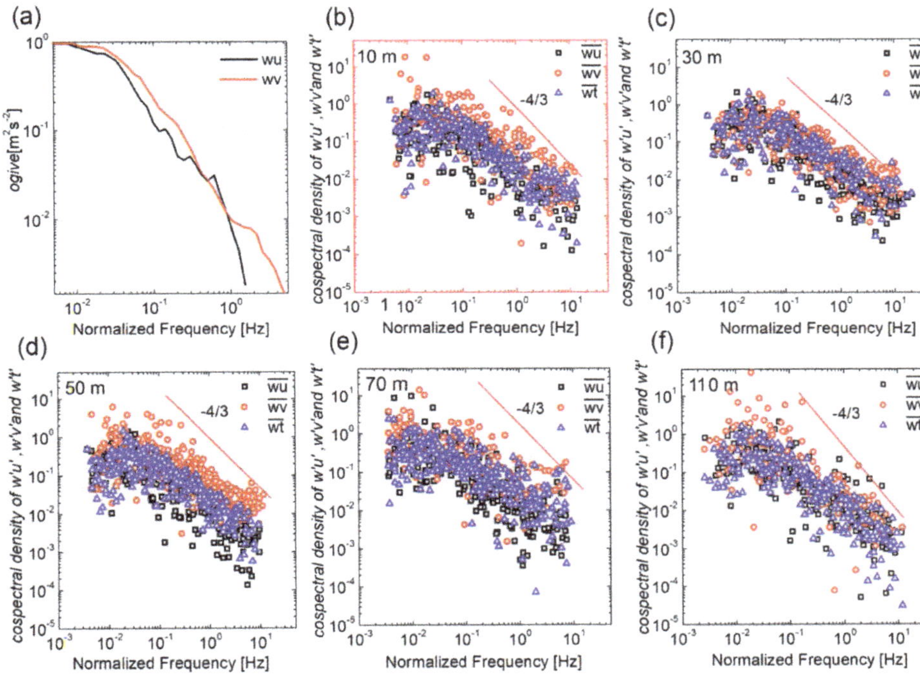

Figure 2. (**a**) Cumulative frequency curves (ogive) of the vertical velocity w with the along−wind u and cross−wind v components. The cospectra densities of the vertical velocity (w) with the along-wind (u) and cross-wind (v) components and ultrasonic virtual temperature (T) at (**b**) 10 m, (**c**) 30 m, (**d**) 50 m, (**e**) 70 m, and (**f**) 110 m. Ten-minute averaged data are for Typhoon Maria (2018) from 0000 local standard time (LST) 10 July to 0000 LST 12 July 2018.

(5) A turbulence (co)spectrum check.

According to the Kolmogorov theory for the inertial subrange [35], a turbulence (co)spectrum of the slope check has been used for turbulence data quality control in previous studies, such as Zhang et al. [10,11], Fortuniak et al. [36], Zhao et al. [21], and the one-dimensional spectrum $S_i(n)$, of any wind component, $i = (u, v, w)$, normalized by the squared friction velocity u_*^2, can be expressed in the form

$$\frac{nS_i(n)}{u_*^2} = \frac{\alpha_i}{(2\pi k)^{\frac{2}{3}}}\varphi_\varepsilon^{2/3}(\zeta)f^{-2/3} = \alpha_i\varphi_\varepsilon^{2/3}(\zeta)f^{-2/3} \tag{11}$$

where n is the natural frequency, f is the non-dimensional frequency, $\alpha_i = (\alpha_u, \alpha_v, \alpha_w)$ are universal Kolmogorov inertial subrange constants and κ is the von Kármán constant, $\varphi_\varepsilon(\zeta)$ is the non-dimensional dissipation rate of TKE (ε). The $\varphi_{(\varepsilon)}$ is a function of the stability parameter, $\zeta = z/L$, where L is the Obukhov length. The velocity spectra normalized by $\varphi_\varepsilon^{2/3}$ and u_*^2, are presented in Figure 3. Our neutral limits from the stable side included the spectra from the narrow stability band $0 < \zeta < 1.2$ (similarly, in the unstable limit) as compared to the relatively wide band ($0 < \zeta < 0.1$) used by Kaimal et al. [37]. In the inertial subrange, the spectra slopes of the horizontal wind speed (u) ranged from -0.61 to -0.65; they follow the $-2/3$ slope quite well.

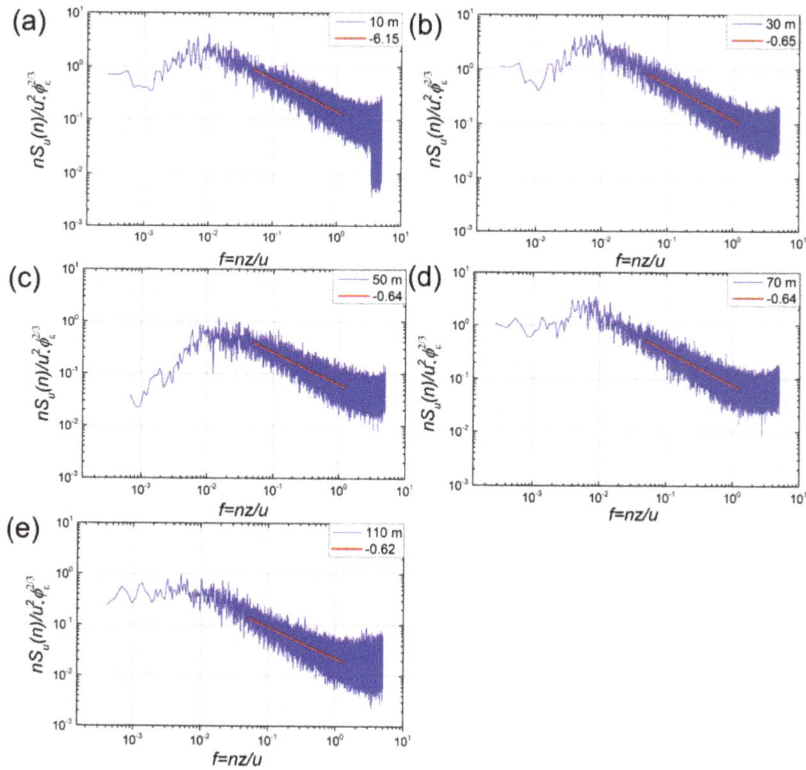

Figure 3. Normalized logarithmic along-wind spectrum plotted against the non-dimensional frequency, $f = n(z)/u$ at (**a**) 10 m, (**b**) 30 m, (**c**) 50 m, (**d**) 70 m, and (**e**) 110 m. Short red solid lines indicate the along-wind spectrum of slopes for the inertial subrange. Twenty hertz raw data were for Typhoon Maria (2018) from 0000 local standard time (LST) 10 July to 0000 LST 12 July 2018.

Cospectra analyses of the measured turbulent fluctuations are a useful tool for testing the reliability of flux data. Figure 2b–f shows examples of cospectra ($nCo_{xw}(n)/u_*^2$) of the longitudinal (u), lateral (v), and ultrasonic virtual temperature (T) averaged from data collected at five heights over three days during the landfall of Typhoon Maria. Previous studies indicated that the empirical cospectra are not as well formed as the empirical spectral functions [36]. So positive values of $nCo_{xw}(n)/u_*^2$ for 0.1 < f < 5 were used to fit the power function to the cospectrum in the inertial subrange, and the −4/3 line is plotted by green line in each panel for the convenience of comparison with the ideal situation. In the inertial sub-range, the cospectra slopes of the horizontal wind speed (u, v) and T ranged from 0.78 to 1.29, and a majority of the three cospectra was close to the ideal slope of –4/3, especially for the longitudinal wind components. The sharp of the three cospectra trends rolled-off at f = 0.01 Hz at each height. However, the slopes deviated from the –4/3 slope more obviously with height. In addition, some low-frequency and high-frequency data deviated from the –4/3 slope on a log-log scale (Figure 2). Ortiz-Suslow et al. [38] and Dorman et al. [39] found that there may be natural deviations in the inertial subrange bandwidth and spectral slope from Kolmogorov's turbulence that occur within ±20% caused by mechanical wind–wave interactions. Compared with the classical 4/3 ratio in the inertial subrange at a homogenous flat surface reported by Kaimal et al. [37] for the Kansas experiment, the sensor noise rising at the high-frequency may be due to the high wind and high wet environment of the typhoon. It may also generate larger-scale eddies during typhoon landfall due to the effect of flow field shear on the turbulence spectrum in the coast process [39].

Verification of the Monin–Obukhov Similarity (MOST) during typhoon landfall: The variations in the normalized standard deviations of turbulence with the dimensionless stability parameter ($\zeta = z/L$) have been widely used to evaluate the applicability of the MOST in a series of previous studies [34,35]. According to MOST, any scaled statistics of turbulence at reference height z are universal functions of the dimensionless stability parameter, z is the observation height and L is the Obukhov length scale (see Equation (6)).

Figures 4–6 show plots of the φ_u, φ_v, and φ_w versus z/L with the left panels (Figures 4, 5 and 6a–e) representing unstable ($z/L < 0$) conditions and the right panels (Figures 4, 5 and 6f–j) representing stable ($z/L > 0$) conditions at five heights during Typhoon Maria's landfall. The blank dashed lines are exponential fitting lines correspond to φ_α. Note that, irrespective of whether the conditions were stable or unstable, σ_u/u_*, σ_v/v_*, and σ_w/w_* exhibited an obvious relationship with z/L, and this is consistent with classical Monin–Obukhov z-less scaling [35]. It is evident that the data collected in the coastal region during the typhoon period were generally in agreement with previous studies that follow the canonical MOST predictions during the non-typhoon period [40,41]. From the perspective of stability, the exponential fitting lines of unstable layers and stable layers were consistent with the –1/3 power law, whereas the formation of stable layers followed the polynomial fitting law from observations at the Downs site. However, our observations during the typhoon period showed a larger scatter of individual data points as compared with the results measured from the North Carolina coastal zone[40]. Additionally, σ_w/w_* showed somewhat greater scatter than σ_u/u_*. The larger scatter of data and poor correspondence of σ_w/w_* with the classical MOST theory may possibly be related to non-local mixing in the complexity of the coastal terrain and typhoon conditions; the shear production rate exceeded the buoyant production rate in the high wind condition. The heights of 10 and 110 m exhibited the smallest and greatest deviation from the MOST theory, respectively.

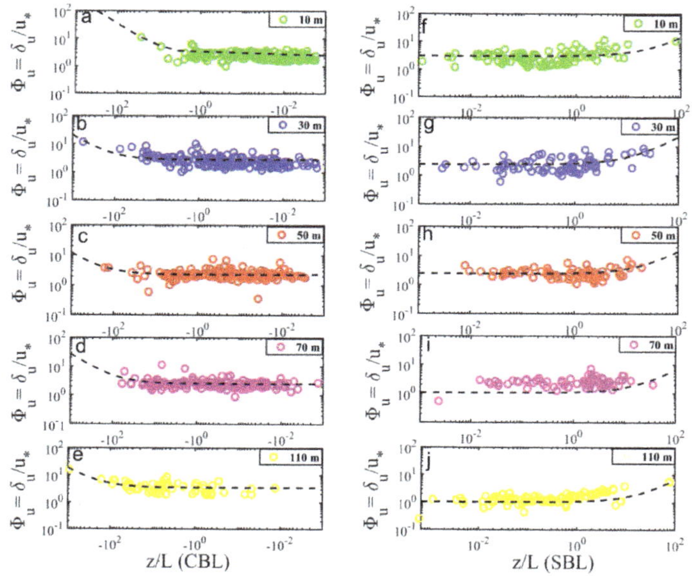

Figure 4. Plots of the normalized standard deviations of the longitudinal wind velocity components (σ_u/u_*) in log-log scales versus the stability parameter (z/L) under unstable (or CBL, <0,) and stable (or UBL, $\zeta > 0$) stratification conditions for the 10-min-averaged onshore data collected during Typhoon Maria's landfall (10–12 July 2018). The black dashed lines correspond to $\varphi_u(\zeta) = 1.1(10 - 1.56\zeta\,)^{1/3}$ for $\zeta < 0$ and $\varphi_u(\zeta) = (1 + 1\zeta\,)^{1/3}$ for $\zeta > 0$ (Kaimal and Finnigan [39]).

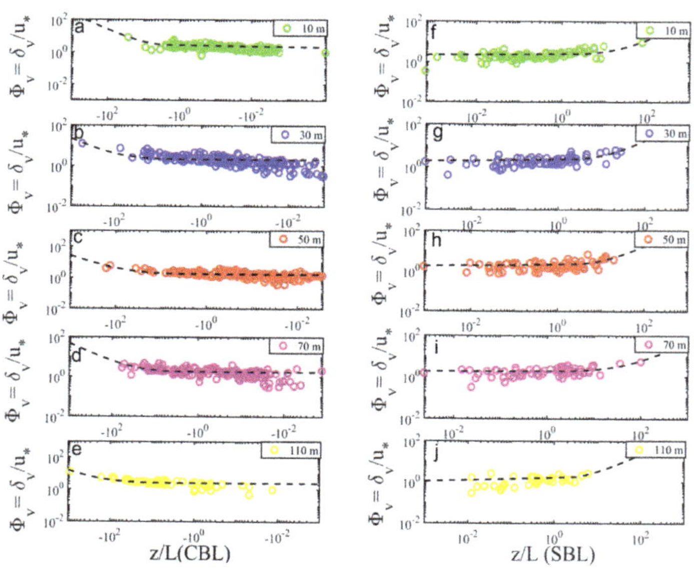

Figure 5. Plots of the normalized standard deviations of the lateral wind velocity components (σ_v/u_*) in log–log scales versus the stability parameter (z/L) under unstable (or CBL, <0,) and stable (or UBL, $\zeta > 0$) stratification conditions for the 10-min-averaged onshore data collected during Typhoon Maria's landfall (10–12 July 2018). The black dashed lines correspond to $\varphi_v(\zeta) = 0.8(10 - 3.5\zeta\,)^{1/3}$ for $\zeta < 0$ and $\varphi_v(\zeta) = 1.7(1 + 0.7\zeta\,)^{1/3}$ for $\zeta > 0$ (Kaimal and Finnigan [39]).

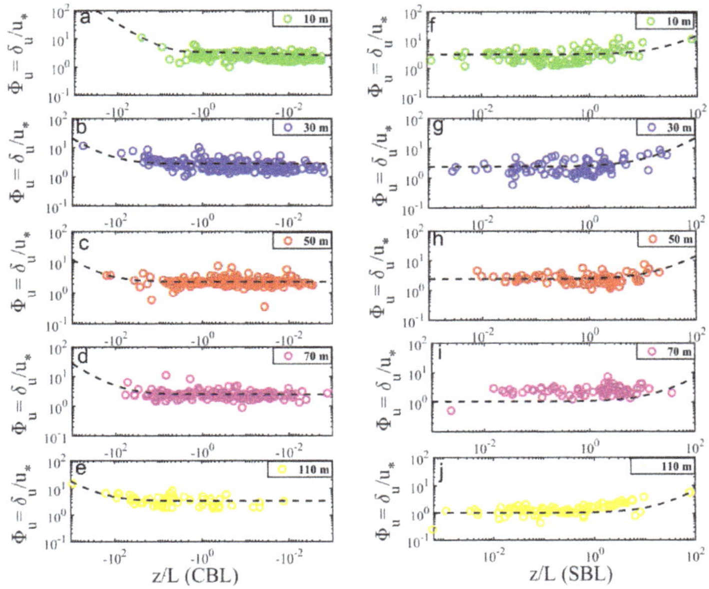

Figure 6. Plots of the normalized standard deviations of the longitudinal wind velocity components (σ_w/u_*) in log–log scales versus the stability parameter (z/L) under unstable (or CBL, <0,) and stable (or UBL, $\zeta > 0$) stratification conditions for the 10-min averaged onshore data collected during Typhoon Maria's landfall (10–12 July 2018). The black dashed lines correspond to $\varphi_w(\zeta) = 0.54(10 - 4.24\zeta\,)^{1/3}$ for $\zeta < 0$ and $\varphi_w(\zeta) = (1 + 1\zeta\,)^{1/3}$ for $\zeta > 0$ (Kaimal and Finnigan [39]).

3. Results and Discussion

3.1. General Meteorological Conditions

3.1.1. Wind Characteristics during the Observation Period

Figure 7 shows 10-min averaged wind directions and speeds from the five heights during Typhoon Maria's landfall. Unfortunately, due to the lack of measurement data from the high tower, there was no wind information from a height of 110 m after landfall. It still can be seen that the wind directions at the five heights exhibited a significant daily cycle from northwest in the daytime (offshore) to northeast in the nighttime (onshore) related to the land–sea breeze. In addition, the wind direction at a height of 110 m on the high tower was slightly different from that of the low tower, which is likely due to the topography. For the wind speeds, compared with the annual 10-min mean u_n of 5–10 m s^{-1}, a maximum 10-min mean u_n of 42.27 m s^{-1} was observed in the front right quadrant of the typhoon at a height of 110 m on the high tower as Typhoon Maria passed through the towers. The u_n became higher and peaked before and after the typhoon passed. With increasing height, the first peaks from the offshore wind direction before the typhoon passed had a u_n of 16.57, 19.05, 22.26, 29.61, and 21.28 m s^{-1}. The second peaks from the onshore wind direction after the typhoon passed had a u_n of 36.51, 36.79, 39.06, 39.67, and 42.27 m s^{-1} (Table 1). In addition, the offshore wind direction changed from approximately southeasterly to southwesterly at the first peak, and from approximately southeasterly to southwesterly during the second peak. This made it possible to conduct wind-direction-dependent analysis of the wind characteristics of this strong typhoon.

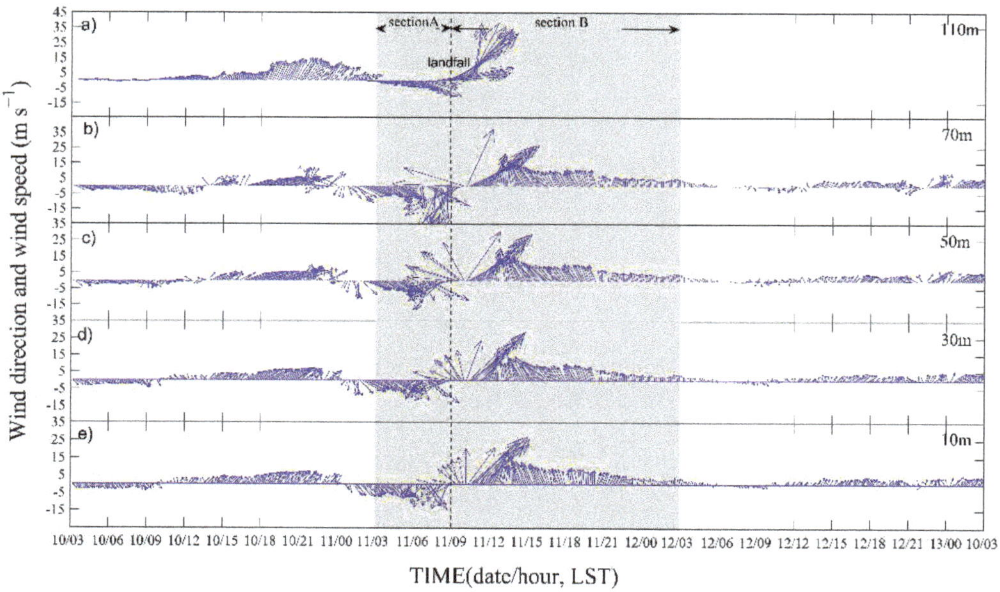

Figure 7. Temporal changes in the 10-min averaged wind speed (u_n) and wind direction (WD) obtained from the sonic anemometers on the five heights [(**a**) 110 m, (**b**) 70 m, (**c**) 50 m, (**d**) 30 m, and (**e**) 10 m of the two towers during Typhoon Maria (2018). Section A is the time period before typhoon landfall on 11 July and Section B is the time period after typhoon landfall.

Table 1. Maximum 10-min mean wind speed and directions during Typhoon Maria in 2018.

	First Peak		Second Peak	
	Onshore (52.5–227.5°)		Offshore (272–5°)	
Instrument Height (m)	Wind Speed (m s^{-1})	Wind Direction (°)	Wind Speed (m s^{-1})	Wind Direction (°)
10	16.57	300.22	36.51	48.28
30	19.05	316.41	36.79	51.17
50	22.26	234.08	39.06	47.72
70	29.61	243.53	39.67	42.17
110	21.28	329.92	42.27	54.50

3.1.2. Meteorological Conditions

To better depict the meteorological conditions during the observation period, Figure 8a shows a time-series of the 10-min averaged TKE values, 1-h averaged air pressure from the Sansha meteorological observatory (located near the EC flux towers), precipitation during landfall (Figure 8b), distance from the typhoon center to the tower during the period 0000 LST 10 July to 0000 LST 12 July, and radar reflectivities (dBz) at Z = 2 km from 0200 to 1200 LST 11 July 2018 obtained by the Ningde weather radar before and after landfall of Typhoon Maria.

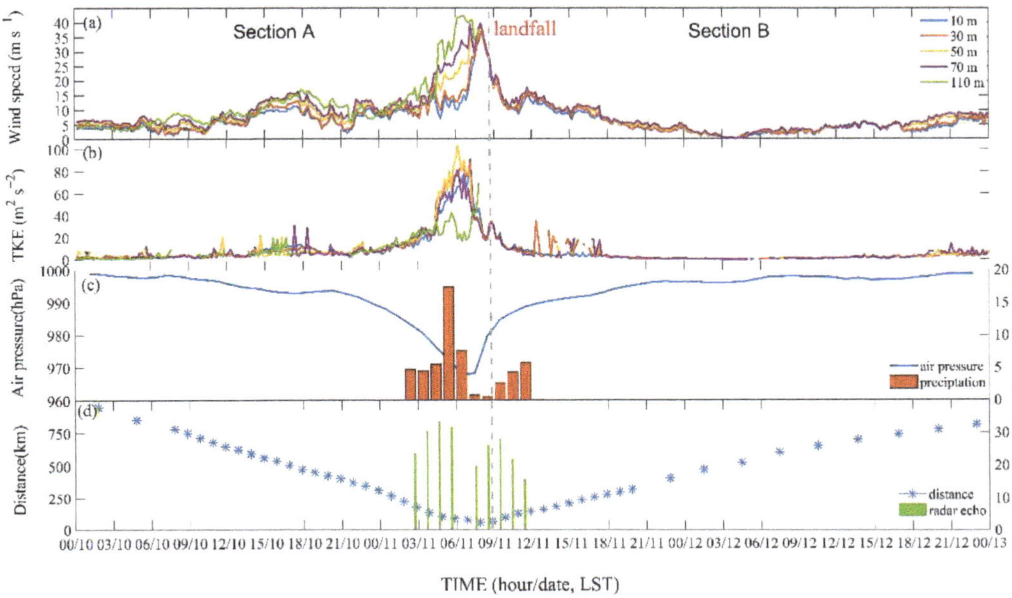

Figure 8. (**a**) Time-series of the 10-min averaged wind speed and (**b**) turbulent kinetic energy (TKE), (**c**) air pressure of the typhoon center and precipitation during landfall, and (**d**) distance of the typhoon center to the tower and radar reflectivities (dBz) observed nearby flux towers from 10 to 12 July (distance data are from http://www.typhoon.gov.cn/, accessed on 1 January 2020).

Section A is the period prior to typhoon landfall and section B is the period after typhoon landfall. Typhoon Maria passed over the observation towers at about 0850 LST 11 July 2018 (Figure 8c). The smallest distance between the observation towers and the center of Typhoon Maria was about the RMW (60 km: Figure 8c). The 10-min averaged TKE on the lower tower increased as Typhoon Maria approached and reached a maximum value of 102.6 m^2 s^{-2} at a 50 m height on the lower tower ahead of landfall, and then decreased rapidly. This may indicate that the inner core of the typhoon increased the local values of the TKE. The 10-min averaged TKE on the high tower exhibited an opposite trend from 0600 to 0800 LST 11 July 2018 before landfall, and the 1-h averaged TKE on the high tower was smaller than on the lower tower during the 3 h (from 0600 to 0800 LST 11 July) (Figure 8a). The main reason for the difference is that the vertical turbulent component (w') decreased abruptly during this 3 h. It is possible that the momentum transport was weak during the period, and the momentum fluxes (τ) were highly correlated with the TKE. This also implies that the momentum transport was closely related to the turbulent intensity. The air pressure at the Sansha meteorological observatory, which is ~1 km from the towers, began to decrease as Typhoon Maria approached, and the air pressure dropped to its lowest value (968 hPa) at this time when the minimum distance between the towers and typhoon center was ~60 km at 0800 LST 11 July 2018.

3.2. Variations in Momentum Fluxes and Turbulent Kinetic Energy with wind Speed at Each Height

Momentum fluxes obtained by the EC method are plotted as a function of $u_{\rm n}$ at the five measurement levels (10, 30, 50, 70, and 110 m) on the two towers during Typhoon Maria (Figure 9a). $|\tau|$ values at all heights exhibited a clear parabolic trend with $u_{\rm n}$, and at low-moderate $u_{\rm n}$ (<30 m s^{-1}), $|\tau|$ initially increased up to 22 m s^{-1}, at which point $|\tau|$ was 11 m^{-1} s^{-2}, then slowly decreased with $u_{\rm n}$ increasing further (>30 m s^{-1}). Moreover, the values of $|\tau|$ in the $u_{\rm n}$ range of 15–30 m s^{-1} was higher than previous observational results from oceanic and inland regions [19,22]. For example, |τ| obtained in this study was five times larger than that at sea at heights of 420–500 m obtained using aircraft observations [22]. Previous studies have mostly attributed this to surface roughness over an island, which is greater than that over the open ocean [22].

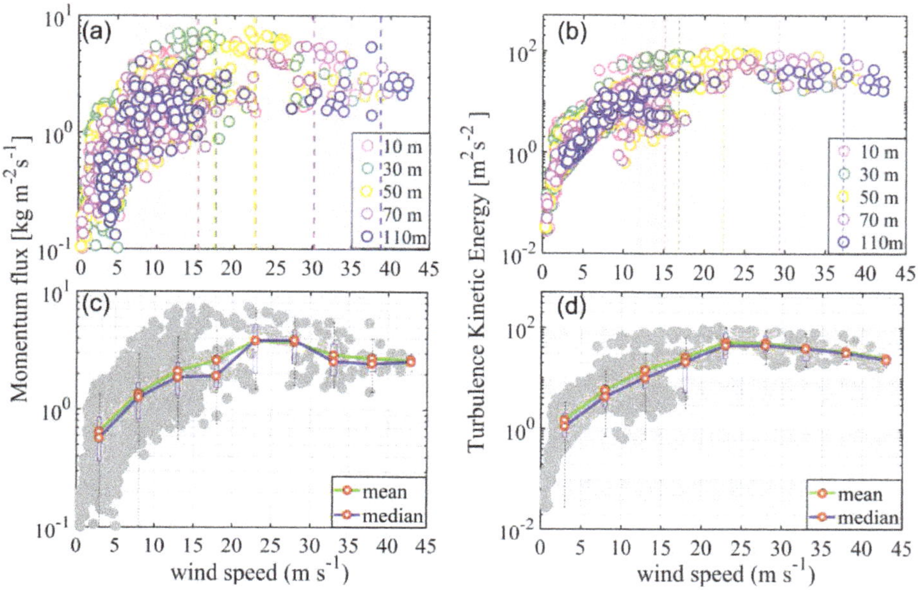

Figure 9. Plots of the (**a**) momentum flux and (**b**) turbulent kinetic energy at the five height levels during Typhoon Maria. Plots of the (**c**) momentum fluxes and (**d**) turbulent kinetic energy as a function of wind speeds at the five heights during landfall of Typhoon Maria. The blue line with the red circles is the bin median average at 5 m s^{-1} intervals of u_n from the observations. The yellow line with red circle is the bin mean average at 5 m s^{-1} intervals of u_n from the observations.

At higher wind speeds (>30 m s^{-1}), |τ| decreased with increasing u_n, this phenomenon is similar to results from some previous studies, which found that |τ| or TKE no longer increase significantly at high wind speed [38–40]. However, this result is inconsistent with the previous observational study by Zhao et al. [42], who found |τ| increased with increasing u_n to 60 m s^{-1} over the ocean.

Furthermore, |τ| peak values varied with height. With increasing height, |τ| peaked at 24.1, 39.0, 40.6, 24.2, and 20.5 m^{-1} s^{-2}, and the corresponding u_n at the |τ| peaks were 12.5, 14.6, 22.1, 32.4, and 37.6 m s^{-1}, respectively. The maximum |τ| value occurred at 50 m height (40.6 m^{-1} s^{-2}), and then decreased with height above 50 m.

The variations in TKE against u_n at the five measurement levels are shown in Figure 9b. In general, the TKE varied similarly to the momentum flux, because the TKE is generated by wind turbulence and the momentum flux is generated from the wind flow. As such, the TKE was highly correlated with |τ| as shown by Equations (1) and (2). The TKE increased significantly at low and moderate wind speeds (<22 m s^{-1}) and slightly decreased at higher wind speeds. In addition, the maximum TKE values also varied with height. With increasing height, the TKE peaked at 24.1, 39.0, 40.6, 24.2, and 20.5 m^{-2} s^{-2}, and the corresponding u_n values at the TKE peaks were 10.0, 16.7, 22.3, 28.6, and 38.0 m s^{-1}, respectively. The maximum TKE peaked at 50 m height with a value of ~100 m^2 s^{-2}. The result showed that the TKE and |τ| peaks increased with height from 10 to 50 m, and then decreased at greater heights with the corresponding wind speeds increasing continue.

3.3. Variations in Momentum Fluxes and Turbulent Kinetic Energy with Distance from TC Center

Although both $|\tau|$ and TKE exhibited parabolic trends with increasing u_n at all five heights (Figure 9c,d), these relationships changed with distance from the TC center. The $|\tau|$ and TKE variations with u_n are plotted as a function of the distance to the TC center normalized by the RMW in Figure 10a,b. All data were classified into seven categories according to the TC radius normalized to the RMW. The first and second categories are at distance outside five times the RMW(5 RMW) from the TC center before and after typhoon landfall, respectively; The third and fourth categories are from the region between the three and five times RMW (3–5 RMW) before and after typhoon landfall, respectively; The fifth and sixth categories are from the region between the two and three times RMW (2–3 RMW) before and after typhoon landfall, respectively; the seventh category is from the region near the RMW. At larger radii (outside 3 times RMW), both $|\tau|$ and TKE increased significantly with increasing u_n at <15 m s^{-1}. With the approach of the typhoon (within three times RMW), both $|\tau|$ and TKE increased slowly with increasing u_n. Most of $|\tau|$ and TKE data were concentrated in the region of two to three times the RMW before the typhoon landfall, and the $|\tau|$ and TKE values were 60 m^{-1} s^{-2} and 10 m^{-2} s^{-2}, respectively, which were five to ten times larger than that in the 2–3 RMW region after typhoon landfall. In the case of the TKE, when the u_n was >15 m s^{-1} in the region of 2–3 RMW before typhoon landfall, the TKE values no longer changed significantly as the u_n further increased. This shows that the TKE was basically constant with height in the 2–3 RMW region before the typhoon landfall, due to adequate vertical turbulent mixing in the surface layer. The constant of TKE with height may reflect that the constant surface layer exists in the 2–3 RMW region before the typhoon landfall. Near the eyewall region at high wind speeds, the $|\tau|$ and TKE values decreased slightly with increasing u_n. The values were smaller than those in the region of 2–3 RMW prior to typhoon landing at similar wind speeds, whereas five times higher than those in the outer typhoon core regions.

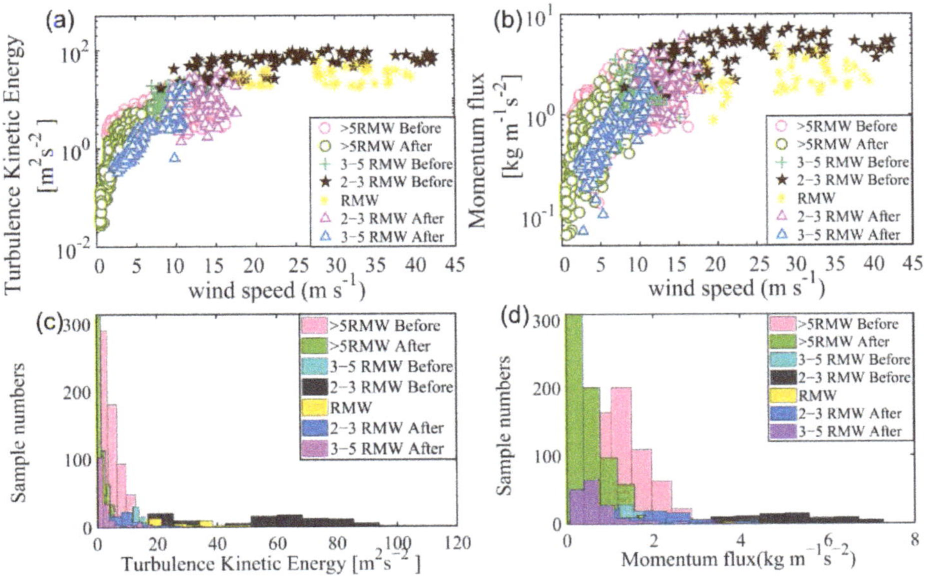

Figure 10. Plots of the (**a**) momentum fluxes and (**b**) turbulent kinetic energy versus wind speed as a function of the TC radius normalized by the maximum wind speed radius (RMW). The data were classified into seven categories according to the TC radius normalized by RMW. Histograms of the (**c**) turbulent kinetic energy and (**d**) momentum fluxes. Each color represents the seven categories defined by the TC radius normalized to RMW.

3.4. Vertical Diffusion Transport of Momentum

Figure 11a shows the bin-averaged K_m values estimated using the eddy covariance method as a function of u_n at all levels during landfall of Typhoon Maria. Seven wind speed bins of 5 m s^{-1} were chosen, ranging up to 45 m s^{-1}.

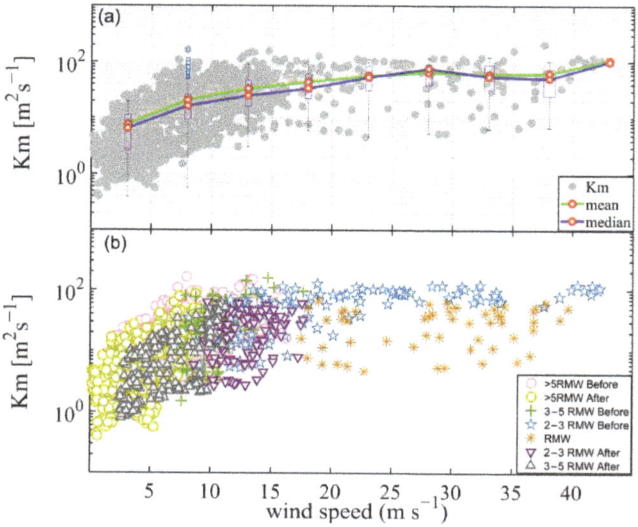

Figure 11. Vertical eddy diffusivity of momentum (K_m) as a function of wind speed (u_n). The blue line with the red circles is the bin median averaged K_m at 5 m s^{-1} intervals of u_n from the observations. The yellow line with the red circles is the bin mean averaged K_m at 5 m s^{-1} intervals of u_n from the observations (**a**). The data were classified into seven categories according to their distance from the typhoon center (**b**).

It was expected that K_m would increase gradually with increasing u_n at <30 m s^{-1}, and then decrease slightly with u_n up to 40 m s^{-1}. K_m increased again when the u_n was >40 m s^{-1}, but at a lower increasing rate than at < 30 m s^{-1}. The spatial patterns and magnitude of the coastal variations in K_m values with u_n are similar to the aircraft observation results of Zhao et al. [22]. The slope of increasing K_m with u_n at <30 m s^{-1} is similar to the results of Zhang and Drenman [5], Tang et al. [24], and Zhao et al. [25]. The decreasing K_m at wind speeds of >30 m s^{-1} is different from the observational results of hurricane eyewalls at 500 m altitude from Zhang et al. [19], which shows that K_m consistently increases with increasing u_n to >40 m s^{-1}. The coastal K_m values obtained in the present study reached uniform values at a slower u_n (28 m s^{-1}) than in the studies of Zhang et al. [19] and Zhao et al. [22] over oceanic regions and Tang et al. [24] over inland areas. This may be due to the larger momentum flux and lower altitude of the present study conditions. The presented observational data support the findings of Zhang and Zhu [11] and suggest that different boundary layer parameterizations should be used for vertical eddy diffusivity over land and oceanic regions in TC simulations and forecasts.

K_m values tended to increase with u_n outside three RMW, but were less dependent on u_n near the eyewall region (Figure 11b). The higher values of K_m were concentrated in the region of 2–3 RMW before typhoon landfall, with an almost constant (100 m^{-2} s^{-1}) high wind speed of 20.0–42.5 m s^{-1}. This indicates that the bulk parameters should be different over different regions when modeling the typhoon PBL.

3.5. Vertical Eddy Diffusivity of the Sensible Heat Flux

Figure 12a shows the bin-averaged K_t values estimated by the EC method as a function of u_n at the five height levels during the landfall of Typhoon Maria. The different colored symbols represent different regions of Typhoon Maria (Figure 12b).

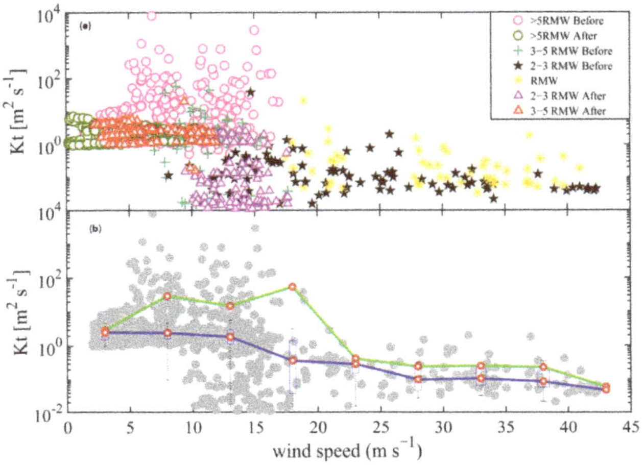

Figure 12. Vertical eddy diffusivity of heat (K_t) as a function of wind speed (u_n). The blue line with red circles is the bin median average K_t at 5 m s^{-1} intervals of u_n from the observations. The yellow line with the red circle represents the bin mean averaged K_t with 5 m s^{-1} interval of u_n from the observations (**a**). The data were classified into seven categories according to their distance from the typhoon center (**b**).

In general, K_t decreased with increasing mean u_n from 5 to 45 m s^{-1}, and most of K_t following the trend were from the region before typhoon landfall. The decreasing trend was opposite to the trend for K_m at moderate u_n. Most K_t values were <100 m^2 s^{-1}, with a small number of large values of K_t (>100 m^2 s^{-1}) in the region outside 5 RMW before typhoon landfall (Figure 12b). From the median line of K_t, all values of K_t were <12 m^2 s^{-1} (Figure 12a). K_t was about one to two orders of magnitude smaller than K_m, possibly because large and small eddies usually transport the sensible heat flux in opposite directions.

The variations in K_t with u_n were different between outside 3 RMW and 1–3 RMW. K_t values showed a clear decrease with increasing wind speed located in the 1–3 RMW, but K_t exhibited little dependence on u_n outside 3 RMW (Figure 12b). K_t from the region outside 3 RMW was 10 times larger than that from the inner core region.

3.6. Variation in Vertical Mixing Length

L_m in models needs to be parameterized by observations [20], and it can be directly estimated from the eddy covariance data using Equation (8). The relationship between L_m and u_n exhibited segmented trends (Figure 13a). At weak and moderate wind speeds (<20 m s^{-1}), there was a weak dependence of L_m on wind speed at each level, due to a large scatter during weak winds outside 5 RMW (Figure 13b). It is consistent with the result of Tang et al. [24], who showed a weak dependence of L_m on the wind speed over land and Zhang et al. [19] using aircraft observations over the ocean. While, at high wind speeds (>30 m s^{-1}) in the region of 1–3 RMW before typhoon landfall, L_m increased with u_n and then began to decrease at >38 m s^{-1}. This indicates that a single PBL scheme with a constant mixing length is inappropriate for hurricane conditions.

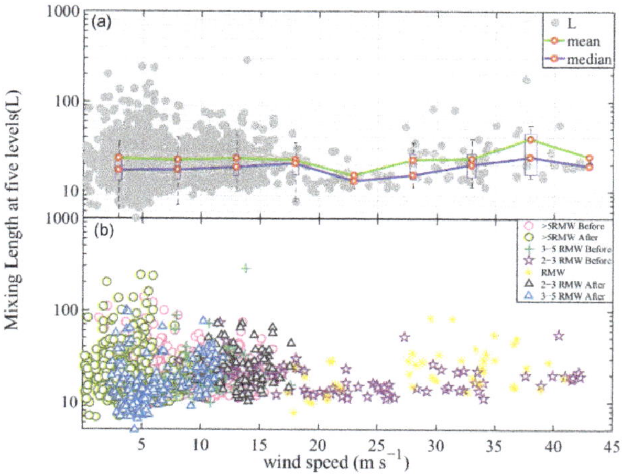

Figure 13. (**a**) Vertical mixing length (L_m) as a function of wind speed (u_n). The data were classified into seven categories according to their distance from the typhoon center. (**b**) The blue line with the red circles is the bin median average K_t at 5 m s^{-1} intervals of u_n from the observations. The green line with red circles is the bin mean average K_t at 5 m s^{-1} intervals of u_n from the observations.

3.7. *Variations in the Strain Ratio*

The strain ratio (S) as a function of u_n with the bin averaged mean and median lines of S are shown in Figure 14a. Meanwhile, the S at different layer with u_n is shown in Figure 14b. Overall, it is evident from Figure 14a that S exhibited a parabolic trend with increasing wind speed from 30 m 70 m heights, while S always increased with increasing u_n at 110 m height, especially when wind speed exceeded 30 m s^{-1}. The present study was similar to the previous study from Tang et al. [24] when wind speed below 30 m s^{-1} under 100 m height. The present study also presented the trend under higher wind up to 110 m height.

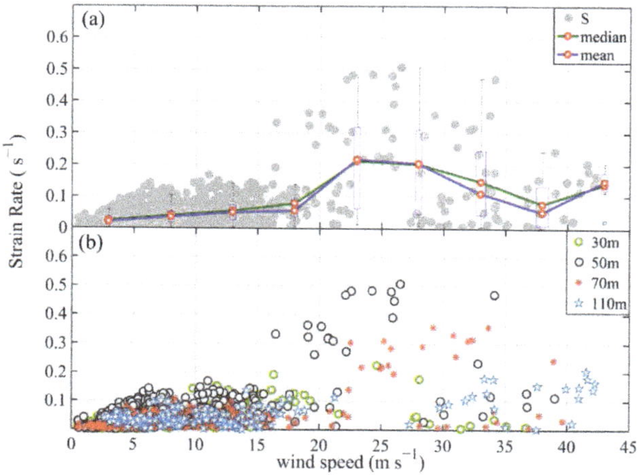

Figure 14. (**a**) S as a function of u_n during landfall of Typhoon Maria. The blue line with the red circles is the bin median average S at 5 m s^{-1} intervals of u_n from the observations. The yellow line with red circles is the bin mean average S at 5 m s^{-1} intervals of u_n from the observations. (**b**) Strain rate (S) at four heights plotted versus wind speed (u_n) during landfall of Typhoon Maria. The strain rate was calculated using Equation (7).

Furthermore, when increasing the height from 30 m to 110 m, S peaked at 0.22, 0.51, 0.38, and 0.21 s^{-1} with the corresponding wind speeds being 24.8, 27.0, 22.3, 33.2, and 41.3 m s^{-1}, respectively. This indicates that the maximum of S increased with increasing altitude below a height of 50 m, and then decreased at heights above 50 m. The bin-averaged values of S was 0.035, 0.066, 0.036, and 0.091 s^{-1} from 30 m to 110 m, respectively. However, as Zhao et al. [25] found S decreased with height up to about 100 m using 356 m tower using landfall, it calls for more observation to be carried out to validate the S trend with altitude.

4. Conclusions and Discussion

Previous studies have focused mainly on TCs over open ocean areas, and few studies have investigated turbulent mixing processes in the high u_n over the land. In this study, high-frequency wind data were examined using eddy–covariance and flux–gradient methods, which were observed at multiple heights (10–110 m) on the two flux towers in a coastal zone. Small-scale turbulence characteristics were analyzed at high wind speeds of up to 42.5 m s^{-1} in the near-surface layer of Super Typhoon Maria. The focus was on investigating the characteristics of the turbulent vertical transport and key parameters ($|\tau|$, TKE, VED, L_m, and S) under high wind speeds (>30 m s^{-1}) at five heights. The main conclusions are as follows:

$|\tau|$, TKE, K_m, and S increased with increasing u_n under low and moderate wind conditions, and these decreased with increasing u_n at high wind speeds (>30 m s^{-1}). The saturated wind speed threshold for the change of $|\tau|$, TKE, and S were about 23 m s^{-1} when regressed from the medium numbers. Furthermore, our results also imply that the saturated wind speed for the change of τ, TKE, and K_m increased with height. The maximum of K_m and S increased from the surface to a maximum value at a height of 50 m, and then decreased with greater heights. Additionally, at any given u_n, $|\tau|$, TKE, and K_m were larger than over oceanic areas, and K_m was about one to two orders of magnitude bigger than K_t, which may be attributable to differences in surface roughness.

Furthermore, the variations of turbulent parameters with radius in landfalling TC were also analyzed. The $|\tau|$, TKE, and K_m values increased with increasing u_n outside 3 RMW from the TC center but decreased slowly with increasing u_n located near 1–3 RMW from the TC center. High $|\tau|$, TKE, and K_m were basically constant with height at high u_n of 20.0–42.5 m s^{-1} within 2–3 RMW from the TC center before typhoon landfall, which indicated turbulent mixing fully suggesting nearly constant flux layers. As compared with K_m, K_t with about one to two orders of magnitude smaller than K_m showed a clear decrease with increasing wind speed at 1–3 RMW, but little dependence on wind speed outside 3 RMW from the TC center.

The spatial variations in L_m values with u_n were different outside 3 RMW and within 1–3 RMW from the TC center. In the region of 1–3 RMW before typhoon landfall, L_m increased with u_n and then began to decrease when u_n exceeded 38 m s^{-1}. With increasing height from 30 m to 130 m, S peaked at 0.22, 0.51, 0.38, and 0.21 with the corresponding wind speeds being 24.8, 27.0, 22.3, 33.2, and 41.3 m s^{-1}, respectively. The relationships between $|\tau|$, TKE, K_m, L_m, and u_n changed with height and radii from the TC center should be accounted for in sub-grid scale physical processes of momentum fluxes in numerical TC models. It will provide guidance for advance the PBL schemes in improving hurricane intensity forecasts on landfalling and their evolution, especially over land.

In our observation, the analyses showed that the magnitudes of the eddy diffusivity for sensible heat flux was much smaller than those for momentum, the shapes of the vertical distributions of the eddy diffusivities were increasing from the surface to a maximum value, then decreasing with height in the surface layer. The results were consistent with those of Zhang et al. [13] above 500 m height by the CBLAST results, however, it was different from that in earlier theoretical and numerical models. It is typically assumed that the value of K_t is equal to that of K_m, or it is calculated using Prandtl and Schmidt numbers (P_r) [42]. In future research, in addition to the momentum flux, more observational data of the heat and moisture fluxes in high winds need to be analyzed, to evaluate the characteristic of Pr to advance our understanding of the air–sea momentum and latent heat exchange under extreme wind conditions [43–48].

Although the EC technique is still the most reliable direct method, clearly, turbulence observations are still too limited in the high-wind boundary layer at this point. The vertical variation in K_m with u_n was also consistent with those of Tang et al. [24] using eddy-covariance methods and Zhao et al. [22] using air-craft observation. However, the magnitude was much smaller than that reported by using sodium radar measurements [49]. The TKE was almost constant at a distance of 2–3 times the RMW before landfall, while the TKE increased with decreasing distance to the typhoon center found in the meteorology radar from Shi et al. [50] before Typhoon Maria's landfall. There still needs to be more turbulence observations and to couple the remotely sensed results to verify the process under extreme wind conditions.

Funding: This research was funded by the Shanghai Municipal Natural Science Foundation with grant number 22ZR1476400, the Key Program for International Science and Technology Cooperation Projects of China with grant number 2017YFE0107700, the National Key Research and Development Program of China with grant number 2020YFE0201900, and the Research Program from the Science Foundation of China with grant numbers 41475060 and 41775065. This research was also supported by the ESCAP/WMO Project (EXOTICCA), the National Natural Science Foundation of China with grant number 42075056, and the Shanghai Science and Technology Research Program with grant number 19dz1200101.

Institutional Review Board Statement: Not applicable.

Informed Consent Statement: Not applicable.

Data Availability Statement: Not applicable.

Acknowledgments: The observational data in this study were from a typhoon experiment observed by Jie Tang, Bingke Zhao, Shuai Zhang and Limin Lin in the Shanghai Typhoon Institute, and we thank the editors and anonymous reviewers for their helpful comments that greatly improved the manuscript.

Conflicts of Interest: The authors declare no conflict of interest.

References

1. Emanuel, K.A. An air–sea interaction theory for tropical cyclones. Part I: Steady-state maintenance. *J. Atmos. Sci.* **1986**, *43*, 585–604. [CrossRef]
2. Smith, R.K.; Montgomery, M.T.; Sang, N.V. Tropical cyclone spin-up revisited. *Q. J. R. Meteorol. Soc.* **2009**, *135*, 1321–1335. [CrossRef]
3. Gopalakrishnan, S.G.; Marks, F.; Zhang, J.A.; Zhang, X.; Bao, J.-W.; Tallapragada, V. A study of the impacts of vertical diffusion on the structure and intensity of tropical cyclones using the high resolution HWRF system. *J. Atmos. Sci.* **2013**, *70*, 524–541. [CrossRef]
4. Gopalakrishnan, S.G.; Koch, D.; Upadhayay, S.; DeMaria, M.; Marks, F.; Rappaport, E.N.; Mehra, A.; Tallapragada, V.; Jung, Y.; Alaka, Jr.; et al. *2019 HFIP R&D Activities Summary: Recent Results and Operational Implementation*; HFIP Technical Report: HFIP 2018-1.2020; National Oceanic and Atmospheric Administration United States Department of Commerce: Washington, DC, USA, 2019. Available online: http://www.hfip.org/documents/HFIP_AnnualReport_FY2019.pdf (accessed on 13 May 2022).
5. Zhang, J.A.; Drennan, W.M. An observational study of vertical eddy diffusivity in the hurricane boundary layer. *J. Atmos. Sci.* **2012**, *69*, 3223–3236. [CrossRef]
6. Zhu, P. Simulation and parameterization of the turbulent transport in the hurricane boundary layer by large eddies. *J. Geophys. Res.* **2008**, *113*, D17104. [CrossRef]
7. Bao, J.-W.; Gopalakrishnan, S.G.; Michelson, S.A.; Marks, F.D.; Montgomery, M.T., Jr. Impact of physics representations in the HWRFX on simulated hurricane structure and pressure–wind relationships. *Mon. Wea. Rev.* **2012**, *140*, 3278–3299. [CrossRef]
8. Kepert, J.D. Choosing a boundary layer parameterization for tropical cyclone modeling. *Mon. Wea. Rev.* **2012**, *140*, 1427–1445. [CrossRef]
9. Zhu, P.; Menelaou, K.; Zhu, Z. Impact of sub-grid scale vertical turbulent mixing on eyewall asymmetric structures and mesovortices of hurricanes. *Q. J. R. Meteorol. Soc.* **2013**, *140*, 416–438. [CrossRef]
10. Zhang, J.A.; Nolan, D.S.; Rogers, R.F.; Tallapragada, V. Evaluating the impact of improvements in the boundary layer parameterizations on hurricane intensity and structure forecasts in HWRF. *Mon. Wea. Rev.* **2015**, *143*, 3136–3155. [CrossRef]
11. Zhang, J.A.; Zhu, P. Effects of vertical eddy diffusivity parameterization on the evolution of landfalling hurricanes. *J. Atmos. Sci.* **2017**, *74*, 1879–1905. [CrossRef]

12. Ming, J.; Zhang, J.A.; Rogers, R.F. Typhoon kinematic and thermodynamic boundary layer structure from dropsonde composites. *J. Geophys. Res. Atmos.* **2015**, *120*, 3158–3172. [CrossRef]
13. Zhang, J.A.; Rogers, R.F.; Tallapragada, V. Impact of parameterized boundary layer structure on tropical cyclone rapid intensification forecasts in HWRF. *Mon. Wea. Rev.* **2017**, *145*, 1413–1426. [CrossRef]
14. Li, X.; Pu, Z. Vertical eddy diffusivity parameterization based on a large-eddy simulation and its impact on prediction of hurricane landfall. *Geophys. Res. Lett.* **2020**, *48*, e2020GL090703. [CrossRef]
15. Zhu, P.; Hazelton, A.; Zhang, Z.; Marks, F.D.; Tallapragada, V. The role of eyewall turbulent transport in the pathway to intensification of tropical cyclones. *J. Geophys. Res. Atmos.* **2021**, *126*, e2021JD034983. [CrossRef]
16. Li, Y.; Zhu, P.; Gao, Z.; Cheung, K.K. Sensitivity of large eddy simulations of tropical cyclone to sub-grid scale mixing parameterization. *Atoms. Res.* **2022**, *265*, 105922. [CrossRef]
17. Nolan, D.S.; Zhang, J.A.; Stern, D.P. Evaluation of planetary boundary layer parameterizations in tropical cyclones by comparison of in-situ data and high resolution simulations of Hurricane Isabel (2003). Part I: Initialization, maximum winds, and outer core boundary layer structure. *Mon. Wea. Rev.* **2009**, *137*, 3651–3674. [CrossRef]
18. Kepert, J.D. The dynamics of boundary layer jets within the tropical cyclone core. Part I: Linear theory. *J. Atmos. Sci.* **2001**, *58*, 2469–2484. [CrossRef]
19. Zhang, J.A.; Marks, F.D.; Montgomery, M.T.; Lorsolo, S. An estimation of turbulent characteristics in the low-level region of intense Hurricanes Allen (1980) and Hugo (1989). *Mon. Wea. Rev.* **2011**, *139*, 1447–1462. [CrossRef]
20. Gopalakrishnan, S.; Hazelton, A.; Zhang, J.A. Improving hurricane boundary layer parameterization scheme based on observations. *Earth Space Sci.* **2021**, *8*, e2020EA001422. [CrossRef]
21. Black, P.G.; Asaro, E.A.D.; Sanford, T.B.; Drennan, W.M.; Zhang, J.A.; French, J.R.; Nüler, P.P.; Terrill, E.J.; Walsh, E.J. Air–sea exchange in hurricanes: Synthesis of observations from the Coupled Boundary Layer Air–Sea Transfer experiment. *B. Meteorol. Soc.* **2007**, *88*, 357–374. [CrossRef]
22. Zhao, Z.K.; Chan, P.W.; Wu, N.; Zhang, J.A.; Hon, K.K. Aircraft observations of turbulence characteristics in the tropical cyclone boundary layer. *Bound. Layer Meteorol.* **2020**, *174*, 493–511. [CrossRef]
23. Katz, J.; Zhu, P. Evaluation of surface layer flux parameterizations using in-situ observations. *Atmos. Res.* **2017**, *194*, 150–163. [CrossRef]
24. Tang, J.; Zhang, J.A.; Aberson, S.D.; Marks, F.D.; Lei, X. Multilevel tower observations of vertical eddy diffusivity and mixing length in the tropical cyclone boundary layer during landfalls. *J. Atmos. Sci.* **2018**, *75*, 3159–3168. [CrossRef]
25. Zhao, Z.; Gao, R.; Zhang, J.A.; Zhu, Y.; Liu, C.; Chan, P.W.; Wan, Q. Observations of Boundary Layer Wind and Turbulence of a Landfalling Tropical Cyclone. 20 August 2021, PREPRINT (Version 1) available at Research Square. Available online: https://www.researchgate.net/publication/354043506_Observations_of_Boundary_Layer_Wind_and_Turbulence_of_a_Landfalling_Tropical_Cyclone (accessed on 28 May 2022).
26. Li, L.; Xiao, Y.; Zhou, H.; Xing, F.; Song, L. Turbulent wind characteristics in typhoon Hagupit based on field measurements. *Int. J. Distrib. Sens. Netw.* **2018**, *14*. [CrossRef]
27. Xia, D.; Dai, L.; Lin, L.; Wang, H.; Hu, H. A Field Measurement based Wind Characteristics Analysis of a Typhoon in Near-ground Boundary Layer. *Atmosphere* **2021**, *12*, 873. [CrossRef]
28. Lee, X.; Massman, W.; Law, B. *Handbook of Micrometeorology: A Guide for Surface Flux Measurement and Analysis*; Kluwer Academic Publishers: New York, NY, USA, 2004; pp. 33–66.
29. Qin, Z.; Xia, D.; Dai, L.; Zheng, Q.; Lin, L. Investigations on wind characteristics for typhoon and monsoon wind speeds based on both stationary and non-stationary models. *Atmosphere* **2022**, *13*, 178. [CrossRef]
30. Ying, M.; Zhang, W.; Yu, H.; Lu, X.; Feng, J.; Fan, Y.; Zhu, Y.; Chen, D. An overview of the China Meteorological Administration tropical cyclone database. *J. Atmos. Ocean Tech.* **2014**, *31*, 287–301. [CrossRef]
31. Bao, X.; Wu, L.; Zhang, S.; Li, Q.; Lin, L.; Zhao, B.; Wu, D.; Xia, W.; Xu, B. Distinct raindrop size distributions of convective inner- and outer-rainband rain in Typhoon Maria (2018). *J. Geophys. Res. Atmos.* **2020**, *125*, e2020JD032482. [CrossRef]
32. Stull, R.B. *An Introduction to Boundary Layer Meteorology*; Kluwer Academic Publishers: Dordrecht, The Netherlands, 1988; p. 670.
33. Schmid, H.P.; Grimmond, C.S.B.; Cropley, F.; Offerle, B.; Su, H.B. Measurements of CO_2 and energy fluxes over a mixed hardwood forest in the mid-western United States. *Agric. Forest Meteorol.* **2000**, *103*, 357–374. [CrossRef]
34. Foken, T.; Göockede; Mauder, M.; Mahrt, L.; Amiro, B.; Munger, W. *Handbook of Micrometeorology: Post-Field Data Quality Control*; Lee, X., Massman, W., Law, B., Eds.; Springer: Dordrecht, The Netherlands, 2005; pp. 181–208.
35. Monin, A.S.; Obukhov, A.M. Basic laws of turbulent mixing in the surface layer of the atmosphere. *Akad. Nauk. SSSR. Geofiz. Inst. Trudy* **1954**, *151*, 163–187.
36. Fortuniak, K.; Pawlak, W. Selected Spectral Characteristics of Turbulence over an Urbanized Area in the Centre of Łód'z, Poland. *Bound. Layer Meteorol.* **2015**, *154*, 137–156. [CrossRef]
37. Kaimal, J.C.; Wyngaard, J.C.; Coté, O.R. Spectral characteristics of surface-layer turbulence. *Q.J.R. Meteorol. Soc.* **1972**, *98*, 563–589. [CrossRef]
38. Ortiz-Suslow, D.G.; Kalogiros, J.; Yamaguchi, R.; Wang, Q. An evaluation of the constant flux layer in the atmospheric flow above the wavy air-sea interface. *J. Geophys. Res. Atmos.* **2021**, *126*, e2020JD032834. [CrossRef]
39. Kaimal, J.C.; Finnigan, J.J. *Atmospheric Boundary Layer Flows: Their Structures and Measurements*; Oxford University Press: Oxford, UK, 1994.

40. Grachev, A.A.; Leo, L.S.; Fernando, H.J.S.; Fairall, C.W.; Creegan, E.; Blomquist, B.W.; Christman, A.J.; Hocut, C.M. Air-sea/land interaction in the coast. *Bound. Layer Meteorol.* **2018**, *167*, 181–210. [CrossRef]
41. Grachev, A.A.; Krishnamurthy, R.; Fernando, H.J.S.; Fairall, C.W.; Bardoel, S.L.; Wang, S. Atmospheric turbulence measurements at a coastal zone with and without Fog. *Bound. Layer Meteorol.* **2021**, *181*, 395–422. [CrossRef]
42. Zhang, J.A.; Rogers, R.F. Effects of parameterized boundary layer structure on hurricane rapid intensification in shear. *Mon. Wea. Rev.* **2019**, *147*, 853–871. [CrossRef]
43. Lükö, G.; Torma, P.; Weidinger, T. Intra-Seasonal and Intra-Annual Variation of the Latent Heat Flux Transfer Coefficient for a Freshwater Lake. *Atmosphere* **2022**, *13*, 352. [CrossRef]
44. Dudorova, N.V.; Belan, B.D. The Energy Model of Urban Heat Island. *Atmosphere* **2022**, *13*, 457. [CrossRef]
45. Yang, L.; Qian, F.; Song, D.-X.; Zheng, K.-J. Research on Urban Heat-Island Effect. *Procedia Eng.* **2016**, *169*, 11–18. [CrossRef]
46. Kalina, E.A.; Biswas, M.K.; Zhang, J.A.; Newman, K.M. Sensitivity of an Idealized Tropical Cyclone to the Configuration of the Global Forecast System–Eddy Diffusivity Mass Flux Planetary Boundary Layer Scheme. *Atmosphere* **2021**, *12*, 284. [CrossRef]
47. Do Nascimento, A.C.L.; Galvani, E.; Gobo, J.P.A.; Wollmann, C.A. Comparison between Air Temperature and Land Surface Temperature for the City of São Paulo, Brazil. *Atmosphere* **2022**, *13*, 491. [CrossRef]
48. Elmarakby, E.; Khalifa, M.; Elshater, A.; Afifi, S. Tailored methods for mapping urban heat islands in Greater Cairo Region. *Ain Shams Eng. J.* **2021**. [CrossRef]
49. Li, J.; Collins, R.; Lu, X.; Williams, B. Lidar observations of instability and estimates of vertical eddy diffusivity induced by gravity wave breaking in the Arctic mesosphere. *J. Geophys. Res. Atmos.* **2021**, *126*, e2020JD033450. [CrossRef]
50. Shi, W.; Tang, J.; Chen, Y.; Chen, N.; Liu, Q.; Liu, T. Study of the Boundary Layer Structure of a Landfalling Typhoon Based on the Observation from Multiple Ground-Based Doppler Wind Lidars. *Remote Sens.* **2021**, *13*, 4810. [CrossRef]

atmosphere

MDPI

Article

Parameterization of Sea Surface Drag Coefficient for All Wind Regimes Using 11 Aircraft Eddy-Covariance Measurement Databases

Zhiqiu Gao [1,2,*], Shaohui Zhou [1], Jianbin Zhang [1], Zhihua Zeng [3] and Xueyan Bi [4]

[1] Climate and Weather Disasters Collaborative Innovation Center, Key Laboratory for Aerosol-Cloud-Precipitation of China Meteorological Administration, School of Atmospheric Physics, Nanjing University of Information Science and Technology, Nanjing 210044, China; 20191203039@nuist.edu.cn (S.Z.); 20201903002@nuist.edu.cn (J.Z.)
[2] State Key Laboratory of Atmospheric Boundary Layer Physics and Atmospheric Chemistry, Institute of Atmospheric Physics, Chinese Academy of Sciences, Beijing 100029, China
[3] Shanghai Typhoon Institute, China Meteorological Administration, Shanghai 200030, China; zengzh@typhoon.org.cn
[4] Guangzhou Institute of Tropical and Marine Meteorology, China Meteorological Administration, Guangzhou 510640, China; xybi@gd121.cn
* Correspondence: zgao@nuist.edu.cn

Citation: Gao, Z.; Zhou, S.; Zhang, J.; Zeng, Z.; Bi, X. Parameterization of Sea Surface Drag Coefficient for All Wind Regimes Using 11 Aircraft Eddy-Covariance Measurement Databases. *Atmosphere* **2021**, *12*, 1485. https://doi.org/10.3390/atmos12111485

Academic Editor: Massimiliano Burlando

Received: 30 August 2021
Accepted: 29 October 2021
Published: 10 November 2021

Abstract: The drag coefficient is essential for calculating the aerodynamic friction between air and sea. In this study, we regress a set of relationships between the drag coefficient and the wind speed for different wind ranges using an observational dataset that consists of 5941 estimates of the mean flow and fluxes from 11 aircraft turbulent measurements over the sea surface. Results show that: (1) the drag coefficient is a power function of wind speed over smooth sea surface when it is no greater than 4.5 ms^{-1}, and the drag coefficient decreases with the increase of wind speed; and (2) for rough sea surface, when the wind speed is greater than 4.5 ms^{-1} and less than or equal to 10.5 ms^{-1}, the drag coefficient increases linearly with the increase of horizontal wind speed; when the wind speed is greater than 10.5 ms^{-1} and less than or equal to 33.5 ms^{-1}, the drag coefficient changes parabolically with the increase of wind speed; when the wind speed is greater than 33.5 ms^{-1}, the drag coefficient is constant. Additionally, regressed from drag coefficient, the saturated wind speed threshold is 23 ms^{-1}. Parameterizations of turbulent heat transfer coefficient (C_h) and water vapor transfer coefficient (C_e) are also investigated.

Keywords: drag coefficient; wind speed; aircraft turbulent measurements; saturated wind speed

1. Introduction

The wind stress acting on the sea surface is important for both the structure of the atmospheric boundary layer over the sea as well as the movements of ocean mixed layer and ocean currents [1]. The wind stress depends not only on the wind speed, but also on the drag coefficient (C_d) or the aerodynamic roughness length (z_0) [2,3]. Over the years, there has been great interest and effort in the community to parameterize z_0 for a wide range of wind conditions [4]. Traditionally, in very low wind environments, sea surface is aerodynamically smooth, and the roughness Reynolds number is approximately 0.11, so that z_0 is given by Equation (1).

$$z_0 = 0.11\upsilon/u_* \tag{1}$$

where υ is the kinematic viscosity of air, and u_* is the surface friction velocity, $u_* = \left[\left(\overline{w'v'}\right)^2 + \left(\overline{w'u'}\right)^2\right]^{1/4}$, where w', v', and u' are the perturbations in vertical velocity, meridional wind, and zonal wind. The roughness Reynolds number $\mathrm{Re}_* \equiv u_* z_0/\upsilon$. When the

10-m wind speed u_{10} is greater than 5.5 ms^{-1}, correspondingly, u_* is greater than 0.23 ms^{-1}, and Re_* is greater than 2, and sea surface becomes fully rough flow while z_0 is given by Equation (2).

$$z_0 = a_c u_*^2 / g \tag{2}$$

where a_c is the Charnock's constant, and a_c = 0.012, and g is the acceleration due to gravity, which characterizes the equilibrium between wind and waves, with the spectrum of gravity waves acting as roughness elements. The transitional regime between smooth and fully rough sea surfaces corresponds to the 10-m wind speed ranging from 2.5 to 5.5 ms^{-1}. Once z_0 is determined from u_*, the neutral drag coefficient (C_{dn}) can be obtained for smooth surface by

$$C_{dn} = k^2 / [\ln(u_* z / 0.11 v)]^2 \tag{3}$$

and for rough surface by

$$C_{dn} = k^2 / \left[\ln\left(zg / a_c u_*^2\right)\right]^2 \tag{4}$$

where k is the von Karman constant, and z is the measurement height.

Much effort has also gone into investigating the variation of C_{dn} with 10-m wind speed, particularly under high wind speed regime. Recently, it was found that drag coefficient increases with the 10-m wind speed under moderate wind speed and decreases with the 10-m wind speed under high wind speed (Table 1). However, the saturated wind speed thresholds obtained by different studies vary greatly, as summarized in Table 1. The lack of consensus on the magnitude of the saturated wind speed poses a challenge in modeling applications.

Table 1. Values of the saturated 10-m wind speed (ms^{-1}) from the literature.

Reference	Saturated Wind Speed (ms^{-1})	Comments
Alamaro et al. [6]	35	Laboratory annular wind wave tank.
Powell et al. [7]	33	Global Positioning System sonde observations in tropical cyclone environments.
Donelan et al. [8]	33	Laboratory extreme wind experiments.
Makin [9]	30–40	Solving the turbulent kinetic energy balance equation for airflow under the limited saturation (by suspended sea-spray droplets) regime.
Black et al. [10]	23	The Coupled Boundary Layer Air-Sea Transfer (CBLAST) Experiment.
Troitskaya et al. [11]	25	Theoretically and laboratory experiment.
Soloviev et al. [12]	30	The unified wave-form and two-phase parameterization model.
Donelan [13]	30	Laboratory extreme wind experiments.
Gao et al. [5]	22–23	Mathematical regression for aircraft eddy-covariance measurements over the tropical Eastern Pacific.

Following Gao et al. [5], we directly used the wind speed measured by aircraft in Equations (3) and (4), and did not convert the wind speed measured by the aircraft to the wind speed at a height of 10 m, since the logarithmic wind profile hypothesis and the constant flux layer hypothesis over the layer may bring additional errors.

Previous studies usually used a single or fewer databases for regression tests where the wind speed usually could not cover the wind speed range from calm, light wind to high wind, and a limited number of samples usually reduce the representativeness of regression results. Unlike most of the prior studies, the motivation of this study is to use a large sample database to obtain more convincing regression results, and to provide a simple parameterization that is more representative and may be used in a fully coupled (atmosphere–wave–ocean) hurricane prediction model. To achieve this goal, we developed a new set of equations of the sea surface drag coefficient (C_d) that are solely dependent on wind speed and over a full range of wind speeds. We also used a large dataset of eddy-covariance turbulence flux measurements over the sea surfaces, which are collected from 11 different experiments from 1992 to 2008 on four different aircrafts. There are 5941 sample data in total.

2. Database

The Oregon State University aircraft marine boundary layer air-sea dataset described in detail by Vickers et al. [14] is used in this study, and the locations of the 11 aircraft experiments are shown in Figure 1. The horizontal averaging segment is 4 km at the typical aircraft speed, and the maximum (4 km average) aircraft altitude is 50 m. This dataset consists of turbulent wind speed, air temperature, and water vapor collected by the following:

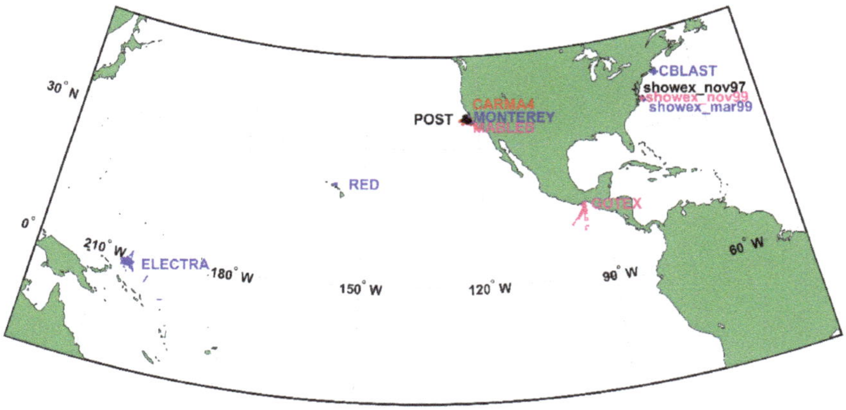

Figure 1. The 11 aircraft experiments' locations.

(1) The Long-EZ aircraft of the National Oceanic and Atmospheric Administration (NOAA) during the four experiments: (1) the pilot program of the Coupled Boundary Layers and Air Sea Transfer experiment (CBLAST Weak Wind) conducted during July–August 2001 over the Atlantic Ocean south of Martha's Vineyard Island, MA [15]; (2) the Shoaling Waves experiment (SHOWEX) over the Atlantic east of the Outer Banks near Duck, NC during November and December 1999 [16]; (3) the SHOWEX pilot study in November 1997; and (4) the SHOWEX pilot study in March 1999.

(2) The Naval Postgraduate School's Center for Interdisciplinary Remotely-Piloted Aircraft Studies (CIRPAS) Twin Otter aircraft in five experiments: (1) outside Monterey Bay off the coast of California during the Cloud-Aerosol Research in the Marine Atmosphere IV experiment (CARMAIV) in August 2007; (2) outside Monterey Bay (Monterey) during April 2008 [17]; (3) the Rough Evaporation Duct experiment (RED) during August–September of 2001 to the east (windward side) of Oahu in the Hawaiian Islands [18]; (4) the Marine Atmospheric Boundary Layer Energy Budget (MABLEB) experiment during April 2007; and (5) the Physics Of Stratocumulus Top during July–August 2008 (POST).

(3) The C-130 Hercules aircraft of the National Center for Atmospheric Research (NCAR) in the Gulf of Tehuantepec Experiment (GOTEX) in February 2004 on the Pacific coast of the Isthmus of Tehuantepec, Mexico [19], and the data collected by the NCAR Electra aircraft in TOGA COARE during November 1992 to February 1993 in the Pacific warm pool [20].

3. Results

3.1. Variation of Friction Velocity (u_*) against Wind Speed

Figure 2 shows the friction velocity against the wind speed (U) for all 11 experiments. u_* is derived from the fast measurements of three-dimensional wind speed. Overall, u_* increased with increasing U. Figure 2 shows that most of the data were collected with a wind speed of less than 17 ms^{-1} during the 11 experiments. Almost all of the data with wind speeds greater than 17 ms^{-1} were collected by the C-130 Hercules aircraft of National Center for Atmospheric Research (NCAR) in February 2004, in the Gulf of Tehuantepec Experiment (GOTEX) which took place near the coast area of Isthmus of Tehuantepec, Mexico [19]. The correlation coefficient between u_* and U is 0.90.

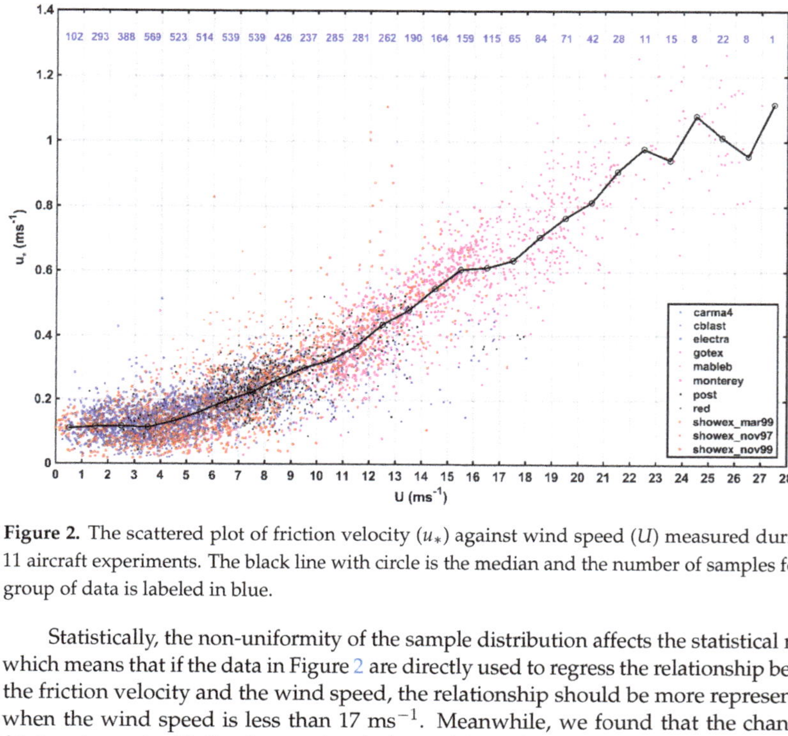

Figure 2. The scattered plot of friction velocity (u_*) against wind speed (U) measured during the 11 aircraft experiments. The black line with circle is the median and the number of samples for each group of data is labeled in blue.

Statistically, the non-uniformity of the sample distribution affects the statistical results, which means that if the data in Figure 2 are directly used to regress the relationship between the friction velocity and the wind speed, the relationship should be more representative when the wind speed is less than 17 ms^{-1}. Meanwhile, we found that the changes in frictional speed with the changes in wind speed from the 11 experiments were consistent, so we combined the observational data from those experiments. With an interval of 1 ms^{-1}, these data were categorized into 28 bins, and the number of samples in each bin was shown at the top of Figure 2.

3.2. Parameterizations of Turbulent Drag Coefficient (C_d)

Figure 3I shows the drag coefficient (C_d) against wind speed (U) for all of the 11 experiments. C_d is obtained by taking the square of the ratio u_* over U. If the wind speed is less than 4.5 ms^{-1}, the air–sea interaction is mainly achieved by viscous motion. We regressed the relationship between C_d and U for all of the 11 experiments in Equation (5). The standard error of the estimate (SEE) of C_d is 1.4×10^{-7}.

$$C_d = 0.0113/U^{1.785} \tag{5}$$

when the wind speed is higher than 4.5 ms^{-1}, C_d increased with increasing U. By using the median numbers of these data bins, the relationship between the drag coefficient C_d and U is regressed as

$$C_d = a(U - b)^2 + c \tag{6}$$

where a, b, and c are regression coefficients. C_d is found to have a parabolic relationship with using these 11 different regression tests:

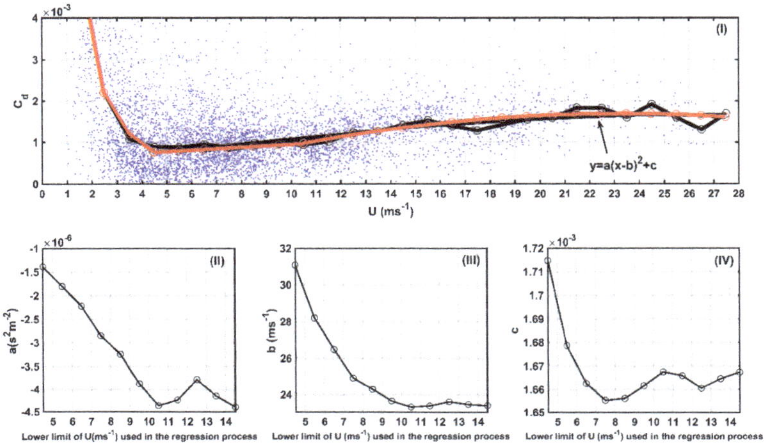

Figure 3. (**I**) Similar to Figure 2, but for drag coefficient. The black line with circles is the median number line. The red lines are the parabolic regression lines for all for all 11 tests. (**II–IV**) are variations of parabolic regression coefficients (a, b, and c) against the lower limit of wind speed in the regression process. The red line with dots is regression line segments.

Test #1: U are median numbers of bins: 4–5 ms^{-1}, 5–6 ms^{-1}, ... , and 27–28 ms^{-1};
Test #2: U are median numbers of bins: 5–6 ms^{-1}, 6–7 ms^{-1}, ... , and 27–28 ms^{-1};
Test #11: U are median numbers of bins: 14–15 ms^{-1}, 16–17 ms^{-1}, ... , and 27–28 ms^{-1}.

The regression coefficients (a, b, and c) obtained during 11 different tests are shown in Figure 3II–IV, respectively. The value of (a) reflects the nonlinearity of the parabola. While the critical (or saturated) wind speed is indicated by the value of b in Equation (6), and at the critical wind speed C_d reaches its local maximum. Figure 3III shows that the values of b range from 31.07 to 23.30 ms^{-1} and gradually converge as the lower limit of the wind speed range corresponding to the regression data continues to increase. The results of b obtained here are not only consistent with the results of many previous studies, but they also explain the great difference in the critical wind speeds from previous studies. The maximum critical wind speed we obtained (corresponding to test #1) is 31.07 ms^{-1}, which is slightly smaller than those from previous studies [6–8]. A possible reason is that the data in this work have wind speed generally smaller than 28 ms^{-1}, while the relatively lower wind speed range may affect the regression and bring in uncertainty. In this study, we calculated the drag coefficient directly from the wind speed measured by aircrafts, and we did not convert the wind speed measured by the aircrafts to the wind speed at a height of 10 m, since the conversion may introduce additional errors. Recently, some studies also mentioned the height conversion of wind speed. For example, Mahrt et al. [21] used the data collected during two Floating Instrument Platform field campaigns and the data collected at the Air–Sea Interaction Tower site, to investigate the relationship between the wind and sea surface stress for contrasting conditions. They found that the sea surface wind stress decreased significantly with height near the surface under thin marine boundary layers and/or enhanced stress divergence close to the sea surface conditions. It is worth noting that b converges when the lower limit of the wind speed range corresponding to the regression process is about 10.5 ms^{-1}. Therefore, 10.5 ms^{-1} may be a threshold value, that is, when the wind speed is less than 10.5 ms^{-1}, C_d is a linear function of wind speed; when the wind speed is greater than 10.5 ms^{-1}, C_d is a parabolic function of wind speed. Figure 3IV shows that the value of c ranges from 1.715×10^{-3} to 1.655×10^{-3}. It is lower than previous values because we calculated the drag coefficient directly from the wind speed measured by the aircraft rather than using the 10-m wind speed. In summary, when the wind speed is less than 10.5 ms^{-1}, we recommend using a linear equation to approximate the relationship between C_d and U. Following Garratt [1], we regressed the relationship between C_d and U for case: 4.5 $ms^{-1} < U \leq 10.5$ ms^{-1} for all of the 11 experiments in Equation (6). The standard error of the estimate (SEE) of C_d is 5.3×10^{-9}. We plotted the values of C_d estimated by Equations (5) and (7) onto Figure 3I (red line with circles).

$$C_d = 3.5 \times 10^{-5} \times U + 0.6 \times 10^{-3} \tag{7}$$

For wind speeds greater than 10.5 ms^{-1}, we recommend using the following equation to approximate it.

$$C_d = -4.4 \times 10^{-6} \times (U - 23)^2 + 1.7 \times 10^{-3} \tag{8}$$

Equation (8) implies that C_d will be negative when $U > 43$ ms^{-1}. Unfortunately, we have never observed such high wind speeds, so it is difficult to define the upper limit of the applicable wind speed scope of Equation (8). For the sake of brevity, and considering the symmetry of parabola, we recommend that modelers use the C_d value (1.20×10^{-3}) when the wind speed is 10.5 ms^{-1} as the C_d value when the wind speed is greater than or equal to 33.5 ms^{-1}. It is obvious that, since the maximum wind speed in the database we used is 27.05 ms^{-1}, the Equation (8) under conditional wind speed greater than 27 ms^{-1} is purely mathematical, lacking the support of observational data.

3.3. Parameterizations of Turbulent Heat Transfer Coefficient (C_h) and Water Vapor Transfer Coefficient (C_e)

We regressed the relationship between C_h and U for the wind speed less than or equal to 4.5 ms^{-1} for all of the 11 experiments in Equation (9). SEE of C_h is 6.78×10^{-8}.

$$C_h = 0.00229/U^{0.96} \tag{9}$$

and a linear relationship between C_h and U for the wind speed higher than 4.5 ms^{-1} and less than or equal to 10.5 ms^{-1} for all of the 11 experiments is expressed in Equation (10). SEE of C_h is 6.36×10^{-8}.

$$C_h = 7.35 \times 10^{-5} \times U + 0.19 \times 10^{-3} \tag{10}$$

We plotted the values of C_h estimated by Equations (9) and (10) onto Figure 4a (red line with circles). When 10.5 $ms^{-1} < U \leq 23$ ms^{-1}, C_h is almost a constant (9.39×10^{-4}) and When $U > 23$ ms^{-1}, C_h is almost a constant (3.25×10^{-4}). We also plot these two values onto Figure 4a.

The variation of turbulent heat transfer coefficient (C_h) and water vapor transfer coefficient (C_e) along with wind speed (U) are presented in Figure 4. It can be seen from Figures 3I and 4a that C_h is distributed more dispersedly than C_d. This is because that not only the dynamic process but also the thermal process affect the turbulent heat transfer, which lead to more complexity and uncertainty. We plotted the values of C_h estimated by Equation (9) onto Figure 4a (red line with circles). Unlike Figure 4a,b shows a more concentrated distribution of turbulent water vapor transport coefficients, which is also a result of the assumed saturated surface water vapor in the C_e calculation. We also regressed the relationship between C_e and U for the wind speed less than 4.5 ms^{-1} for all of the 11 experiments in Equation (10). SEE of C_e is 2.66×10^{-9}. We plotted the values of C_e estimated by Equation (11) onto Figure 4b (red line with circles)

$$C_e = 0.0008/U^{0.76} \tag{11}$$

When $U > 4.5$ ms^{-1}, C_e is almost a constant (3.4×10^{-4}), which is also shown in Figure 4b.

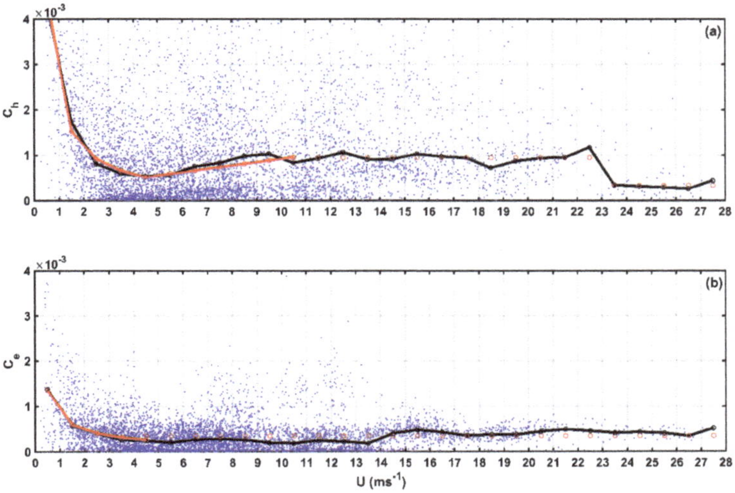

Figure 4. Similar to Figure 3I, but for turbulent heat transfer coefficient (**a**) and turbulent water vapor transfer coefficient (**b**). The red line with circles is regression results.

Alamaro et al. [6] concluded that the value of C_k/C_d strongly affected hurricane intensity. Strictly speaking, C_k is not the sum of C_h and C_e, but since we only have sensible heat flux and latent heat flux data, and lack turbulence original observation data, here, we roughly take C_k as the sum of C_h and C_e. We illustrate enthalpy transfer coefficient C_k ($\equiv C_h + C_e$) and C_d in Figure 5a, and the variations of C_k/C_d with wind speed are shown in Figure 5b. It can be seen from Figures 4 and 5 that when U reaches at 23 ms^{-1} C_h suddenly decreases and keeps almost invariant after. However, variations of C_h with wind speed (especially a sudden drop) under strong wind conditions are rarely shown in previous studies. Figure 5b shows $C_k/C_d > 1$ under moderate wind speed and $C_k/C_d < 1$ under low and high wind speed.

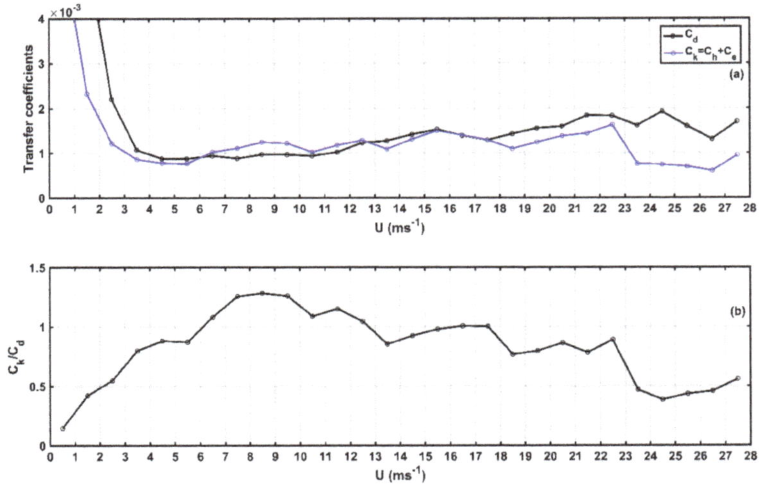

Figure 5. (**a**) The median number variations of drag coefficient (C_d) and enthalpy transfer coefficient C_k against wind speed (U) measured during the 11 aircraft experiments; (**b**) The value of C_k/C_d against wind speed (U) measured during the 11 aircraft experiments.

4. Conclusions

We derived a suite of functions of drag coefficient, heat transfer coefficient, and water vapor transfer coefficient with respect to wind speed under all wind conditions as follows:

$$C_d = \begin{cases} 0.0113/U^{1.785} & U \leq 4.5\,\text{ms}^{-1} \\ 3.5 \times 10^{-5} \times U + 0.6 \times 10^{-3} & 4.5\,\text{ms}^{-1} < U \leq 10.5\,\text{ms}^{-1} \\ -4.4 \times 10^{-6} \times (U-23)^2 + 1.7 \times 10^{-3} & 10.5\,\text{ms}^{-1} < U \leq 33.5\,\text{ms}^{-1} \\ 1.20 \times 10^{-3} & U \geq 33.5\,\text{ms}^{-1} \end{cases}$$

We also found a power relationship between heat and water vapor transfer coefficients and wind speed using the turbulent data collected from aircraft measurements conducted in 11 experiments. A saturated wind speed of 23 ms^{-1} is found based on the regression of wind speed and drag coefficient. Since the maximum wind speed in the database we used is 27.05 ms^{-1}, the expression of C_d under conditional wind speed greater than 27 ms^{-1} is only the product of mathematical derivation, lacking the support of observational data. Therefore, we provide the expression of C_d for the full wind speed range for use in models. In this study, since these 11 experiments did not measure these parameters, we did not consider the dependence of C_d on wave parameters (such as wave age and wave height). Therefore, the equations we have derived in this study are useful for atmospheric models, and they could be further improved in the future with advanced measurements of other parameters. Parameterizations of turbulent heat transfer coefficient (C_h) and water vapor transfer coefficient (C_e) are also investigated. It is found that they are almost independent of wind speed under strong wind conditions: C_h is 9.39×10^{-4} when 10.5 $\text{ms}^{-1} < U \leq 23$ ms^{-1}; while C_e is 3.4×10^{-4} when $U > 4.5$ ms^{-1}.

Author Contributions: Conceptualization, methodology, investigation, data curation, S.Z., J.Z. and Z.Z.; Formal analysis and writing–review, X.B.; Editing and supervision, Z.G. All authors have read and agreed to the published version of the manuscript.

Funding: This study was funded by the National Key Research and Development Program of Ministry of Science and Technology of China (2018YFC1506405), and by National Natural Science Foundation of China (Grants 41275022 and 41975067).

Institutional Review Board Statement: Not applicable.

Informed Consent Statement: Not applicable.

Data Availability Statement: The data presented in this study are available on request from the corresponding references in this paper.

Acknowledgments: All the dedicated scientists who collected and made available the fast response aircraft data are greatly appreciated. The database used in this work was created by the North West Research Associates under the U.S. Office of Naval Research (ONR) supports. We are very grateful to three anonymous reviewers for their careful review and valuable comments, which lead to substantial improvement of this manuscript.

Conflicts of Interest: The authors declare no conflict of interest.

References

1. Garratt, J.R. *The Atmospheric Boundary Layer*; Cambridge University Press: New York, NY, USA, 1992; p. 316.
2. Nosov, V.V.; Lukin, V.P.; Nosov, E.V.; Torgaev, A.V. Turbulence Scales of the Monin-Obukhov Similarity Theory in the Anisotropic Mountain Boundary Layer. *Russ. Phys. J.* **2020**, *63*, 244–249. [CrossRef]
3. Shikhovtsev, A.; Kovadlo, P.; Lukin, V. Temporal variations of the turbulence profiles at the Sayan solar observatory site. *Atmosphere* **2019**, *10*, 499. [CrossRef]
4. Deskos, G.; Carre, A.; Palacios, R. Assessment of low-altitude atmospheric turbulence models for aircraft aeroelasticity. *J. Fluids Struct.* **2020**, *95*, 102981. [CrossRef]
5. Gao, Z.; Peng, W.; Gao, C.Y.; Li, Y. Parabolic dependence of the drag coefficient on wind speed from aircraft eddy-covariance measurements over the tropical Eastern Pacific. *Sci. Rep.* **2020**, *10*, 1805. [CrossRef] [PubMed]

6. Alamaro, M.; Emanuel, K.A.; McGillis, W.R. Experimental investigation of air-sea transfer of momentum and enthalpy at high wind speed. In *Preprints, Proceedings of the 25th Conference on Hurricane and Tropical Meteorology, San Diego, CA, USA, 28 April–3 May 2002*; American Meteorological Society: San Diego, CA, USA, 2002; pp. 667–668.
7. Powell, M.D.; Vickery, P.J.; Reinhold, T.A. Reduced drag coefficient for high wind speeds in tropical cyclones. *Nature* **2003**, *422*, 279–283. [CrossRef] [PubMed]
8. Donelan, M.A.; Haus, B.K.; Reul, N.; Plant, W.J.; Stiassnie, M.; Graber, H.C.; Brown, O.B.; Saltzman, E.S. On the limiting aerodynamic roughness of the ocean in very strong winds. *Geophys. Res. Lett.* **2004**, *31*, L18306. [CrossRef]
9. Makin, V.K. A note on drag of the sea surface at Hurricane winds. *Bound. Layer Meteorol.* **2005**, *115*, 169–176. [CrossRef]
10. Black, P.G.; Dasaro, E.A.; Drennan, W.M.; French, J.R.; Niiler, P.P.; Sanford, T.B.; Terrill, E.; Walsh, E.J.; Zhang, J.A. Air–Sea Exchange in Hurricanes: Synthesis of Observations from the Coupled Boundary Layer Air-Sea Transfer Experiment. *Bull. Am. Meteorol. Soc.* **2007**, *88*, 357–374. [CrossRef]
11. Troitskaya, Y.I.; Sergeev, D.A.; Kandaurov, A.A.; Baidakov, G.A.; Vdovin, M.A.; Kazakov, V.I. Laboratory and theoretical modeling of air-sea momentum transfer under severe wind conditions. *J. Geophys. Res.* **2012**, *117*, C00J21. [CrossRef]
12. Soloviev, A.V.; Lukas, R.; Donelan, M.A.; Haus, B.K.; Ginis, I. The air-sea interface and surface stress under tropical cyclones. *Sci. Rep.* **2014**, *4*, 5306. [CrossRef]
13. Donelan, M.A. On the decrease of the oceanic drag coefficient in high winds. *J. Geophys. Res. Ocean.* **2018**, *123*, 1485–1501. [CrossRef]
14. Vickers, D.; Mahrt, L.; Andreas, E.L. Estimates of the 10-m neutral sea surface drag coefficient from aircraft eddy-covariance measurements. *J. Phys. Oceanogr.* **2013**, *43*, 301–310. [CrossRef]
15. Edson, J.B.; Crawford, T.; Crescenti, J.; Farrar, T.; Frew, N.; Gerbi, G.; Helmis, C.; Hristov, T.; Khelif, D.; Jessup, A.; et al. The coupled boundary layers and air-sea transfer experiment in low winds. *Bull. Am. Meteorol. Soc.* **2007**, *88*, 341–356. [CrossRef]
16. Sun, J.; Vandemark, D.; Mahrt, L.; Vickers, D.; Crawford, T.; Vogel, C. Momentum transfer over the coastal zone. *J. Geophys. Res.* **2001**, *106*, 12437–12448. [CrossRef]
17. Khelif, D.; Burns, S.P.; Friehe, C.A. Improved wind measurements on research aircraft. *J. Atmos. Ocean. Technol.* **1999**, *16*, 860–875. [CrossRef]
18. Anderson, K.; Brooks, B.; Caffrey, P.; Clarke, A.; Cohen, L.; Crahan, K.; Davidson, K.; De Jong, A.; De Leeuw, G.; Dion, D.; et al. The RED Experiment: An assessment of boundary layer effects in a trade winds regime on microwave and infrared propagation over the sea. *Bull. Am. Meteorol. Soc.* **2014**, *85*, 1355–1366. [CrossRef]
19. Raga, G.; Abarca, S. On the parameterization of turbulent fluxes over the tropical eastern Pacific. *Amer. Chem. Phys.* **2007**, *7*, 635–643. [CrossRef]
20. Vickers, D.; Esbensen, S.K. Subgrid surface fluxes in fair weather conditions during TOGA COARE: Observational estimates and parameterization. *Mon. Weather Rev.* **1998**, *126*, 620–633. [CrossRef]
21. Mahrt, L.; Scott, M.; Tihomir, H.; James, E. On Estimating the Surface Wind Stress over the Sea. *J. Phys. Oceanogr.* **2018**, *48*, 1533–1541. [CrossRef]

atmosphere

Article

Surface Layer Drag Coefficient at Different Radius Ranges in Tropical Cyclones

Lei Ye [1,2], Yubin Li [1,2,*] and Zhiqiu Gao [1,3]

1 Collaborative Innovation Center on Forecast and Evaluation of Meteorological Disasters, Key Laboratory for Aerosol-Cloud-Precipitation of China Meteorological Administration, School of Atmospheric Physics, Nanjing University of Information Science and Technology, Nanjing 210044, China; 20191103014@nuist.edu.cn (L.Y.); zgao@mail.iap.ac.cn (Z.G.)
2 Southern Marine Science and Engineering Guangdong Laboratory (Zhuhai), Zhuhai 519080, China
3 State Key Laboratory of Atmospheric Boundary Layer Physics and Atmospheric Chemistry, Institute of Atmospheric Physics, Chinese Academy of Sciences, Beijing 100029, China
* Correspondence: liyubin@nuist.edu.cn

Abstract: Using dropsonde data and a flux-profile method, this study investigates the drag coefficient (C_d)–wind speed relationship within different radius ranges. The results show a systematic decrease of friction velocity u_* from the range of R/RMW > 1.05 to that of R/RMW < 0.95 (R is the radial location of a dropsonde profile, and RMW is the radius of maximum wind), and the reduction is 5~25% for different wind speeds. Further, within the ranges of either R/RMW > 1.05 or R/RMW < 1.05, a clear feature of "roll-off" at about 35 m s^{-1} can be obtained. However, the roll feature becomes vague in the ranges of R/RMW < 0.95, R/RMW < 0.85, and R/RMW < 0.75, indicating the TC dynamics within and near RMW play a role in affecting the flux-profile relationship. Even more, C_d of R < 0.75RMW deviates significantly from the C_d of R < 0.85RMW and R < 0.95RMW, while the deviation between R < 0.85RMW and R < 0.95RMW is much smaller. Especially when 10 m winds exceed 40 m s^{-1}, u_* of R < 0.75RMW is significantly larger than that of R < 0.85RMW. This phenomenon is also linked to the TC dynamics (e.g., the large radial gradients of winds and the drastic vertical variation of the bulk Richardson number), but the speculation needs to be verified in future study.

Keywords: dropsonde; drag coefficient; wind speed; friction velocity; radial position

Citation: Ye, L.; Li, Y.; Gao, Z. Surface Layer Drag Coefficient at Different Radius Ranges in Tropical Cyclones. *Atmosphere* **2022**, *13*, 280. https://doi.org/10.3390/atmos13020280

Academic Editor: Eric A. Hendricks

Received: 6 December 2021
Accepted: 1 February 2022
Published: 8 February 2022

1. Introduction

Tropical cyclone (TC) is one of the most devastating weather systems that cause huge losses of lives and properties [1–4]. Its enormous energy mostly comes from and dissipates near the surface, which is expressed as enthalpy flux and surface drag [5,6], and the theory is well accepted that the intensity development of TC strongly depends on the ratio of the drag coefficient (C_d) and enthalpy flux transfer coefficient (C_k) (e.g., [7–10]).

With field dropsondes observation and the flux-profile method, Powell et al. [11] first found that C_d, under strong wind conditions, peaked at a 10-m wind speed of around 38 m s^{-1}. Donelan et al. [12] also found a similar variation pattern with laboratory measurements, only that C_d tended to level off rather than decline after it reached the peak value. While, by using ocean current velocity profiles, Jarosz et al. [13] calculated the drag coefficient with a bottom-up method and also found C_d first increased, but later when the 10-m wind speed was larger than about 32 m s^{-1}, it declined with wind speed. Bi et al. [14] used offshore tower observation and found C_d peaked at 18 m s^{-1} and decreased afterward, but leveled off when larger than 27 m s^{-1}. Other studies with various methods generally also found the same feature that C_d decreased or leveled off with wind speed when it was larger than around 30 m s^{-1} (e.g., [15–24]).

The observation of hurricanes by dropsondes has been carried out for several decades and thousands of profiles have been obtained (e.g., [25,26]), which leads to the hope to thoroughly understand the relationship between drag coefficient and wind speed. Nevertheless, Richter et al. [27] concluded from virtual dropsondes of a numerical simulation that, caused by the sensitivity of this method to the regression procedure, the calculation of C_d with dropsondes observation and the flux-profile method was accurate to within approximately 50%. Richter et al. [28] further showed that the flux-profile method had an inherent underestimation inclination of C_d at hurricane-force winds, which was a result of the uncertainty in the vertical position of the sonde near the surface and the non-monotonic profile of wind speed with height, but these might be mitigated by selecting the regression profile within about 20–150 m. Through virtual dropsondes, Richter et al. [28] also showed that within the radius of maximum wind (RMW), where the traditional Monin-Obukhov similarity theory was not applicable anymore, C_d was drastically underestimated. However, this characteristic was not observed in their study with real dropsondes.

Based on the findings of Richter et al. [28], this study takes a further step to understand how the C_d–wind speed relationship is within the RMW based on the real dropsondes. Especially, under the same 10-m wind speed, how the C_d calculated from the flux-profile method varies with RMW. This paper is organized as follows. Section 2 describes the data and the methods used in this paper. Section 3 presents the results, including the variation of friction velocity (u_*) against R/RMW in the real dropsonde observations. And this is followed by a summary and conclusions in Section 4.

2. Data and Methods

2.1. Data

Dropsonde data used in this study are from the Long-Term NOAA Dropsonde Hurricane Archive released by NOAA's National Hurricane Center and Hurricane Research Division (https://data.eol.ucar.edu/dataset/542.001 (accessed on 5 December 2021)). The data in this archive include over 13,600 Global Positioning System (GPS) dropsonde profiles from 120 TCs in the Atlantic and East Pacific basins from 1996 to 2012 [26]. The GPS dropsondes are released from aircrafts flying into the TC system, and aircraft types include P3 (flying at 1–5 km altitude in the inner and outer core of TC) and G-IV (flying at 14–15 km altitude in the outside environment of TC). The data quality control of the sounding data is processed by Atmospheric Sounding Processing Environment (ASPEN) software, and various visualization tools and statistical methods are used to evaluate data products to identify and correct data quality problems caused by various errors and deviations. The wind speed is observed at 4 Hz frequency, corresponding to 3–8 m vertical resolution, but prior to 2010, the frequency of wind data was measured at 2 Hz. More detailed information about the dataset can be found in the data archive and Wang et al. [26].

In this work, the radial locations of the profiles (R) are also given by the Dropsonde Hurricane Archive introduced above, while the RMW data are from the Extended Best Track (EBT) dataset [29], which includes 6-h interval RMW data of the corresponding TCs in the Dropsonde Hurricane Archive. The EBT RMW data are then interpolated linearly to fit the R data of the dropsonde profiles.

2.2. Method

Following Powell et al. [11], Holthuijsen et al. [16], and Richter et al. [27], we use the flux-profile method to estimate the drag coefficient C_d, which is based on the logarithmic profile of wind speed within the hurricane boundary layer. According to Monin-Obukhov (MO) similarity theory, the wind profile in the neutral condition can be described as

$$U = \frac{u_*}{k} \ln\left(\frac{z}{z_0}\right), \tag{1}$$

$$lnz = \left(\frac{k}{u_*}\right) U + lnz_0, \tag{2}$$

where U is the wind speed at altitude z and the Von Karman constant k is taken as 0.4. The friction velocity u_* and the roughness length z_0 can be estimated through logarithmic linear regression. Then C_d can be derived from

$$\tau = \rho u_*^2 = \rho C_d U_{10}^2, \tag{3}$$

where U_{10} is the 10-m wind speed and ρ is air density.

Following the approach used in Powell et al. [11] and Richter et al. [28], the vertical profiles of wind speed are analyzed in a composite sense, as a function of the mean boundary layer (MBL, here defined as the height of 10 to 500 m) wind speed. That is, the dropsonde profiles are grouped by their MBL mean wind speeds. This study considers only dropsonde profiles with MBL wind speed larger than 20 m s^{-1}, and 3661 dropsonde profiles meet this criterion. Figure 1 presents the number of profiles in each 5 m s^{-1} MBL mean wind speed bin, which shows a decreasing trend with increasing wind speed. The data are also categorized by R/RMW to analyze the C_d–wind speed relationship at different R/RMW ranges. Figure 1 also presents the number of profiles in the ranges of R > 1.05RMW, R < 1.05RMW, and R < 0.75RMW. Here, the value of 1.05 of R/RMW is chosen to divide the profiles within and outside of the RMW, while the selection of 0.75 of R/RMW as a dividing point is determined by the fact that in the range of R < 0.75RMW there are still relatively large quantity observations. It also can be seen from Figure 1 that the sample numbers are less in high wind speed 5 m s^{-1} bins, and to maintain a relatively large quantity of samples in each MBL mean wind speed group, the wind speed intervals are chosen as 20–25 m s^{-1}, 25–30 m s^{-1}, 30–35 m s^{-1}, 35–40 m s^{-1}, 40–45 m s^{-1}, 45–50 m s^{-1}, 50–60 m s^{-1}, 60–70 m s^{-1}, and 70–90 m s^{-1} as in Table 1. Richter et al. [28] used both real and virtual dropsondes to examine the flux-profile method, and they found an underestimation of C_d through the flux-profile method with virtual dropsondes, but not with real dropsondes. By using the virtual dropsondes from a full physics simulation, a large number of profiles can be obtained, and Richter et al. [28] divided these virtual samples by an interval of 0.26RMW (Figures 10 and 11 in Richter et al. [28]). However, for the real dropsondes, they used an interval of 1 RMW due to the limited samples (Figure 3 in Richter et al. [28]). Here, we follow Richter et al. [28] using 1.05RMW as a watershed, and 0.75RMW as the lowest limit to guarantee relatively enough samples (more than ~30 for each 10 m s^{-1} bin), and between 1.05RMW to 0.75RMW we use an interval of 0.05RMW to see how the flux-profile relationship varied from 0.75 to 1.05 RMW (Table 1).

Next, each measurement in each profile is further binned into 10-m height intervals, and the measurements within each height bin are collected and averaged together. Variability associated with mesoscale, convective, and under-sampled turbulent scales is then removed by averaging all profiles in a given wind speed group [11,30]. Finally, we choose the 20 to 160 m height range over which the fit to the mean profiles will be made, based on the suggestion of Powell [31] that 20–160 m surface layer is more representative for the lowest levels. Meanwhile, Richter et al. [28] and Jiang et al. [32] also indicated that under 20 m height, the vertical position of dropsondes might be contaminated; besides, the regressed C_d began to decrease monotonically once the upper height bound exceeded roughly 150 m. Byrne and Zhang [33] found that there was a transition of the flow from 3-D to 2-D turbulence in the hurricane boundary layer occurring around 150 m, suggesting that above 150 m MO similarity theory does not apply.

The 20–160 m of each MBL mean wind speed group are plotted in Figure 2 with a semi-log coordinate, and a least-squares linear regression fit is used to estimate the roughness length z_0 (as intercept), friction velocity u_* (k/u_* as slope), and 10-m wind speed U_{10}, then drag coefficient C_d was computed from Equation (3). It can be seen that the regression coefficient of each group is above 99%, indicating the sample size of each MBL mean wind speed group is large enough, and this guarantees the validity of the fitting results.

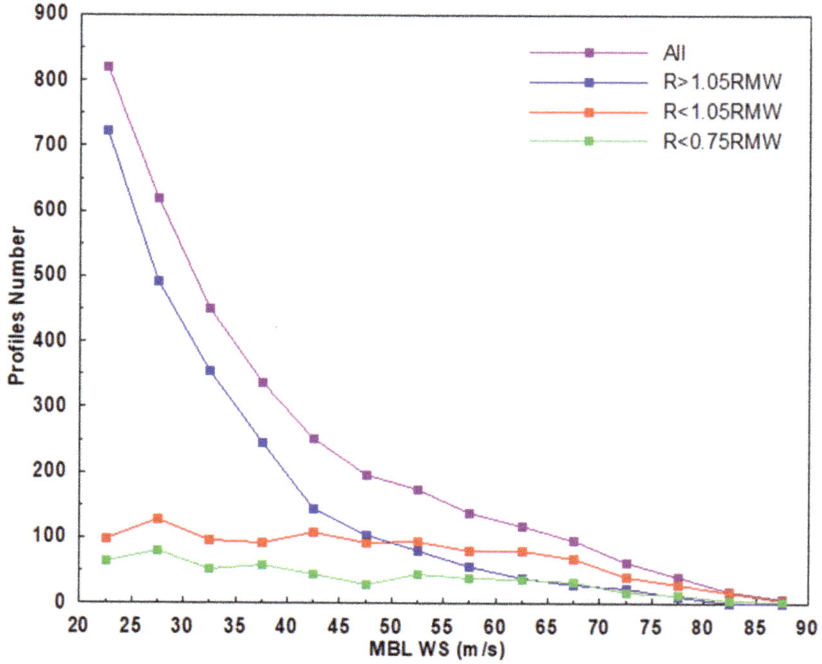

Figure 1. The profile number in each 5 m s^{-1} MBL mean wind speed bin.

Table 1. The profile number in each MBL mean wind speed group for different R/RMW ranges.

MBL WS Group (Unit: m s^{-1})	R/RMW Range								
	All	**>1.05**	**<1.05**	**<1.00**	**<0.95**	**<0.90**	**<0.85**	**<0.80**	**<0.75**
20–25	820	722	98	92	85	77	73	70	64
25–30	619	491	128	117	113	107	100	88	80
30–35	450	355	95	92	87	83	73	64	53
35–40	336	245	91	86	79	76	71	67	59
40–45	251	143	108	92	82	75	68	57	44
45–50	196	104	92	84	74	65	52	39	29
50–60	310	135	175	160	144	128	113	96	83
60–70	213	67	146	139	127	117	107	87	70
70–90	127	35	92	79	70	61	51	41	35
Sum	3322	2297	1025	941	861	789	708	609	517

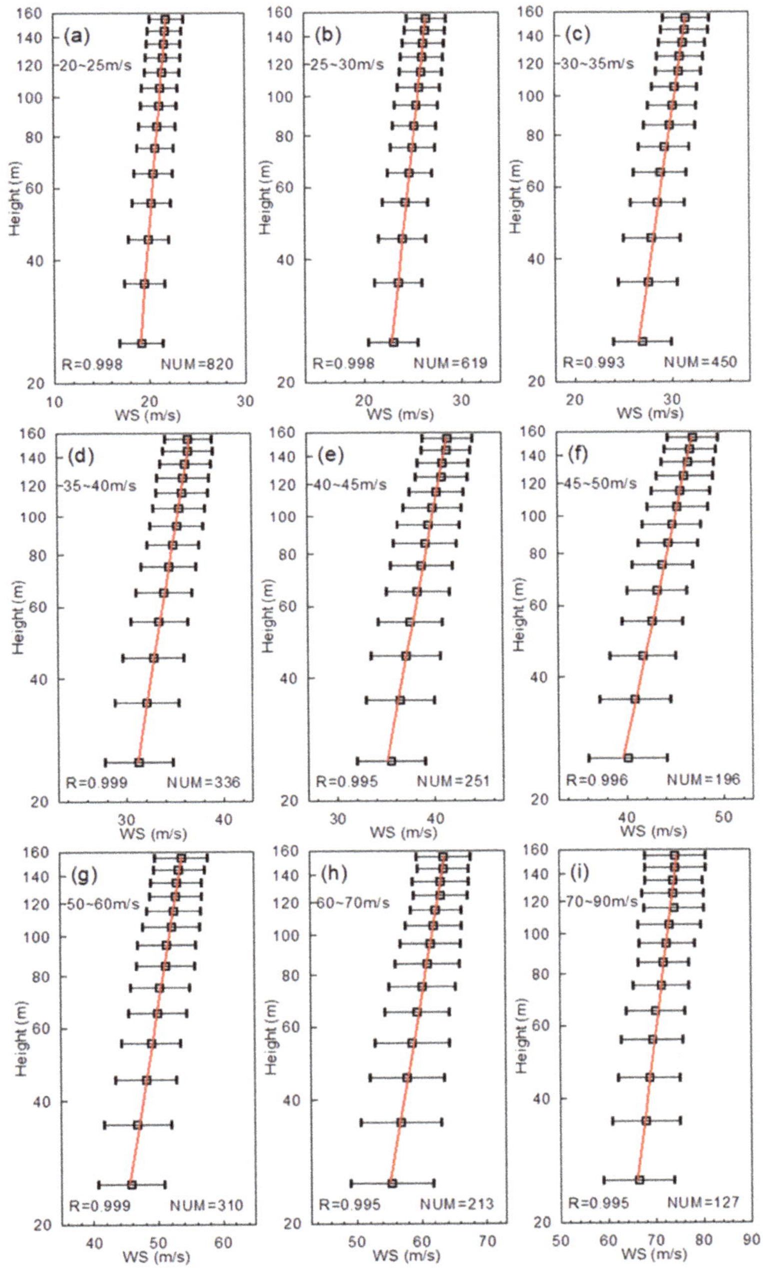

Figure 2. Wind profiles of each MBL mean wind speed group between 20 and 90 m s^{-1}: (**a**) 20–25 m s^{-1}, (**b**) 25–30 m s^{-1}, (**c**) 30–35 m s^{-1}, (**d**) 35–40 m s^{-1}, (**e**) 40–45 m s^{-1}, (**f**) 45–50 m s^{-1}, (**g**) 50–60 m s^{-1}, (**h**) 60–70 m s^{-1}, and (**i**) 70–90 m s^{-1}. The black squares and horizontal bars represent the mean and standard deviation for each height bin. Red lines represent the least square fitting within 20–160 m height. R is the regression coefficient. NUM is the number of dropsonde profiles.

3. Results

Figure 3 depicts the u_* obtained in this study, but within different R/RMW ranges, and also the u_* derived by previous studies. It can be found that the u_* obtained from the various studies generally shows the same increasing trend when the 10-m wind speed is less than 40 m s^{-1}, but levels off (or presents a smaller increasing trend) at higher wind speeds. Specifically, the data points obtained from the range of R/RMW > 1.05 and R/RMW < 1.05 bear very little difference, except that only at the 10-m wind speed of about 38 m s^{-1} does the R/RMW < 1.05 group show an evidently smaller value of u_* than the R/RMW > 1.05 group, which is even smaller in the group of R/RMW < 0.75. By using virtual dropsondes, Richter et al. [28] showed that with a decreasing R/RMW, the u_* obtained by the flux-profile method tends to be underestimated (Figures 10 and 11 in Richter et al. [28]), which are seen here in the 10-m wind speed of about 38 m s^{-1}. At other wind speed ranges, either no significant difference is observed (e.g., at a 10-m wind speed range of 40–50 m s^{-1}), or the results from the group of R/RMW < 0.75 are even larger than from the group of R/RMW > 1.05 (e.g., at a 10-m wind speed ranges of 50–70 m s^{-1}).

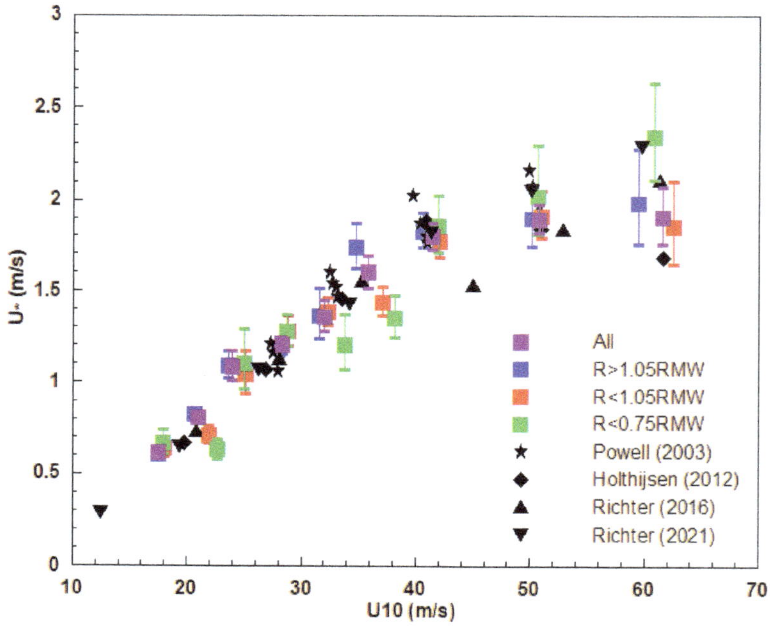

Figure 3. Variation of the friction velocity (u_*, unit: m s^{-1}) with 10-m wind speed (U_{10}, unit: m s^{-1}). The squares with 4 different colors represent the mean value in each wind speed bin from the 4 different R/RMW groups, and vertical bars represent the 95% confidence interval. Black symbols with different shapes represent the results from previous studies including Powell et al. [11], Holthuijsen et al. [16], Richter et al. [27], and Richter et al. [28].

Richter et al. [28] also draw a similar conclusion that the different R/RMW ranges generally do not affect the C_d–10-m wind speed relationship (Figure 3), which is a result of the various profiles from many TCs with different sizes and intensities.

To further look into the u_*–U_{10} relationship under different R/RMW conditions, we plot the variation of u_* with R/RMW in the range of >1.05, <1.05, and later on at 0.05 interval to <0.75 (Figure 4). When comparing the u_* in the ranges of R/RMW > 1.05 and <0.95, a systematic decrease of u_* with R/RMW is generally found for all MBL mean wind speed groups, except the 20–25 m s^{-1} and 35–40 m s^{-1} MBL mean wind speed groups. Richter et al. [28] showed by virtual dropsondes, that with a U_{10} of about 65 m s^{-1}

and R/RMW of about 0.85 (and U_{10} of about 40 m s^{-1} and R/RMW of about 0.7), u_* was underestimated by about 10~20% percent (Figure 10c,d in Richter et al. [28]). Here, although the real dropsondes do not repeat these values exactly, a reduction of u_* by 5~25% from R/RMW > 1.05 to <0.95 is observed in the 7 out of 9 MBL mean wind speed groups.

Nonetheless, an unexpected increase of u_* with R/RMW from R/RMW < 0.85 to <0.75 is found in the groups of an MBL mean wind speed larger than 50 m s^{-1}. This contradicts the findings of Richter et al. [28]. To examine whether this is caused by some extreme values that could bias the mean of the distribution when sample sizes are small, the wind profiles of R < 0.85RMW and R < 0.75RMW in all MBL mean wind speed groups are presented here as Figure 5. It can be found that for all the wind profiles, a logarithmic profile is generally kept, but when the mean MBL mean wind speed is larger than 50 m s^{-1}, an increase of u_* (i.e., decrease of the slope of the profile) is found from R < 0.85RMW to R < 0.75RMW. This phenomenon may be caused by the drastic radial gradient of winds and temperature inside the RMW, as shown in previous studies (e.g., Kepert and Wang [34]; Bell and Montgomery [35]; Zhang et al. [36]). Zhang et al. [36] also found that the depth of the non-strongly stable layer (i.e., the layer with a bulk Richardson number < 0.25) decreased significantly from ~700 m at the RMW to ~200 m at 0.5 RMW (Figure 9 in Zhang et al. [36]), which is less than half of our 500 m MBL layer depth. A bulk Richardson number is an important parameter to indicate the turbulent mixing intensity, and in the atmospheric models, 0.25 is usually used as the marker of the boundary layer height. Combining these findings, it is speculated that when it comes to about < 0.75RMW, the logarithmic wind profile is affected by the TC dynamics (e.g., the large radial gradients of winds and the drastic vertical variation of the bulk Richardson number). This phenomenon will be further investigated in the future through a high-resolution simulation study.

Figure 6 shows the C_d–10-m wind speed relationship under different R/RMW ranges. Generally, when considering all the data or the data with R/RMW > 1.05, a clear feature of "roll-off" at about 35 m s^{-1} can be obtained. Even, if considering the whole range of R/RMW < 1.05, such a feature may still roughly exist. However, at the ranges of R/RMW < 0.95, R/RMW < 0.85, or R/RMW < 0.75, the roll feature becomes vague, indicating the TC dynamics within and near the RMW play a role in affecting the flux-profile relationship. On one hand, the drift of dropsondes with the tangential and radial winds may play a role. Real dropsondes are advected with the horizontal wind, and, in the region near the RMW, that advection can move the sondes 10–20 km downwind as the sondes descend to the surface, violating the flux-profile assumptions. Richter et al. [28] concluded that the effects of a dropsonde drift were negligible, and the features of C_d–U_{10} relationship with and without drift effects were similar to each other. Nevertheless, some differences are still seen in the C_d–U_{10} with and without drift effects, especially, the C_d–U_{10} at 0.78 < R/RMW $\leq$ 1.04 is significantly different: with/without drift effect, C_d is about 0.0018/0.0012, and U_{10} is about 65/52 m s^{-1} (Figures 10 and 11 in Richter et al. [28]). Therefore, the drift of dropsondes within and near RMW indeed plays a role in the C_d–U_{10} relationship. On the other hand, the assumption of the constant flux layer may not be valid anymore within and near the RMW, as pointed by Richter et al. [28], even though a logarithmic layer is persistent throughout all the different ranges of R/RMW. The violation of the constant flux layer may be caused by the large radial gradient of winds and the transport of the momentum flux (e.g., Kepert [37]; Zhang et al., [38]; Bryan et al., [39]).

In addition, in Figure 6b, the C_d of R < 0.75RMW (i.e., green square) deviates significantly from the C_d of R < 0.85RMW and R < 0.95RMW (i.e., brown and pink squares), while the deviation between R < 0.85RMW (brown squares) and R < 0.95RMW (pink squares) is much smaller, especially when 10 m winds exceed 40 m s^{-1}. As discussed previously, when it comes to about < 0.75RMW, the logarithmic wind profile may be altered by the drift, the large radial gradients, and the advection of the momentum flux here.

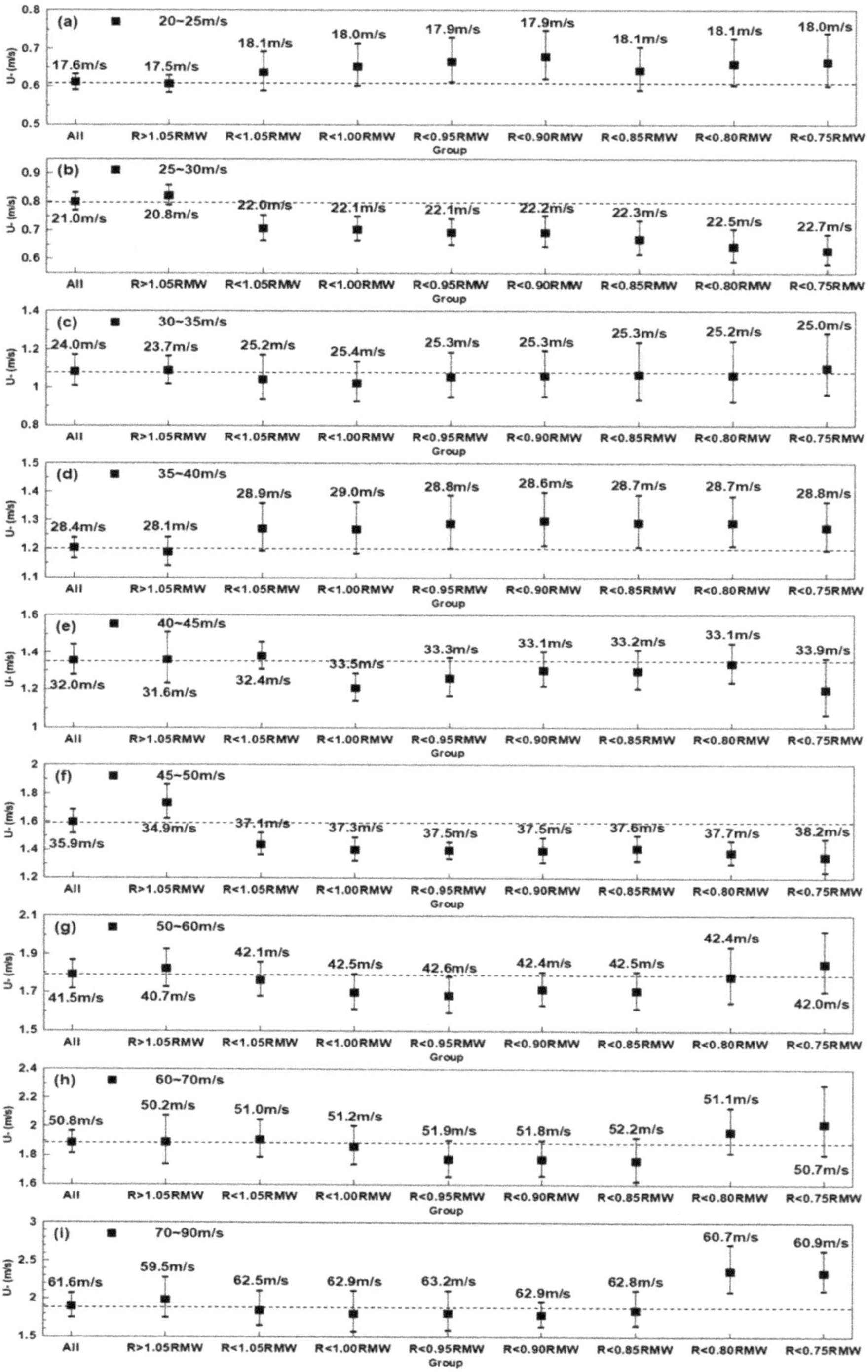

Figure 4. Friction velocity (u_*, unit: m s^{-1}) at every wind speed bin from different R/RMW groups (including all data, and data with R > 1.05RMW, R < 1.05RMW, R < 1.00RMW, R < 0.95RMW, R < 0.90RMW,

R < 0.85RMW, R < 0.80RMW, and R < 0.75RMW): (**a**) 20–25 m s^{-1}, (**b**) 25–30 m s^{-1}, (**c**) 30–35 m s^{-1}, (**d**) 35–40 m s^{-1}, (**e**) 40–45 m s^{-1}, (**f**) 45–50 m s^{-1}, (**g**) 50–60 m s^{-1}, (**h**) 60–70 m s^{-1}, and (**i**) 70–90 m s^{-1}. The squares represent the mean value in each wind speed bin, vertical bars represent the 95% confidence interval, and the number above each square is the mean 10 m wind speed of the data. Dashed lines represent the mean value of all data group at each wind speed bin.

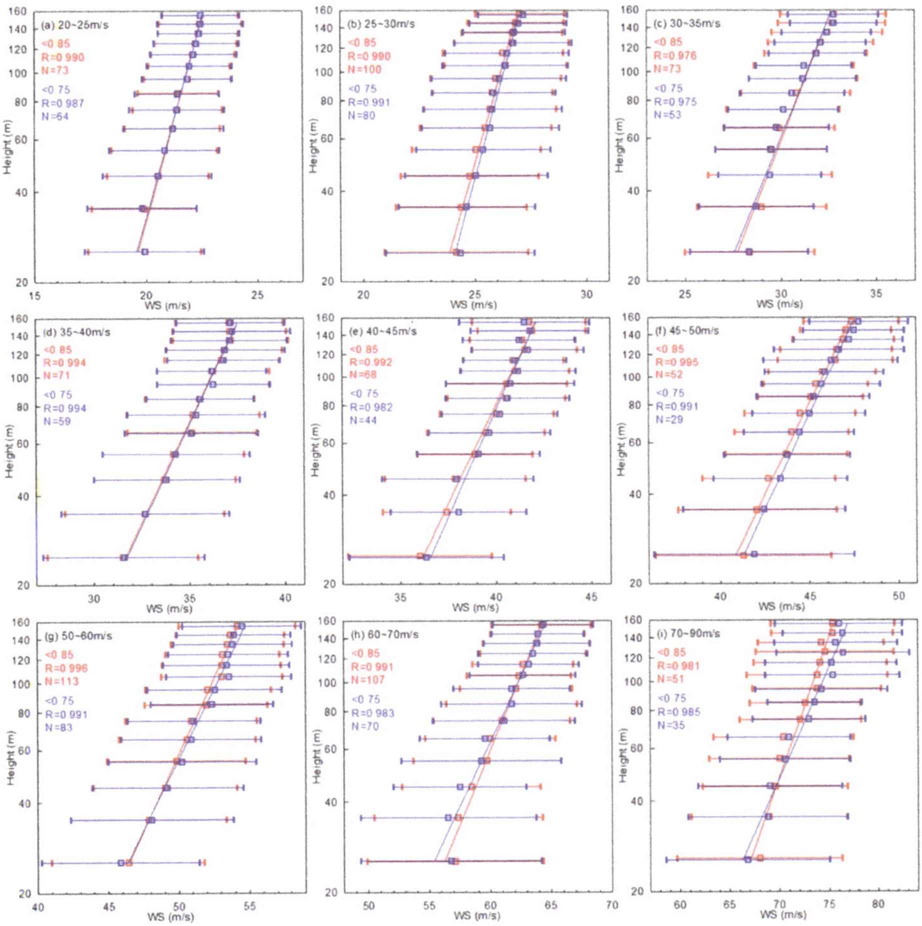

Figure 5. Wind profiles of R < 0.85RMW (in red) and R < 0.75RMW (in blue) in different MBL mean wind speed groups, (**a**) 20–25 m s^{-1}, (**b**) 25–30 m s^{-1}, (**c**) 30–35 m s^{-1}, (**d**) 35–40 m s^{-1}, (**e**) 40–45 m s^{-1}, (**f**) 45–50 m s^{-1}, (**g**) 50–60 m s^{-1}, (**h**) 60–70 m s^{-1}, and (**i**) 70–90 m s^{-1}. The squares and horizontal bars represent the mean and standard deviation for each height bin. Lines from top to bottom represent the least square fitting within 20–160 m height. R is the regression coefficient. N is the number of dropsonde profiles.

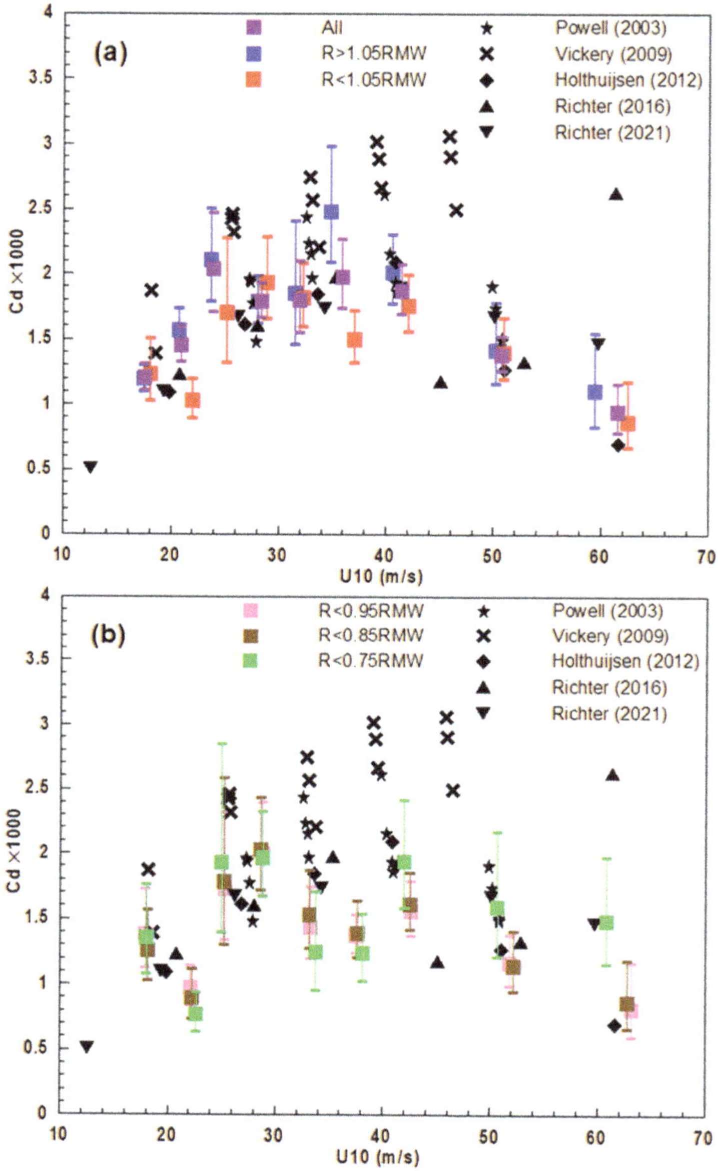

Figure 6. Same as Figure 3, but for the variation of the surface drag coefficient ($C_d \times 10^3$) with 10-m wind speed (U_{10}, unit: m s^{-1}). (**a**) All, R/RMW > 1.05 and R/RMW < 1.05, (**b**) R/RMW < 0.95, R/RMW < 0.85 and R/RMW < 0.75. The various black symbols are the result from Powell et al. [11], Vickery et al. [30], Holthuijsen et al. [16], Richter et al. [27], and Richter et al. [28].

4. Summary and Conclusions

Using dropsonde data and flux-profile method, in this study, we investigated the C_d–wind speed relationship within different R/RMW ranges, and also the variation of C_d with R/RMW under the same wind speed bins. The main findings and conclusions are as follows:

(1) From the range of R/RMW > 1.05 to that of R/RMW < 0.95, a systematic decrease of 5~25% in friction velocity u_* is found for 7 out of the 9 MBL mean wind speed groups.

(2) A clear feature of "roll-off" at about 35 m s^{-1} can be obtained for the ranges of either R/RMW > 1.05 or R/RMW < 1.05. However, in the ranges of R/RMW < 0.95, R/RMW < 0.85, and R/RMW < 0.75, the roll feature becomes vague, indicating the TC dynamics within and near the RMW play a role in affecting the flux-profile relationship.
(3) When U_{10} exceeds 40 m s^{-1}, u_* of R < 0.75RMW is significantly larger than that of R < 0.85RMW. This phenomenon is not caused by the extreme values in the relatively small sample size, and it is speculated to link with the TC dynamics (e.g., the large radial gradients of winds and the drastic vertical variation of the bulk Richardson number).

At last, these results may be affected by the limited data quantity in this study, and future work will be carried out with more observations and simulations to further understand how the TC dynamics within and near the RMW affect the wind profiles in TCs.

Author Contributions: Conceptualization, Y.L.; methodology, L.Y. and Y.L.; software, L.Y.; validation, L.Y.; formal analysis, L.Y. and Y.L.; investigation, L.Y. and Y.L.; resources, L.Y.; data curation, L.Y.; writing—original draft preparation, L.Y.; writing—review and editing, L.Y., Y.L. and Z.G.; visualization, L.Y.; supervision, Y.L. and Z.G.; project administration, Y.L. and Z.G.; funding acquisition, Y.L. and Z.G. All authors have read and agreed to the published version of the manuscript.

Funding: This study was supported by the National Key Projects of the Ministry of Science and Technology of China (2018YFC1506405); National Natural Science Foundation of China (42075072, 41675009, 41475014); The Innovation Group Project of Southern Marine Science and Engineering Guangdong Laboratory (Zhuhai) (311021008).

Institutional Review Board Statement: Not applicable.

Informed Consent Statement: Not applicable.

Data Availability Statement: Not applicable.

Acknowledgments: The authors thank the three anonymous reviewers for their constructive comments. The numerical calculations in this paper have been done on the supercomputing system in the Supercomputing Center of Nanjing University of Information Science & technology.

Conflicts of Interest: The authors declare no conflict of interest.

References

1. Pielke, R.A., Jr.; Gratz, J.; Landsea, C.W.; Collins, D.; Saunders, M.A.; Musulin, R. Normalized hurricane damage in the United States: 1900–2005. *Nat. Hazards Rev.* **2008**, *9*, 29–42. [CrossRef]
2. Schmidt, S.; Kemfert, C.; Höppe, P. The impact of socio-economics and climate change on tropical cyclone losses in the USA. *Reg. Environ. Chang.* **2010**, *10*, 13–26. [CrossRef]
3. Wang, Y.; Wen, S.; Li, X.; Thomas, F.; Su, B.; Wang, R.; Jiang, T. Spatiotemporal distributions of influential tropical cyclones and associated economic losses in China in 1984–2015. *Nat. Hazards* **2016**, *84*, 2009–2030. [CrossRef]
4. Wang, H.; Xu, M.; Onyejuruwa, A.; Wang, Y.; Wen, S.; Gao, A.E.; Li, Y. Tropical cyclone damages in Mainland China over 2005–2016: Losses analysis and implications. *Environ. Dev. Sustain.* **2019**, *21*, 3077–3092. [CrossRef]
5. Zhang, J.A. Estimation of dissipative heating using low-level in situ aircraft observations in the hurricane boundary layer. *J. Atmos. Sci.* **2010**, *67*, 1853–1862. [CrossRef]
6. Heng, J.; Wang, Y. Nonlinear response of a tropical cyclone vortex to prescribed eyewall heating with and without surface friction in TCM4: Implications for tropical cyclone intensification. *J. Atmos. Sci.* **2016**, *73*, 1315–1333. [CrossRef]
7. Emanuel, K.A. Sensitivity of tropical cyclones to surface exchange coefficients and a revised steady-state model incorporating eye dynamics. *J. Atmos. Sci.* **1995**, *52*, 3969–3976. [CrossRef]
8. Emanuel, K.A. Self-stratification of tropical cyclone outflow. Part II: Implications for storm intensification. *J. Atmos. Sci.* **2012**, *69*, 988–996. [CrossRef]
9. Rogers, R.; Aberson, S.; Aksoy, A.; Annane, B.; Black, M.; Cione, J.; Dorst, N.; Dunion, J.; Gamache, J.; Goldenberg, S.; et al. NOAA'S Hurricane Intensity Forecasting Experiment: A Progress Report. *Bull. Amer. Meteor. Soc.* **2013**, *94*, 859–882. [CrossRef]
10. Soloviev, A.V.; Lukas, R.; Donelan, M.A.; Haus, B.K.; Ginis, I. The air-sea interface and surface stress under tropical cyclones. *Sci. Rep.* **2014**, *4*, 5306. [CrossRef]
11. Powell, M.D.; Vickery, P.J.; Reinhold, T.A. Reduced drag coefficient for high wind speeds in tropical cyclones. *Nature* **2003**, *422*, 279–283. [CrossRef] [PubMed]
12. Donelan, M.A.; Haus, B.K.; Reul, N.; Plant, W.J.; Stiassnie, M.; Graber, H.C.; Brown, O.B.; Saltzman, E.S. On the limiting aerodynamic roughness of the ocean in very strong winds. *Geophys. Res. Lett.* **2004**, *31*, L18306. [CrossRef]

13. Jarosz, E.; Mitchell, D.A.; Wang, D.W.; Teague, W.J. Bottom-up determination of air-sea momentum exchange under a major tropical cyclone. *Science* **2007**, *315*, 1707–1709. [CrossRef] [PubMed]
14. Bi, X.; Gao, Z.; Liu, Y.; Liu, F.; Song, Q.; Huang, J.; Huang, H.; Mao, W.; Liu, C. Observed drag coefficients in high winds in the near offshore of the South China Sea. *J. Geophys. Res. Atmos.* **2015**, *120*, 6444–6459. [CrossRef]
15. Sanford, T.B.; Price, J.F.; Girton, J.B. Upper-ocean response to Hurricane Frances (2004) observed by profiling EM-APEX floats. *J. Phys. Oceanogr.* **2011**, *41*, 1041–1056. [CrossRef]
16. Holthuijsen, L.H.; Powell, M.D.; Pietrzak, J.D. Wind and waves in extreme hurricanes. *J. Geophys. Res.* **2012**, *117*, C09003. [CrossRef]
17. Takagaki, N.; Komori, S.; Suzuki, N.; Iwano, K.; Kuramoto, T.; Shimada, S.; Kurose, R.; Takahashi, K. Strong correlation between the drag coefficient and the shape of the wind sea spectrum over a broad range of wind speeds. *Geophys. Res. Lett.* **2012**, *39*, L23604. [CrossRef]
18. Potter, H.; Graber, H.C.; Williams, N.J.; Collins, C.O.; Ramos, R.J.; Drennan, W.M. In situ measurements of momentum fluxes in typhoons. *J. Atmos. Sci.* **2015**, *72*, 104–118. [CrossRef]
19. Komori, S.; Iwano, K.; Takagaki, N.; Onishi, R.; Kurose, R.; Takahashi, K.; Suzuki, N. Laboratory measurements of heat transfer and drag coefficients at extremely high wind speeds. *J. Phys. Oceanogr.* **2018**, *48*, 959–974. [CrossRef]
20. Hsu, J.Y.; Lien, R.C.; D'Asaro, E.A.; Sanford, T.B. Scaling of drag coefficients under five tropical cyclones. *Geophys. Res. Lett.* **2019**, *46*, 3349–3358. [CrossRef]
21. Peng, W.; GAO, Z.; Li, Y.; Zhang, R.; Tong, B.; Wu, B.; He, Y. Testing drag coefficient approaches by using the buoy data collected in moderate to high wind under following, crossing and opposing swell conditions. *J. Trop Meteor.* **2019**, *25*, 336–343. [CrossRef]
22. Gao, Z.; Peng, W.; Gao, C.Y.; Li, Y. Parabolic dependence of the drag coefficient on wind speed from aircraft eddy-covariance measurements over the tropical Eastern Pacific. *Sci. Rep.* **2020**, *10*, 1805. [CrossRef] [PubMed]
23. Troitskaya, Y.I.; Sergeev, D.A.; Vdovin, M.; Kandaurov, A.; Ermakova, O.; Takagaki, N. Alaboratory study of the effect of surface waves on heat and momentum transfer at high wind speeds. *J. Geophys. Res. Oceans* **2020**, *125*, e2020JC016276. [CrossRef]
24. Gao, Z.; Zhou, S.; Zhang, J.; Zeng, Z.; Bi, X. Parameterization of Sea Surface Drag Coefficient for All Wind Regimes Using 11 Aircraft Eddy-Covariance Measurement Databases. *Atmosphere* **2021**, *12*, 1485. [CrossRef]
25. Aberson, S.D.; Franklin, J.L. Impact on hurricane track and intensity forecasts of GPS dropwindsonde observations from the first-season flights of the NOAA Gulfstream-IV jet aircraft. *Bull. Amer. Meteor. Soc.* **1999**, *80*, 421–428. [CrossRef]
26. Wang, J.; Young, K.; Hock, T.; Lauritsen, D.; Behringer, D.; Black, M.; Black, P.G.; Franklin, J.; Halverson, J.; Molinari, J.; et al. A Long-Term, High-Quality, High-Vertical-Resolution GPS Dropsonde Dataset for Hurricane and Other Studies. *Bull. Amer. Meteor. Soc.* **2015**, *96*, 961–973. [CrossRef]
27. Richter, D.H.; Bohac, R.; Stern, D.P. An assessment of the flux profile method for determining air-sea momentum and enthalpy fluxes from dropsonde data in tropical cyclones. *J. Atmos. Sci.* **2016**, *73*, 2665–2682. [CrossRef]
28. Richter, D.H.; Wainwright, C.; Stern, D.P.; Bryan, G.H.; Chavas, D. Potential Low Bias in High-Wind Drag Coefficient Inferred from Dropsonde Data in Hurricanes. *J. Atmos. Sci.* **2021**, *78*, 2339–2352. [CrossRef]
29. Demuth, J.; DeMaria, M.; Knaff, J.A. Improvement of Advanced Microwave Sounding Unit tropical cyclone intensity and size estimation algorithms. *J.J. Appl. Meteor.Meteor. Climatol.* **2006**, *45*, 1573–1581. [CrossRef]
30. Vickery, P.J.; Wadhera, D.; Powell, M.D.; Chen, Y. A hurricane boundary layer and wind field model for use in engineering applications. *J.J. Appl. Meteor.Meteor. Climatol.* **2009**, *48*, 381–405. [CrossRef]
31. Powell, M.D. *Final Report on the NOAA Joint Hurricane Testbed: Drag Coefficient Distribution and Wind Speed Dependence in Tropical Cyclones*; NOAA/AOML: Miami, FL, USA, 2007.
32. Jiang, W.; Wu, L.; Liu, Q. High-Wind Drag Coefficient Based on the Tropical Cyclone Simulated With the WRF-LES Framework. *Front. Earth Sci.* **2021**, *9*, 694314. [CrossRef]
33. Byrne, D.; Zhang, J.A. Height-dependent transition from 3-D to 2-D turbulence in the hurricane boundary layer. *Geophys. Res. Lett.* **2013**, *40*, 1439–1442. [CrossRef]
34. Kepert, J.; Wang, Y. The Dynamics of Boundary Layer Jets within the Tropical Cyclone Core. Part II: Nonlinear Enhancement. *J. Atmos. Sci.* **2001**, *58*, 2485–2501. [CrossRef]
35. Bell, M.M.; Montgomery, M.T. Observed Structure, Evolution, and Potential Intensity of Category 5 Hurricane Isabel (2003) from 12 to 14 September. *Mon. Wea. Rev.* **2008**, *136*, 2023–2046. [CrossRef]
36. Zhang, J.A.; Rogers, R.F.; Nolan, D.S.; Marks, F.D. On the Characteristic Height Scales of the Hurricane Boundary Layer. *Mon. Wea. Rev.* **2011**, *139*, 2523–2535. [CrossRef]
37. Kepert, J.D. The dynamics of boundary layer jets within the tropical cyclone core. Part I: Linear theory. *J. Atmos. Sci.* **2001**, *58*, 2469–2484. [CrossRef]
38. Zhang, J.A.; Drennan, W.M.; Black, P.G.; French, J.R. Turbulence Structure of the Hurricane Boundary Layer between the Outer Rainbands. *J. Atmos. Sci.* **2009**, *66*, 2455–2467. [CrossRef]
39. Bryan, G.H.; Worsnop, R.P.; Lundquist, J.K.; Zhang, J.A. A simple method for simulating wind profiles in the boundary layer of tropical cyclones. *Boundary-Layer Meteorl.* **2017**, *162*, 475–502. [CrossRef]

atmosphere

MDPI

Article

Evaluation of Air–Sea Flux Parameterization for Typhoon Mangkhut Simulation during Intensification Period

Lei Ye [1,2], Yubin Li [1,2,*] and Zhiqiu Gao [1,3]

1 Collaborative Innovation Center on Forecast and Evaluation of Meteorological Disasters, Key Laboratory for Aerosol-Cloud-Precipitation of China Meteorological Administration, School of Atmospheric Physics, Nanjing University of Information Science and Technology, Nanjing 210044, China
2 Southern Marine Science and Engineering Guangdong Laboratory (Zhuhai), Zhuhai 519000, China
3 State Key Laboratory of Atmospheric Boundary Layer Physics and Atmospheric Chemistry, Institute of Atmospheric Physics, Chinese Academy of Sciences, Beijing 100029, China
* Correspondence: liyubin@nuist.edu.cn

Abstract: Using the Advanced Research Weather Research and Forecasting (WRF) model, a series of numerical experiments are conducted to examine the sensitivity of the Typhoon Mangkhut intensification simulation to different air–sea flux parameterization schemes (*isftcflx* option), including option 0 (OPT0), option 1 (OPT1), and option 2 (OPT2). The results show that three schemes basically reproduce tropical cyclone (TC) track and intensity of observation, and the simulated exchange coefficient of three schemes is consistent with theoretical results. Using the same upper limit of C_d as OPT0 and OPT2, OPT1 has much larger C_k than the other two options, which leads to larger latent heat (and sensible heat) flux and produces stronger inflow (within boundary layer) and updrafts (around eyewall), and thus stronger TC intensity. Meanwhile, the results that larger C_k/C_d corresponds with stronger TC in the mature stage are consistent with Emanuel's potential intensity theory. The fact that C_k in OPT1 is evidently larger than the C_k from previous studies leads to produce a better TC intensity simulation. Generally, we should use more reasonable air–sea flux parameterization based on observation to improve TC intensity simulation.

Keywords: air–sea flux parameterization; Typhoon Mangkhut; intensification period; tropical cyclone; *isftcflx* option

Citation: Ye, L.; Li, Y.; Gao, Z. Evaluation of Air–Sea Flux Parameterization for Typhoon Mangkhut Simulation during Intensification Period. *Atmosphere* **2023**, *13*, 2133. https://doi.org/10.3390/atmos13122133

Academic Editor: Anthony R. Lupo

Received: 30 November 2022
Accepted: 16 December 2022
Published: 19 December 2022

1. Introduction

As one of the most destructive weather systems, a tropical cyclone (TC) causes severe loss of lives and property damage around the world [1]. Over the past decades, with the development of numerical weather prediction models, TC track prediction has steadily improved, however, TC intensity forecast, especially accurate prediction of the TC intensification period, remains a challenge and problem [2].

The exchange of heat, moisture, and momentum at the air–sea interface is believed to be one of the most important physical processes determining TC intensity change [3–5]. The air–sea surface flux exchanges of moist enthalpy (the sum of latent and sensible heat fluxes) act as the primary energy source of TC, whereas the surface momentum flux is the sink of TC development [6]. An accurate description of the air–sea interaction processes is important to improve the forecast of TC evolution [7].

In current numerical weather prediction models, air–sea fluxes cannot be resolved directly, and therefore must be accounted for through subgrid-scale parameterizations of surface exchange coefficients [4,5]. The surface fluxes of momentum τ, sensible heat SH, and latent heat LH are expressed as

$$\tau = -\rho u_*^2 = -\rho C_d (\Delta U)^2, \tag{1}$$

$$SH = -\rho c_p u_* \theta_* = -(\rho c_p) C_h U \Delta\theta, \tag{2}$$

$$LH = -\rho L_v u_* q_* = -(\rho L_v) C_q U \Delta q, \tag{3}$$

where ρ is the air density, c_p is the specific heat capacity of air at constant pressure, L_v is the latent heat of vaporization, u_* is the friction velocity, and θ_* and q_* are the surface layer temperature and moisture scales, respectively; $\Delta(U, \theta, q)$ are the differences in wind speed, temperature, and water vapor between a reference height z_{ref} (often 10 m) and the bottom of the surface layer ($U = 0$), respectively. C_d, C_h, and C_q are the bulk exchange coefficients for drag, sensible heat, and latent heat, respectively.

Based on the Monin–Obukhov similarity theory and observations indicating a neutrally stable surface layer within the TC eyewall (e.g., [8]), u_*, θ_*, and q_* in neutral stability are given as follows:

$$u_* = \frac{kU}{\ln\left(\frac{Z_{ref}}{Z_0}\right)}, \tag{4}$$

$$\theta_* = \frac{k\Delta\theta}{\ln\left(\frac{Z_{ref}}{Z_t}\right)}, \tag{5}$$

$$q_* = \frac{k\Delta q}{\ln\left(\frac{Z_{ref}}{Z_q}\right)}, \tag{6}$$

The drag coefficient C_d, sensible heat exchange coefficient C_h, and latent heat exchange coefficient C_q under neutral stability conditions are calculated from

$$C_{d,n} = \frac{k^2}{\left[\ln\left(\frac{z_{ref}}{z_0}\right)\right]^2}, \tag{7}$$

$$C_{h,n} = \frac{k^2}{\ln\left(\frac{z_{ref}}{z_0}\right) \times \ln\left(\frac{z_{ref}}{z_t}\right)} = C_{d,n}^{1/2} \times \frac{k}{\ln\left(\frac{z_{ref}}{z_t}\right)}, \tag{8}$$

$$C_{q,n} = \frac{k^2}{\ln\left(\frac{z_{ref}}{z_0}\right) \times \ln\left(\frac{z_{ref}}{z_q}\right)} = C_{d,n}^{1/2} \times \frac{k}{\ln\left(\frac{z_{ref}}{z_q}\right)}, \tag{9}$$

where k is the von Karman constant, z_{ref} is the reference height (often 10 m), the subscript n represents neutrally stable conditions, and z_0, z_t, and z_q are the roughness lengths for momentum, sensible heat, and latent heat (moisture), respectively. Additionally, the enthalpy exchange coefficient C_k is defined as the combination of the latent and sensible heat exchange [9].

Emanuel's well-known potential intensity (PI) theory states that the potential intensity for a steady-state (or mature) TC is proportional to the ratio of surface bulk transfer coefficients for enthalpy and momentum, C_k/C_d, and suggested the ratio C_k/C_d in real hurricanes lies in the range 0.75–1.5 [10,11]. Clearly, the ratio C_k/C_d is an important factor in the TC intensity of a mature storm.

Using numerical simulations, many studies have shown that TC intensity simulation is affected by the choice of air–sea flux (or surface roughness) parameterization schemes. Green and Zhang [5] examined the impact of surface flux on the intensity and structure of tropical cyclones using different air–sea flux parameterization schemes available in Weather Research and Forecasting (WRF) version 3.4 through simulations of Hurricane Katrina (2005). Chen et al. [12] focused on Hurricane Katrina (2005) through a coupled atmosphere–ocean modeling system and examined the combined impacts of air–sea flux parameterizations and ocean cooling on TC evolution. Additionally, Kueh et al. [6] investigated the effects of horizontal resolution and surface flux formulas on typhoon intensity and structure simulations through the case study of Super Typhoon Haiyan (2013). Additionally, Greeshma et al. [13] and Alimohammadi et al. [14] investigated the impact of air–sea flux parametrization schemes on the simulation of TC over the North Indian Ocean and the Arabian Sea, respectively. Furthermore, Green and Zhang [5] and Nystrom et al. [15,16] focused on the impact of surface exchange coefficient uncertainty on TC simulation through varying parameters in the surface exchange coefficient formulas.

Based on previous studies and the update of the air–sea flux parameterization scheme in the numerical model, in this study, we further evaluate and analyze the difference of air–sea flux parameterization schemes and their impacts on TC intensification simulation with WRF version 4.0 in order to understand the impact of air–sea flux on TC intensification, which can be beneficial to improve TC intensity forecast. This study takes Typhoon Mangkhut in 2018 as an example and compares different air–sea flux parameterization schemes. Their different TC simulation results are analyzed, and the associated mechanisms are investigated.

The rest of the paper is organized as follows. Section 2 introduces Typhoon Mangkhut, the model setup, and the experimental design. Section 3 presents the simulation results and analyzes the mechanism. The summary and conclusions are provided in Section 4.

2. Numerical Simulations of Typhoon Mangkhut

2.1. Overview of Typhoon Mangkhut

Mangkhut was the 22nd named Tropical Cyclone of the 2018 Pacific typhoon season [17]. It was an extremely powerful and catastrophic tropical cyclone that caused extensive damage [18].

Mangkhut formed over the western North Pacific about 2330 km east of Guam on 7 September, moved westwards quickly, and intensified gradually in the following few days [17,19]. Mangkhut developed into a super typhoon on 11 September. It turned to the northwest on 14 September, reaching its peak intensity before making landfall over Luzon with an estimated maximum sustained wind of 250 km h^{-1} near the center. Mangkhut weakened after crossing the northern part of Luzon and continued to track northwestwards quickly across the northern part of the South China Sea towards the coast of Guangdong. Mangkhut weakened into a severe typhoon on the morning of 16 September and made landfall in the vicinity of Taishan in Guangdong before dusk. It then moved into the western part of Guangdong and weakened further. Mangkhut degenerated into an area of low pressure over Guangxi the next night. In this study, we focus on the intensification period: from 0000 UTC (Universal Time Coordinated) 7 September to 0000 UTC 13 September 2018, and the track and intensity in this period can be found in Figure 1.

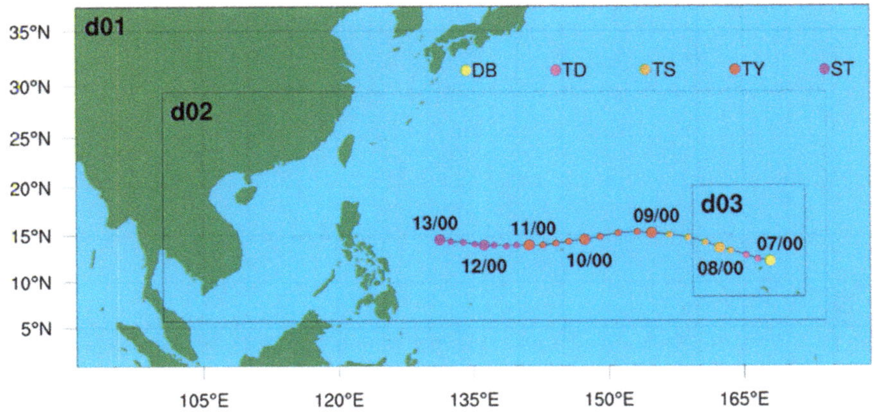

Figure 1. The location of model-simulated domains and the best track of Mangkhut from JTWC during the simulated period from 0000 UTC 7 to 0000 UTC 13 September 2018. Colored dots are the TC track, and colors indicate the categories of TC intensity (DB: disturbance; TD: tropical depression; TS: tropical storm; TY: typhoon; ST: super typhoon).

2.2. Experimental Design

The Advanced Research WRF model version 4.0 developed by the US National Center of Atmospheric Research (NCAR) was used for Typhoon Mangkhut simulation. A two-way interactive, three-level nested grid configuration with a track-following moving-nest option was employed. Figure 1 shows the setting of model domains.

The domains have 27, 9, and 3 km horizontal resolution and 342 × 153, 856 × 292, and 433 × 424 grid points, respectively. A total of 33 vertical levels were configured, and the pressure at the uppermost layer was 50 hPa. The initial and lateral boundary conditions were obtained from the fifth-generation European Centre for Medium-Range Weather Forecasts (ECMWF) atmospheric reanalysis of the global climate (ERA5) dataset at a resolution of 0.25° × 0.25°. ERA5 is a widely used dataset for the initial and boundary condition in numerical simulation because of its high resolution of 0.25° × 0.25°, which can provide ideal initial and boundary conditions for numerical simulations [20–22]. Additionally, ERA5's 0.25° resolution is close to the 27 km resolution of the outermost domain in our simulation, which minimizes the error during the reguiding of the initial and boundary condition preparation. Therefore, ERA5 is used in this study. The ERA5 reanalysis data of wind, specific humidity, temperature, and surface flux were also used to assess the model performance. The TC best track from the Joint Typhoon Warning Center (JTWC) was used for evaluating the simulated TC track and intensity.

All the experiments started at 0000 UTC 7 September and ended at 0000 UTC 13 September 2018, covering the intensification period of Typhoon Mangkhut. The first 6 h of integration (i.e., from 0000 UTC to 0006 UTC 7th September 2018) served as the spin-up period, and the results in this period are not investigated.

The model's physics options were the same for all three domains, except that cumulus parameterization was only used in the 27 km horizontal resolution domain. Physical parameterization options included the Eta (Ferrier) microphysics scheme [23] and the Kain–Fritsch cumulus parameterization scheme [24], the Unified Noah land surface model [25], the Rapid Radiative Transfer longwave radiation scheme (RRTM) [26], the Dudhia shortwave radiation scheme [27], the Mellor–Yamada–Nakanishi–Niino Level 2.5 (MYNN2) [28,29], and the Revised MM5 surface layer scheme [30].

It is noted that the choice of the planetary boundary layer (PBL) and surface layer parameterization scheme is based on our recent work (Ye et al., in preparation), where we found that among the 7 most widely used PBL parameterization schemes, the MYNN2 produced better simulation for the distribution and variation of PBL eddy diffusivity, which is closer to observation, and here, the MYNN2 PBL scheme is coupled to the Revised MM5 surface layer scheme.

2.3. Description of Air–Sea Flux Parameterization in WRF Version 4.0

In Advanced Research WRF, it allows users to change air–sea fluxes formulation through the "namelist" option *isftcflx*. The revised MM5 surface layer scheme in WRF version 4.0 uses three options for calculating the roughness lengths and air–sea flux. In this study, we conducted a series of sensitivity experiments with 3 different air–sea flux parameterization schemes, respectively. The description is considered comprehensively in WRF version 4.0 code and detailed description is in Green and Zhang [4], Kueh et al. [6], and Alimohammadi et al. [14].

2.3.1. Option 0 (*Isftcflx* = 0)

Option 0 (*isftcflx* = 0) is the default option for air–sea flux (or surface roughness) parameterization in WRF Version 4.0. In this option, the momentum roughness length over water is based upon Charnock [31] and adds an upper limit, as in option 1 and 2.

$$z_0 = min\left(\alpha\frac{u_*^2}{g} + \frac{0.11 \times 1.5 \times 10^{-5}}{u_*}, 2.85 \times 10^{-3}\right), \quad (10)$$

where α = 0.0185 is the Charnock coefficient, u_* is the frictional velocity, and g is the gravitational acceleration. Additionally, the constant value of 1.5×10^{-5} is used for the kinematic viscosity of air. The Charnock expression relates the aerodynamic roughness length to friction velocity for describing the gross effect of wavy sea surface induced by wind stress, and the viscous term describes roughness behavior under smooth flow conditions [6]. In the case of the momentum roughness length monotonic increasing with wind speed, option 0 adds an upper limit as in options 1 and 2 from WRF version 3.9.

The heat and moisture roughness lengths for option 0 over water are based upon Fairall et al. [32] and expressed as a function of the roughness Reynolds number, with a lower limit of 2.0×10^{-9} and an upper limit of 1.0×10^{-4}:

$$z_t = z_q = max\left[2.0 \times 10^{-9}, min\left(1.0 \times 10^{-4}, 5.5 \times 10^{-5} \times Re_*^{-0.6}\right)\right], \tag{11}$$

where $Re_* = z_0 u_* / \nu$ is the roughness Reynolds number, and ν is the kinematic viscosity of air as a function of air temperature for the Re_* calculation [6]. The heat roughness length z_t is set to equal to moisture roughness length z_q. The upper limit of 1.0×10^{-4} keeps the heat and moisture roughness length constant at extreme wind speeds, whereas the lower limit of 2.0×10^{-9} is implemented to prevent the model from blowing up.

2.3.2. Option 1 (*Isftcflx* = 1)

Recent field observations and laboratory experiments showed that C_d is saturated at hurricane wind speed [8,33]. For option 1 (*isftcflx* = 1), the momentum roughness length is given by a blend of two roughness length formulas [4]:

$$z_0 = max\left\{1.27 \times 10^{-7}, min\left[z_w z_2 + (1 - z_w) z_1, 2.85 \times 10^{-3}\right]\right\}, \tag{12}$$

$$z_w = min\left[1.0, \left(\frac{u_*}{1.06}\right)^{0.3}\right], \tag{13}$$

$$z_1 = 0.011 \times \frac{u_*^2}{g} + \frac{0.11 \times 1.5 \times 10^{-5}}{u_*}, \tag{14}$$

$$z_2 = \frac{10}{\exp\left(9.5 u_*^{-1/3}\right)} + \frac{0.11 \times 1.5 \times 10^{-5}}{max(u_*, 0.01)}, \tag{15}$$

where z_1 is again the Charnock [31] expression plus a viscous term, with Charnock coefficient α = 0.011. z_2 is the exponential expression from Davis et al. [9] plus a viscous term, with a constant kinematic viscosity $\nu = 1.5 \times 10^{-5}$. The two roughness length formulas are combined using a weight function z_w, with a lower limit of 2.0×10^{-9} and an upper limit of 1.0×10^{-4} on z_0. The lower and upper limits on z_0 are adapted from Davis et al. [9], with a slightly different value of the lower limit. The upper limit is used to prevent a monotonic increase in z_0 and C_d at high wind conditions. Additionally, leveling-off of C_d at high wind speed suggests a decline in the efficiency of the exchange of momentum across the air–sea interface.

For the heat and moisture roughness lengths, based on Large and Pond [34], option 1 sets z_t and z_q as constants of 10^{-4} m for all wind speeds, as:

$$z_t = z_q = 10^{-4}, \tag{16}$$

2.3.3. Option 2 (*Isftcflx* = 2)

For the momentum roughness length, option 2 (*isftcflx* = 2) uses the same formulation as option 1. Additionally, the heat and moisture roughness lengths are expressed based on the formula proposed by Brutsaert [35]:

$$z_t = z_0 \exp\left[-k\left(7.3 Re_*^{1/4} Pr^{1/2} - 5\right)\right], \tag{17}$$

$$z_q = z_0 \exp\left[-k\left(7.3Re_*^{1/4}Sc^{1/2} - 5\right)\right], \tag{18}$$

where $Re_* = z_0 u_* / \nu$ is the roughness Reynolds number, and ν is the kinematic viscosity of air. k is the von Karman constant, Pr is the Prandtl number, and Sc is the Schmidt number, and Garratt [36] is used with values of k = 0.40, Pr = 0.71, and Sc = 0.60.

A brief description of the sensitivity experiments is in Table 1, and the relationship between roughness length (or exchange coefficient) and wind speed for different options is in Figure 2. Additionally, these 3 air–sea flux parameterization schemes in revised MM5 stay unchanged from WRF version 3.9 to 4.3, released in 2022.

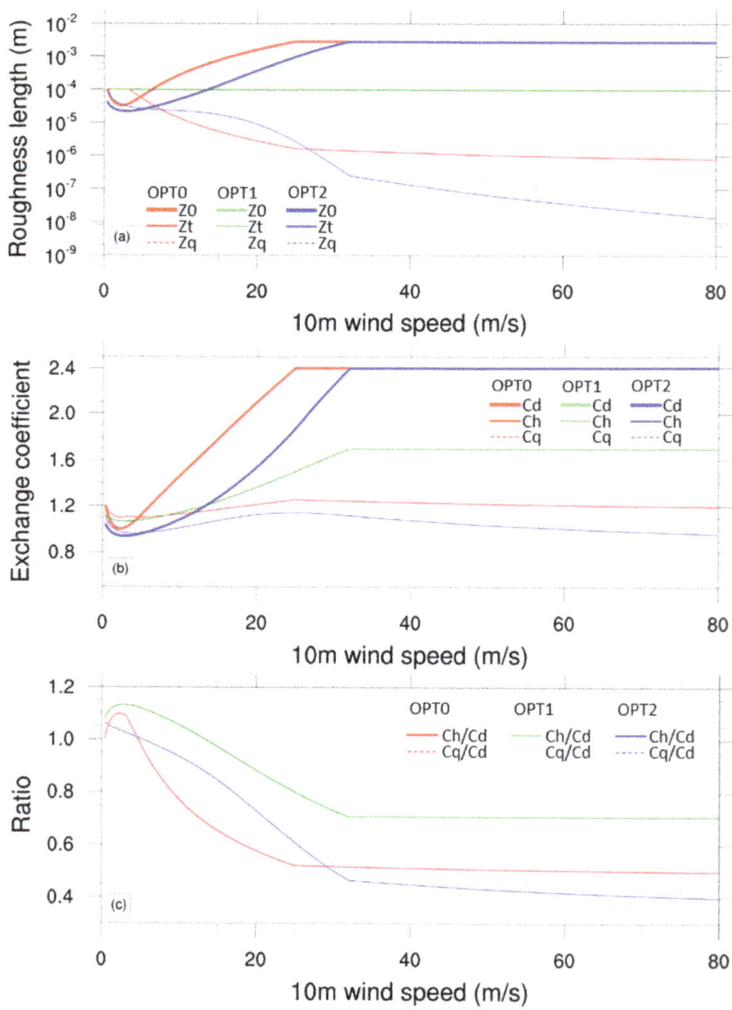

Figure 2. Plots as functions of 10 m wind speed of (**a**) roughness lengths z_0 (thick solid), z_t (thin solid), and z_q (dashed); (**b**) exchange coefficients C_d (thick solid), C_h (thin solid), and C_q (dashed); (**c**) exchange coefficient ratios C_h/C_d (thin solid) and C_q/C_d (dashed), for neutral stability condition. Air–sea flux schemes option 0, option 1, and option 2, colored red, green, and blue, respectively. As stated in the text, some curves are identical: for option 0, $z_t = z_q$ and $C_h = C_q$; for option 1, $z_t = z_q$ and $C_h = C_q$; and between option 1 and option 2, z_0 (and thus C_d) is the same.

Table 1. Brief description of the sensitivity numerical experiments.

Experiment Name	Surface Layer Scheme	Air–Sea Flux Scheme	C_d	C_k
OPT0	Revised MM5	*isftcflx* = 0	Charnock [31], Donelan et al. [33]	Fairall et al. [32]
OPT1	Revised MM5	*isftcflx* = 1	Donelan et al. [33]	Large and Pond [34]
OPT2	Revised MM5	*isftcflx* = 2	Donelan et al. [33]	Garratt [36]

Figure 2 compared the theoretical results of the roughness length and surface exchange coefficient from 3 surface flux parameterization schemes in the WRF version 4.0 model. For the roughness length in Figure 2a, specifically for the momentum roughness length z_0 (thick solid line), OPT1 is the same as OPT2, and all of three have the same upper limit; for the thermal roughness length z_t (thin solid line) and the vapor roughness length z_q (dashed line), OPT1 is greater than OPT0 and OPT2. The surface exchange coefficient is calculated from the roughness length, and for the surface exchange coefficient in Figure 2b, OPT1 is the same as OPT2 for the drag coefficient C_d (thick solid line), and the 3 all have the same upper limit. For the heat exchange coefficient C_h (thin solid line) and water vapor exchange coefficient C_q (dashed line), OPT1 is greater than OPT0 and OPT2. Thus, for the exchange coefficient ratio in Figure 2c, OPT1 is greater than OPT0 and OPT2 for the ratio of C_h/C_d (thin solid line) and C_q/C_d (dashed line), and in the high wind speed, OPT0 is greater than OPT2.

3. Results

3.1. TC Track and Intensity

In this subsection, we focused on the results of the simulated TC track and intensity for three different air–sea flux schemes for Typhoon Mangkhut during the TC intensification to verify the simulation.

Figure 3a,b shows that the simulated tracks of the three air–sea flux schemes are similar and generally reproduce the westward movement of Typhoon Mangkhut from the JTWC best track. Specifically, from the 7th to 9th September, the simulated tracks of the three schemes were very close, and then the track errors of the three schemes began to gradually increase. Overall, the simulated track was very close. It indicates that the simulated track is not very sensitive to the different air–sea flux scheme, which is consistent with the results in Green and Zhang [4]. It might be related to using the same PBL scheme and surface layer scheme.

For simulated TC intensity, Figure 3c,d further compares the time series of the minimum central sea level pressure (MSLP) and the maximum sustained wind speed (VMAX) of 3 schemes. Compared with the JTWC best track data, the three numerical simulations captured the variation trends of MSLP and VMAX, although an underestimation and difference were found in simulated TC intensity for the three schemes. In the first 3 days until 0000 UTC 10 September, the simulation difference of MSLP and VMAX in the experiments was very small. After that, differences started to show. Seen from the parameter of MSLP, it can be found that OPT1 generally captured the rapid intensification processes from 1800 UTC 9 September to 0600 UTC 11 September, followed by OPT1, while OPT2 showed weaker intensification rates. While looking at the parameter of VMAX, a similar pattern can be found. During the rapid intensification processes at the end of the simulation, among the three simulations, OPT1 had the largest VMAX, followed by OPT0, whereas OPT2 was smaller than the JTWC's VMAX. The simulated results of TC intensity demonstrate the evident sensitivity of intensity to the air–sea flux scheme, which is consistent with the conclusion in Green and Zhang [4]. According to theoretical results of exchange coefficient in Figure 2, at high wind speed, OPT1 has the larger C_h/C_d and C_q/C_d with the same C_d and larger C_h and C_q. The ratio of C_h/C_d and C_q/C_d correspond to the TC intensity. In the next subsection, we further study the relationship between TC intensity and exchange coefficient based on the simulated results in the three simulations.

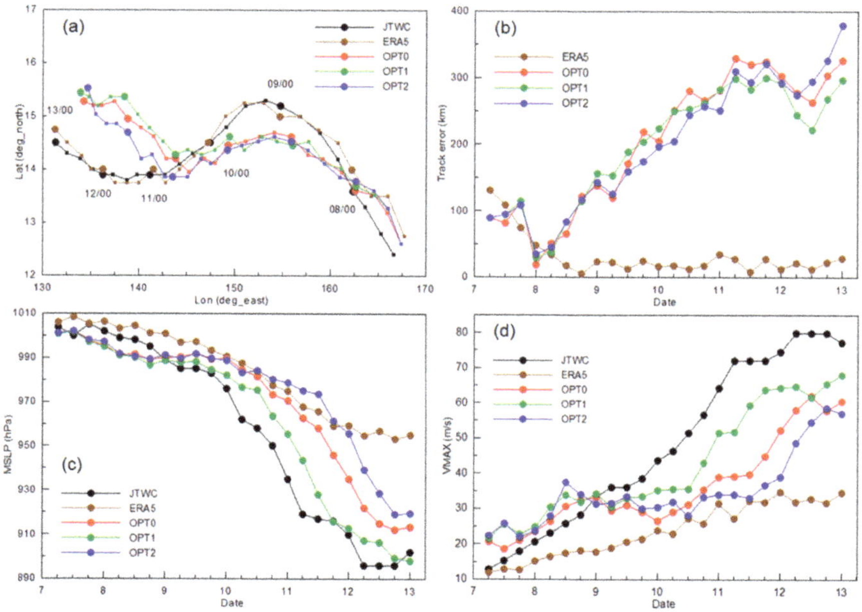

Figure 3. (**a**) TC track, (**b**) time series of track error (unit: km), (**c**) minimum central sea level pressure (MSLP) (unit: hPa), and (**d**) maximum sustained wind speed (VMAX) (unit: m s^{-1}) from the JTWC best track data; ERA5 reanalysis data and the numerical simulations from 0600 UTC 7 to 0000 UTC 13 September 2018.

3.2. *Surface Exchange Coefficient*

The intensity variations among different air–sea flux (or surface roughness) schemes become evident in the TC intensification stage because of the dependence of wind speed on surface exchanges of momentum and moisture enthalpy [37]. In this subsection, we analyze the variation of the simulated surface exchange coefficient with the 10 m wind speed among the three schemes at the TC peak intensity of 1200 UTC 12th September, at which TC reached mature stage.

In Figure 4, the simulated values for C_d and C_k are similar with the theoretical values in Figure 2. In Figure 4a, the C_d of the three schemes all increased initially with increasing wind speed until high wind, and the C_d of OPT0 reached saturation slightly earlier than the C_d of OPT1 and OPT2. Additionally, the C_d of the three schemes share the same upper limit (about 2.4×10^{-3}) under high wind speed (beyond about 33 m s^{-1}).

In Figure 4b, the C_k of OPT0 is nearly unchanged with increasing wind speed. Similar to C_d, the C_k of OPT1 increased initially with increasing wind speed below 33 m s^{-1} and saturated thereafter. Additionally, the C_k of OPT2 increased slowly with increasing wind speed below 33 m s^{-1} and thereafter tended to decrease slightly. Among the three simulations, the C_k of OPT1 is significantly larger than the C_k of OPT0 and OPT2, and in high wind speed, the C_k of OPT0 is slightly larger than the C_k of OPT2.

Due to the same upper limit in C_d for all three schemes, the difference in C_k affects the differences in C_k/C_d. In Figure 4c, the C_k/C_d of OPT1 is the largest, and the C_k/C_d of OPT0 is slightly larger than the C_k/C_d of OPT2 when wind speed exceeds 33 m s^{-1}. Emanuel [11] suggested that in real hurricanes C_k/C_d lies in the range 0.75–1.5. Among the three schemes, only OPT1 follows this rule (Figure 4c). It is noted that the C_k of OPT1 is significantly larger than observation results from previous studies [38–40].

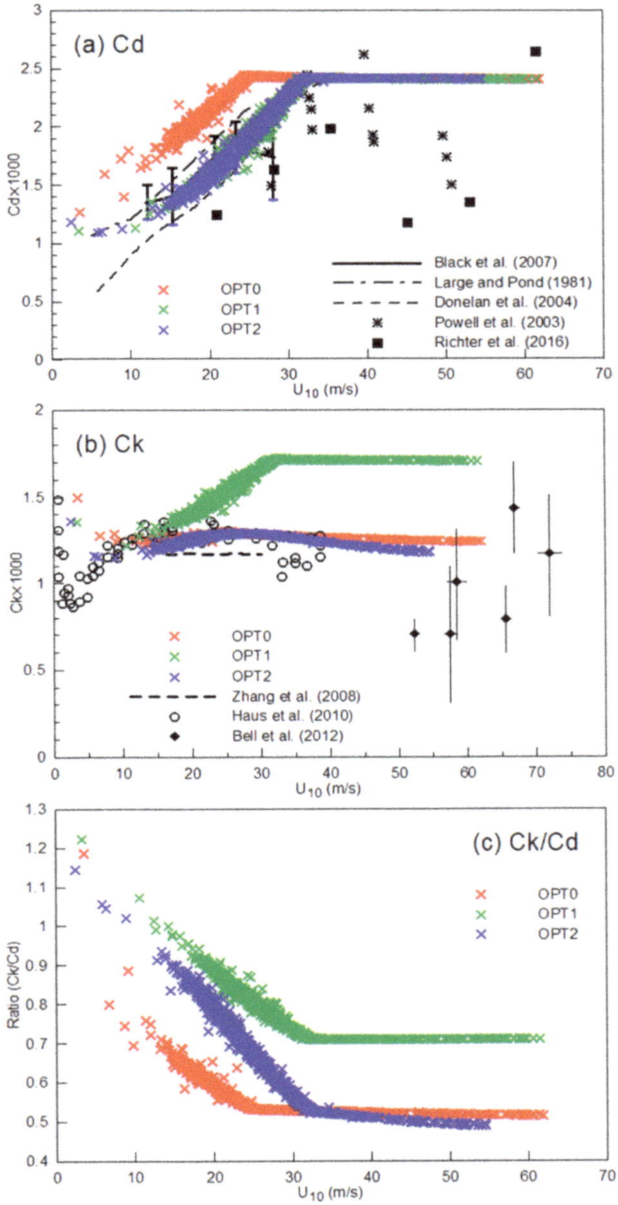

Figure 4. The variation of surface exchange coefficient for (**a**) C_d, (**b**) C_k, and (**c**) ratio of C_k/C_d with 10 m wind speed at 1200 UTC 12 September 2018; numerical simulations by OPT0 (in red), OPT1 (in green), and OPT2 (in blue) compared with data from Large and Pond [38], Powell et al. [8], Donelan et al. [33], Black et al. [3], Zhang et al. [39], Haus et al. [40], Bell et al. [41], and Richter et al. [42] (in different black marks).

Since maximum wind speed is strongly sensitive to C_k/C_d [11], the decrease in C_d (generally tends to reduce the energy loss) and increase in C_k (thus more energy gain) have greater potential to yield a stronger storm [6]. Correspondingly, in this study, the magnitude of C_k/C_d is consistent with the simulated TC intensity relationship for the three schemes because differences in C_d and C_k affect surface energy transport and influence the storm intensification by WISHE type of feedback [4,13,37].

3.3. Surface Flux

As mentioned above, different air–sea flux (or surface roughness) schemes produce different C_d and C_k values, and C_k and C_d determine the surface enthalpy flux and momentum flux, respectively [43]. The surface enthalpy flux is the main source of energy for TC development, while surface momentum flux is the main sink of TC energy. In this subsection, we look into the simulated result of surface fluxes in numerical experiments.

Figures 5–7 plot the distribution of surface latent heat, sensible heat, and momentum flux simulated by different sea–air flux schemes within a radius of 500 km at 1200 UTC 12, September 2018, which represents mature TC. Additionally, ERA5 reanalysis data are used here for comparison. Consistent with TC intensity results in Figure 3, surface fluxes of ERA5 are significantly smaller than simulated surface fluxes from three options, although the distributions of surface fluxes from ERA5 and simulations are similar. Figures 5 and 6 show that sensible heat fluxes are smaller than latent heat fluxes, which is consistent with those derived from field measurements (e.g., [4,38]). Additionally, whether latent and sensible heat flux or momentum flux, surface flux is the weakest in the TC center and gradually increases outward from the TC eye, reaching its maximum near the eyewall.

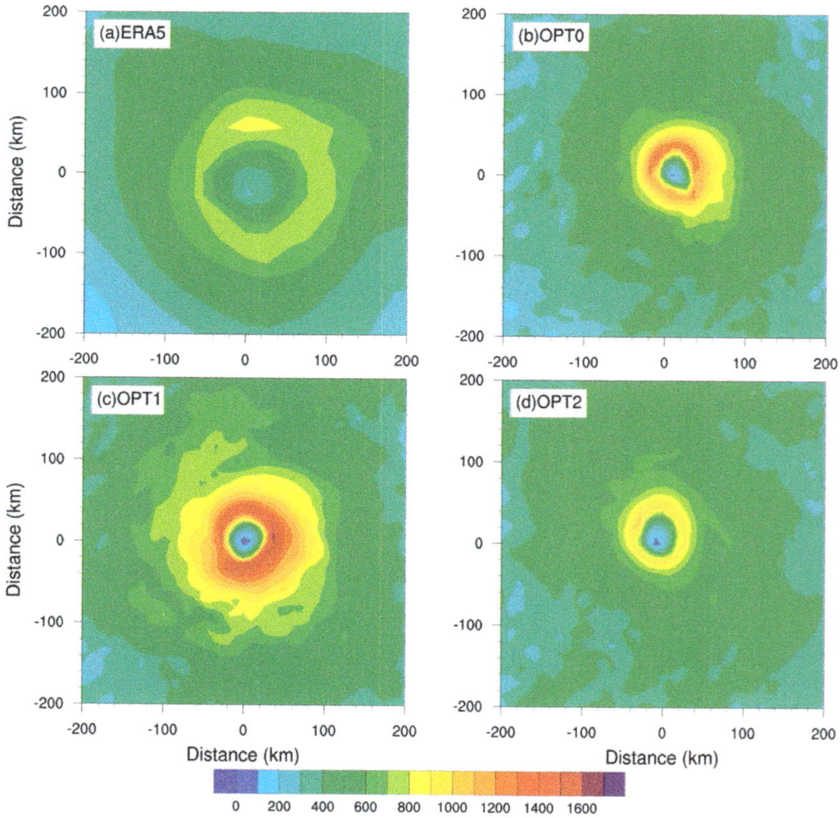

Figure 5. Surface latent heat flux (unit: W m^{-2}) of Typhoon Mangkhut from (**a**) ERA5, (**b**) OPT0, (**c**) OPT1, and (**d**) OPT2 at 1200 UTC 12 September 2018.

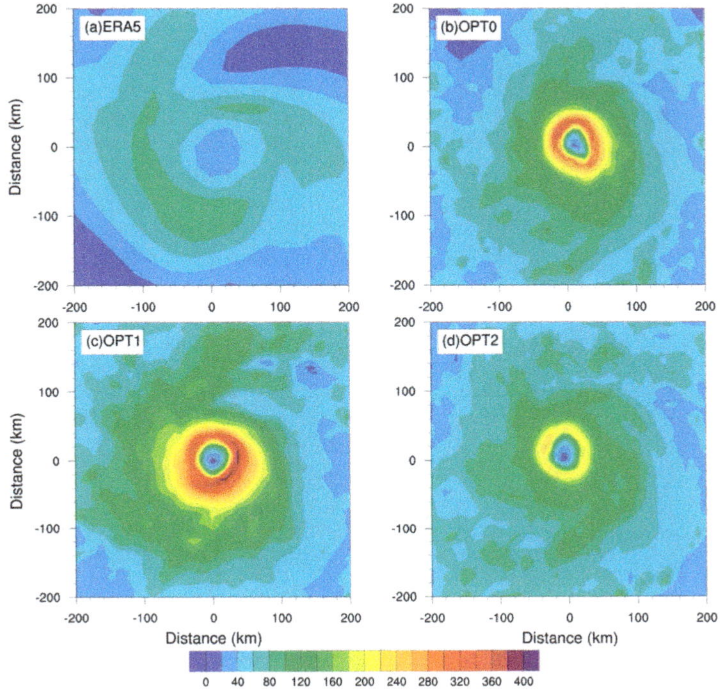

Figure 6. As in Figure 5 but for surface sensible heat flux (unit: W m^{-2}).

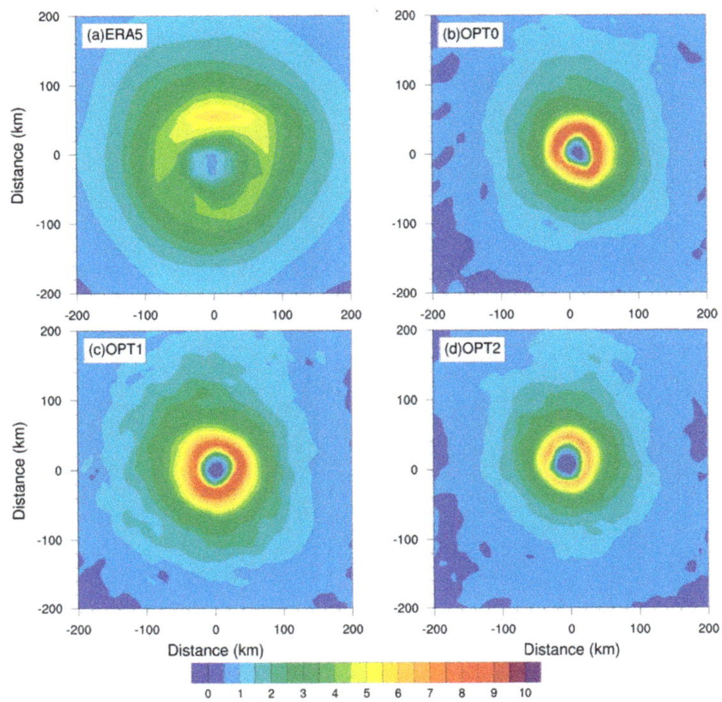

Figure 7. As in Figure 5 but for surface momentum flux (unit: kg m^{-1} s^{-2}).

Consistent with the magnitude relationship of TC intensity, OPT1 produced the largest latent flux and sensibility, followed by OPT0 and OPT2. This better simulation of TC intensity by OPT0 can be attributed to the steady increase in latent flux for high wind speed. The larger latent heat flux transfer results in stronger moist convection, thereby leading to stronger TC intensity for simulation through thermo-dynamical interactions [37]. Similarly, OPT1 produced larger momentum flux than OPT0 and OPT2, since momentum flux is influenced by wind speed. Overall, the difference in momentum is relatively small due to the same peak C_d for the three schemes.

In general, although sharing the same upper limit in C_d as OPT0 and OPT2, OPT1 has much larger C_k than the other two options, which leads to larger enthalpy flux, and especially latent heat flux, and thus stronger TC intensity.

3.4. TC Wind Field

In this subsection, the height–radius distribution of azimuthally averaged tangential wind, radial wind, and upward vertical wind within 200 km from the TC center 1200 UTC 12 September simulated by the three schemes are analyzed. Similar with the last subsection, 1200 UTC 12 September is chosen to represent the TC mature stage when the simulated TC reached its peak intensity. Additionally, ERA5 reanalysis data are used as a reference to compare the simulation results. Although ERA5 generally underestimates the TC intensity, it provides an effective reference for TC size [44]. Consistent with TC intensity results in Figure 3, the wind speed of ERA5 is significantly weaker than the simulated wind speed from the three options. For TC size (i.e., radius of maximum tangential wind speed) in Figure 8, the three options simulate a similar TC size, which is smaller than ERA5's TC size.

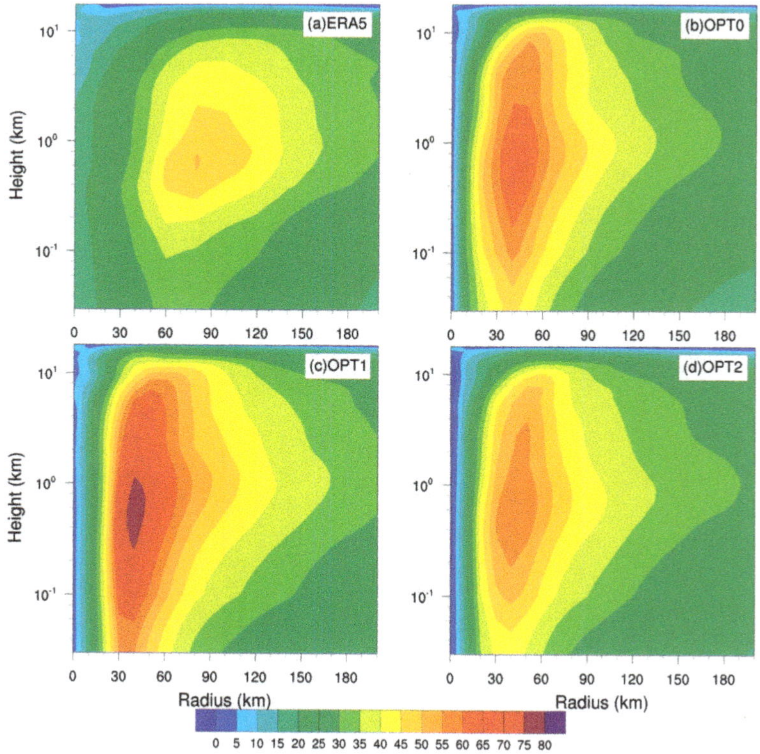

Figure 8. Height–radius distribution of azimuthally averaged tangential wind (unit: m s^{-1}) from (**a**) ERA5, (**b**) OPT0, (**c**) OPT1, and (**d**) OPT2 at 1200 UTC 12 September 2018.

Radius–height plots of the tangential, radial, and vertical winds at 1200 UTC 12 September (Figures 8–10) indicate that the different air–sea flux schemes yield the same general wind structure, which is consistent with results in Green and Zhang [4], albeit with some differences in the magnitude of peak wind speed.

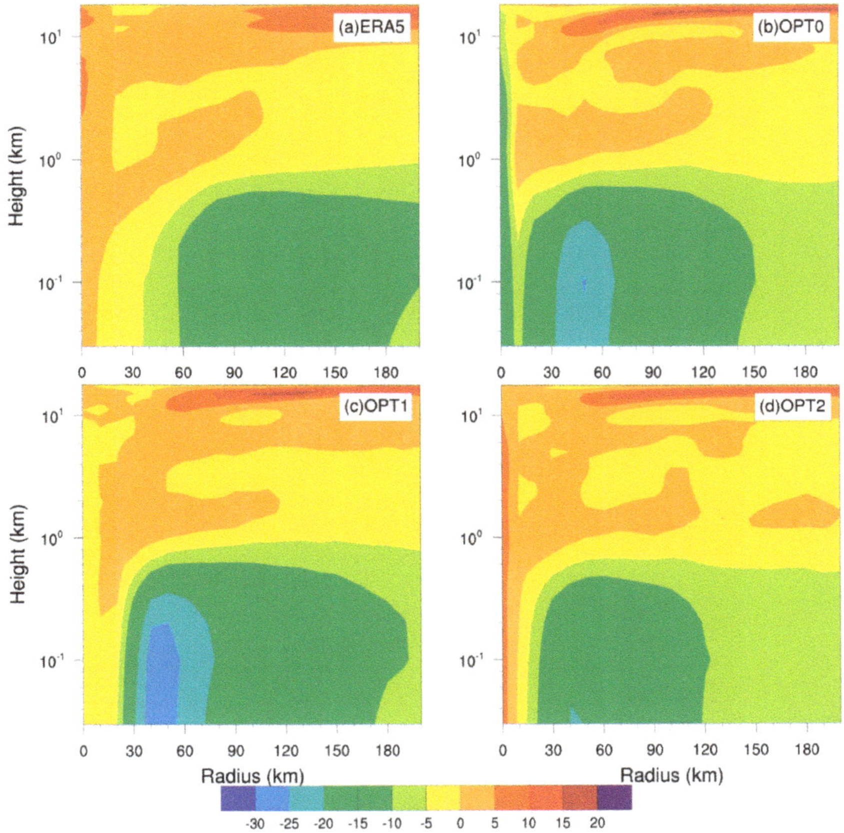

Figure 9. As in Figure 8 but for radial wind (unit: m s^{-1}).

OPT1 has the strongest wind speed, including a peak tangential wind speed greater than 75 m s^{-1}, peak radial inflow velocity (within boundary layer) greater than 25 m s^{-1}, and peak upward velocity (along eyewall) greater than 2.2 m s^{-1}. Next, OPT0 has stronger wind than OPT2, with peak tangential wind speed greater than 65 m s^{-1}, peak radial inflow velocity (within boundary layer) about 25 m s^{-1}, and peak upward velocity (along eyewall) about 1.6 m s^{-1}. Greater tangential wind corresponds to stronger TC, whereas stronger inflow within the boundary layer and updrafts around the eyewall lead to stronger TC. The magnitude relationship of peak wind speed corresponds to the TC intensity, C_k, and surface latent flux.

For better TC intensification simulation by OPT1, although sharing the same upper limit for C_d, the larger C_k of OPT1 produces greater latent flux (and sensible flux) and stronger wind field, including stronger radial inflow (boundary layer) and updrafts (around eyewall), which leads to stronger storms.

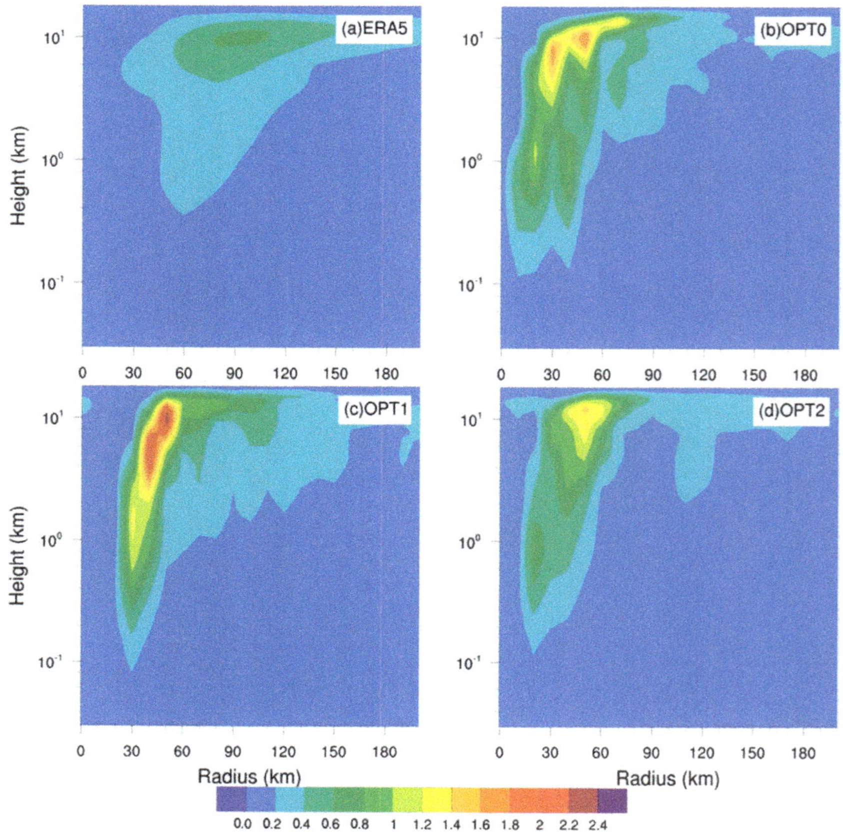

Figure 10. As in Figure 8 but for upward vertical wind (unit: m s^{-1}).

4. Conclusions

In this study, using the Advanced Research WRF version 4.0 model, a series of numerical experiments for Typhoon Mangkhut are performed to examine the sensitivity of TC intensification simulation with three different air–sea flux (or surface roughness) parameterization schemes (*isftcflx* options), including OPT0, OPT1, and OPT2. The major findings can be summarized as follows:

(1) 3 Air–sea schemes simulate and basically reproduce the TC track and intensity of observation, and simulated exchange coefficients are consistent with theoretical results.

(2) Although using the same upper limit of C_d as OPT0 and OPT2, OPT1 has much larger C_k than the other two options, which leads to larger latent heat (and sensible heat) flux and produces stronger inflow (within boundary layer) and updrafts (around eyewall), and thus stronger TC intensity. Meanwhile, the results that larger C_k/C_d corresponds with stronger TC in the mature stage are consistent with Emanuel's PI theory.

(3) Although producing the better TC intensity simulation, as a matter of fact, the C_k in OPT1 is evidently larger than the C_k from previous studies. In brief, more reasonable air–sea flux parameterization can improve TC intensity simulation to a certain extent.

In future, we need to conduct further work on air–sea flux parameterization and observation of TC intensification, in order to produce more accurate intensity forecasts and accurate exchange coefficient simulation under TC condition.

Author Contributions: Conceptualization, L.Y. and Y.L.; methodology, L.Y.; software, L.Y.; validation, L.Y.; formal analysis, L.Y. and Y.L.; investigation, L.Y. and Y.L.; resources, L.Y.; data curation, L.Y.; writing—original draft preparation, L.Y.; writing—review and editing, L.Y., Y.L. and Z.G.; visualization, L.Y.; supervision, Y.L. and Z.G.; project administration, Y.L. and Z.G.; funding acquisition, Y.L. All authors have read and agreed to the published version of the manuscript.

Funding: This study was supported by the National Natural Science Foundation of China (42075072), the Innovation Group Project of Southern Marine Science and Engineering Guangdong Laboratory (Zhuhai) (311021008), and Postgraduate Research and Practice Innovation Program of Jiangsu Province (KYCX22_1152).

Institutional Review Board Statement: Not applicable.

Informed Consent Statement: Not applicable.

Acknowledgments: The numerical calculations in this paper have been done on the supercomputing system in the Supercomputing Center of Nanjing University of Information Science & technology.

Conflicts of Interest: The authors declare no conflict of interest.

References

1. Glenn, S.M.; Miles, T.N.; Seroka, G.N.; Xu, Y.; Forney, R.K.; Yu, F.; Roarty, H.; Schofield, O.; Kohut, J. Stratified coastal ocean interactions with tropical cyclones. *Nat. Commun.* **2016**, *7*, 10887. [CrossRef] [PubMed]
2. Seroka, G.; Miles, T.; Xu, Y.; Kohut, J.; Schofield, O.; Glenn, S. Hurricane Irene Sensitivity to Stratified Coastal Ocean Cooling. *Mon. Weather Rev.* **2016**, *144*, 3507–3530. [CrossRef]
3. Black, P.G.; D'Asaro, E.A.; Drennan, W.M.; French, J.R.; Niiler, P.P.; Sanford, T.B.; Terrill, E.J.; Walsh, E.J.; Zhang, J.A. Air–sea exchange in hurricanes: Synthesis of observations from the Coupled Boundary Layer Air–Sea Transfer experiment. *Bull. Am. Meteorol. Soc.* **2007**, *88*, 357–374. [CrossRef]
4. Green, B.W.; Zhang, F. Impacts of air–sea flux parameterizations on the intensity and structure of tropical cyclones. *Mon. Weather Rev.* **2013**, *141*, 2308–2324. [CrossRef]
5. Green, B.W.; Zhang, F. Sensitivity of tropical cyclone simulations to parametric uncertainties in air–sea fluxes and implications for parameter estimation. *Mon. Weather Rev.* **2014**, *142*, 2290–2308. [CrossRef]
6. Kueh, M.T.; Chen, W.M.; Sheng, Y.F.; Lin, S.C.; Wu, T.R.; Yen, E.; Tsai, Y.L.; Lin, C.Y. Effects of horizontal resolution and air-sea flux parameterization on the intensity and structure of simulated Typhoon Haiyan. *Nat. Hazards Earth Syst. Sci.* **2019**, *19*, 1509–1539. [CrossRef]
7. Rizza, U.; Canepa, E.; Miglietta, M.M.; Passerini, G.; Morichetti, M.; Mancinelli, E.; Virgili, S.; Besio, G.; De Leo, F.; Mazzino, A. Evaluation of drag coefficients under medicane conditions: Coupling waves, sea spray and surface friction. *Atmos. Res.* **2021**, *247*, 105207. [CrossRef]
8. Powell, M.D.; Vickery, P.J.; Reinhold, T. Reduced drag coefficient for high wind speeds in tropical cyclones. *Nature* **2003**, *422*, 279–283. [CrossRef]
9. Davis, C.; Wang, W.; Chen, S.S.; Chen, Y.; Corbosiero, K.; DeMaria, M.; Dudhia, J.; Holland, G.; Klemp, J.; Michalakes, J.; et al. Prediction of landfalling hurricanes with the Advanced Hurricane WRF model. *Mon. Weather Rev.* **2008**, *136*, 1990–2005. [CrossRef]
10. Emanuel, K.A. An air–sea interaction theory for tropical cyclones. Part I: Steady-state maintenance. *J. Atmos. Sci.* **1986**, *43*, 585–604. [CrossRef]
11. Emanuel, K.A. Sensitivity of tropical cyclones to surface exchange coefficients and a revised steady-state model incorporating eye dynamics. *J. Atmos. Sci.* **1995**, *52*, 3969–3976. [CrossRef]
12. Chen, Y.; Zhang, F.; Green, B.W.; Yu, X. Impacts of ocean cooling and reduced wind drag on hurricane Katrina (2005) based on numerical simulations. *Mon. Weather Rev.* **2018**, *146*, 287–306. [CrossRef]
13. Greeshma, M.; Srinivas, C.V.; Hari Prasad, K.B.R.R.; Baskaran, R.; Venkatraman, B. Sensitivity of tropical cyclone predictions in the coupled atmosphere–ocean model WRF-3DPWP to surface roughness schemes. *Meteorol. Appl.* **2019**, *26*, 324–346. [CrossRef]
14. Alimohammadi, M.; Malakooti, H.; Rahbani, M. Comparison of momentum roughness lengths of the WRF-SWAN online coupling and WRF model in simulation of tropical cyclone Gonu. *Ocean Dyn.* **2020**, *70*, 1531–1545. [CrossRef]
15. Nystrom, R.G.; Chen, X.; Zhang, F.; Davis, C.A. Nonlinear impacts of surface exchange coefficient uncertainty on tropical cyclone intensity and air-sea interactions. *Geophys. Res. Lett.* **2020**, *47*, e2019GL085783. [CrossRef]
16. Nystrom, R.G.; Rotunno, R.; Davis, C.A.; Zhang, F. Consistent impacts of surface enthalpy and drag coefficient uncertainty between an analytical model and simulated tropical cyclone maximum intensity and storm structure. *J. Atmos. Sci.* **2020**, *77*, 3059–3080. [CrossRef]
17. He, Y.C.; He, J.Y.; Chen, W.C.; Chan, P.W.; Fu, J.Y.; Li, Q. Insights from Super Typhoon Mangkhut (1822) for wind engineering practices. *J. Wind Eng. Ind. Aerodyn.* **2020**, *203*, 104238. [CrossRef]

18. Yang, J.; Li, L.; Zhao, K.; Wang, P.; Wang, D.; Sou, I.M.; Yang, Z.; Hu, J.; Tang, X.; Mok, K.M.; et al. A Comparative Study of Typhoon Hato (2017) and Typhoon Mangkhut (2018)–Their Impacts on Coastal Inundation in Macau. *J. Geophys. Res. Oceans* **2019**, *124*, 9590–9619. [CrossRef]
19. He, Q.; Zhang, K.; Wu, S.; Zhao, Q.; Wang, X.; Shen, Z.; Li, L.; Wan, M.; Liu, X. Real-Time GNSS-Derived PWV for Typhoon Characterizations: A Case Study for Super Typhoon Mangkhut in Hong Kong. *Remote Sens.* **2020**, *12*, 104. [CrossRef]
20. Li, D.; Staneva, J.; Bidlot, J.R.; Grayek, S.; Zhu, Y.; Yin, B. Improving Regional Model Skills During Typhoon Events: A Case Study for Super Typhoon Lingling Over the Northwest Pacific Ocean. *Front. Mar. Sci.* **2021**, *8*, 613913. [CrossRef]
21. Gevorgyan, A. Convection-permitting simulation of a heavy rainfall event in Armenia using the WRF model. *J. Geophys. Res. Atmos.* **2018**, *123*, 11,008–11,029. [CrossRef]
22. Wu, J.F.; Xue, X.H.; Hoffmann, L.; Dou, X.K.; Li, H.M.; Chen, T.D. A case study oftyphoon-induced gravity wavesand the orographic impactsrelated to Typhoon Mindulle (2004) over Taiwan. *J. Geophys. Res. Atmos.* **2015**, *120*, 9193–9207. [CrossRef]
23. Rogers, E.; Black, T.; Ferrier, B.; Lin, Y.; Parrish, D.; DiMego, G. Changes to the NCEP Meso Eta analysis and forecast system: Increase in resolution, new cloud microphysics, modified precipitation assimilation, and modified 3DVAR analysis. *NWS Tech. Proced. Bull.* **2001**, *488*, 15.
24. Kain, J.S. The Kain–Fritsch convective parameterization: An update. *J. Appl. Meteorol.* **2004**, *43*, 170–181. [CrossRef]
25. Tewari, M.; Chen, F.; Wang, W.; Dudhia, J.; LeMone, M.A.; Mitchell, K.; Ek, M.; Gayno, G.; Wegiel, J.; Cuenca, R.H. Implementation and verification of the unified NOAH land surface model in the WRF model. In Proceedings of the 20th Conference on Weather Analysis and Forecasting/16th Conference on Numerical Weather Prediction, Seattle, WA, USA, 14 January 2004; pp. 11–15.
26. Mlawer, E.J.; Taubman, S.J.; Brown, P.D.; Iacono, M.J.; Clough, S.A. Radiative transfer for inhomogeneous atmospheres: RRTM, a validated correlated–k model for the longwave. *J. Geophys. Res.* **1997**, *102*, 16663–16682. [CrossRef]
27. Dudhia, J. Numerical study of convection observed during the Winter Monsoon Experiment using a mesoscale two–dimensional model. *J. Atmos. Sci.* **1989**, *46*, 3077–3107. [CrossRef]
28. Nakanishi, M.; Niino, H. An improved Mellor–Yamada level 3 model: Its numerical stability and application to a regional prediction of advecting fog. *Bound. -Layer Meteorol.* **2006**, *119*, 397–407. [CrossRef]
29. Nakanishi, M.; Niino, H. Development of an improved turbulence closure model for the atmospheric boundary layer. *J. Meteorol. Soc. Jpn.* **2009**, *87*, 895–912. [CrossRef]
30. Jimenez, P.A.; Dudhia, J.; Gonzalez-Rouco, J.F.; Navarro, J.; Montavez, J.P.; Garcia-Bustamante, E. A revised scheme for the WRF surface layer formulation. *Mon. Weather Rev.* **2012**, *140*, 898–918. [CrossRef]
31. Charnock, H. Wind stress on a water surface. *Q. J. R. Meteor. Soc.* **1955**, *81*, 639–640. [CrossRef]
32. Fairall, C.W.; Bradley, E.F.; Hare, J.E.; Grachev, A.A.; Edson, J.B. Bulk parameterization of air–sea fluxes: Updates and verification for the COARE algorithm. *J. Clim.* **2003**, *16*, 571–591. [CrossRef]
33. Donelan, M.A.; Haus, B.K.; Reul, N.; Plant, W.J.; Stiassnie, M.; Graber, H.C.; Brown, O.B.; Saltzman, E.S. On the limiting aerodynamic roughness of the ocean in very strong winds. *Geophys. Res. Lett.* **2004**, *31*, L18306. [CrossRef]
34. Large, W.G.; Pond, S. Sensible and latent heat flux measurements over the ocean. *J. Phys. Oceanogr.* **1982**, *12*, 464–482. [CrossRef]
35. Brutsaert, W. A theory for local evaporation (or heat transfer) from rough and smooth surfaces at ground level. *Water Resour. Res.* **1975**, *11*, 543–550. [CrossRef]
36. Garratt, J.R. *The Atmospheric Boundary Layer*; Cambridge University Press: Cambridge, UK, 1992; p. 316.
37. Nellipudi, N.R.; Yesubabu, V.; Srinivas, C.V.; Vissa, N.K.; Sabique, L. Impact of surface roughness parameterizations on tropical cyclone simulations over the Bay of Bengal using WRF-OML model. *Atmos. Res.* **2021**, *262*, 105779. [CrossRef]
38. Large, W.G.; Pond, S. Open Ocean momentum flux measurements in moderate to strong winds. *J. Phys. Oceanogr.* **1981**, *11*, 324–336. [CrossRef]
39. Zhang, J.A.; Black, P.G.; French, J.R.; Drennan, W.M. First direct measurements of enthalpy flux in the hurricane boundary layer: The CBLAST results. *Geophys. Res. Lett.* **2008**, *35*, L14813. [CrossRef]
40. Haus, B.K.; Jeong, D.; Donelan, M.A.; Zhang, J.A.; Savelyev, I. Relative rates of sea-air heat transfer and frictional drag in very high winds. *Geophys. Res. Lett.* **2010**, *37*, L07802. [CrossRef]
41. Bell, M.M.; Montgomery, M.T.; Emanuel, K.A. Air–sea enthalpy and momentum exchange at major hurricane wind speeds observed during CBLAST. *J. Atmos. Sci.* **2012**, *69*, 3197–3222. [CrossRef]
42. Richter, D.H.; Bohac, R.; Stern, D.P. An Assessment of the Flux Profile Method for Determining Air-Sea Momentum and Enthalpy Fluxes from Dropsonde Data in Tropical Cyclones. *J. Atmos. Sci.* **2016**, *73*, 2665–2682. [CrossRef]
43. Gopalakrishnan, S.G.; Marks, F.; Zhang, J.A.; Zhang, X.; Bao, M.J.W.; Tallapragada, V. A study of the impacts of vertical diffusion on the structure and intensity of the tropical cyclones using the high-resolution HWRF system. *J. Atmos. Sci.* **2013**, *70*, 524–541. [CrossRef]
44. Bian, G.F.; Nie, G.Z.; Qiu, X. How well is outer tropical cyclone size represented in the era5 reanalysis dataset? *Atmos. Res.* **2021**, *249*, 105339. [CrossRef]

atmosphere

Article

Correction to a Simple Biosphere Model 2 (SiB2) Simulation of Energy and Carbon Dioxide Fluxes over a Wheat Cropland in East China Using the Random Forest Model

Shiqi Zhang [1,†], Zexia Duan [1,*,†], Shaohui Zhou [1] and Zhiqiu Gao [1,2]

[1] Climate and Weather Disasters Collaborative Innovation Center, Key Laboratory for Aerosol-Cloud-Precipitation of China Meteorological Administration, School of Atmospheric Physics, Nanjing University of Information Science and Technology, Nanjing 210044, China
[2] State Key Laboratory of Atmospheric Boundary Layer Physics and Atmospheric Chemistry, Institute of Atmospheric Physics, Chinese Academy of Sciences, Beijing 100029, China
* Correspondence: dzx@nuist.edu.cn
† These authors contributed equally to this work.

Abstract: Modeling the heat and carbon dioxide (CO_2) exchanges in agroecosystems is critical for better understanding water and carbon cycling, improving crop production, and even mitigating climate change, in agricultural regions. While previous studies mainly focused on simulations of the energy and CO_2 fluxes in agroecosystems on the North China Plain, their corrections, simulations and driving forces in East China are less understood. In this study, the dynamic variations of heat and CO_2 fluxes were simulated by a standalone version of the Simple Biosphere 2 (SiB2) model and subsequently corrected using a Random Forest (RF) machine learning model, based on measurements from 1 January to 31 May 2015–2017 in eastern China. Through validation with direct measurements, it was found that the SiB2 model overestimated the sensible heat flux (H) and latent heat flux (LE), but underestimated soil heat flux (G_0) and CO_2 flux (F_c). Thus, the RF model was used to correct the results modeled by SiB2. The RF model showed that disturbances in temperature, net radiation, the G_0 output of SiB2, and the F_c output of SiB2 were the key driving factors modulating the H, LE, G_0, and F_c. The RF model performed well and significantly reduced the biases for H, LE, G_0, and F_c simulated by SiB2, with higher R^2 values of 0.99, 0.87, 0.75, and 0.71, respectively. The SiB2 and RF models combine physical mechanisms and mathematical correction to enable simulations with both physical meaning and accuracy.

Keywords: turbulent flux; CO_2 flux; SiB2 model; RF algorithm; wheat field; machine learning model

Citation: Zhang, S.; Duan, Z.; Zhou, S.; Gao, Z. Correction to a Simple Biosphere Model 2 (SiB2) Simulation of Energy and Carbon Dioxide Fluxes over a Wheat Cropland in East China Using the Random Forest Model. *Atmosphere* **2022**, *13*, 2080. https://doi.org/10.3390/atmos13122080

Academic Editor: Paul V. Doskey

Received: 4 November 2022
Accepted: 1 December 2022
Published: 10 December 2022

1. Introduction

Land surface processes modulate the weather and climate primarily through the exchange of energy, momentum, water, and carbon dioxide (CO_2) across the atmospheric boundary layer [1–5]. Climate simulations are especially sensitive to the temporal characteristics in the energy partitioning of available energy into sensible heat (H) and latent heat (LE) fluxes [6–8]. Therefore, investigating the diurnal and seasonal variations of land-atmosphere interactions is important for improving boundary layer parameterization schemes and the precision of weather predictions [7].

To deepen our understanding of the surface–atmosphere exchanges of water, surface energy, and CO_2 fluxes, numerous measurement methods [e.g., the Bowen ratio–energy balance method, the eddy covariance (EC) method, and the scintillometer method] have been applied [2]. Among them, the EC technique is considered to be the most direct and trustworthy method to obtain data on soil–plant–atmosphere carbon, water, and energy fluxes [9,10]. However, rainy days and power outages can cause data losses [11]. Accordingly, a powerful way to obtain accurate flux variations is to model the fluxes using reliable surface models, while evaluating the outputs against observed data [12,13].

Land surface process models have broadly experienced four main stages of development [14–16]. To begin with, the simple "bucket" model was proposed by Manabe [17], in which all the land surface parameters were set to fixed values. Next, the biosphere transfer scheme [18] and the simple biosphere (SiB) model [19] were proposed, which took into account the role of vegetation in the land surface process. Then, version two of the SiB model (SiB2) was developed [20,21] which introduced a vegetation biochemical process, including photosynthesis and respiration. Finally, in the most recent stage of development, models that can simulate the process of dynamic vegetation changes have emerged [22], such as the dynamic land ecosystem model [23], albeit there have as yet been relatively fewer applications of them in these types of models owing to their reliance on extremely highly complex physical inner mechanisms [16]. Furthermore, there are also models used to investigate and evaluate the environmental impact of energy production or consumption processes, such as life cycle assessment [24,25]. SiB2, a widely used land surface model, is expected to continue to gain importance as a surface flux simulation method. Previous studies have used SiB2 to investigate the water, energy, and CO_2 fluxes over different underlying surfaces, and verified them with observations [4,7,12,14,26–30]. Most such investigations found that the processes of dynamic vegetation changes could not be reflected. There has also been some research carried out into revising SiB2 by correcting its fixed parameters [2,3] and physical equations [31] to address biases that arise from the model's complexity and diversity of study area and vegetation. However, few studies have used machine learning algorithms to correct the outputs of SiB2.

With the development of machine learning, more and more researchers have used this approach to correct the biases of models [32]. For example, the European Center for Medium-Range Weather Forecasts model [33–36], the Weather Research and Forecasting model [37], land surface models such as ORCHIDEE (organising carbon and hydrology in dynamic ecosystems) [38], and that of Abramowitz et al. [39]. Recently, the random forest (RF) model has become a popular machine learning technique owing to its success in selecting and ranking numerous predictor variables [40]. Several works have investigated land surface processes with the RF model, such as the exchange of energy [41] and CO_2 [42,43], and obtained satisfactory results. Accordingly, there is reason to believe that the RF model could also be a promising method to correct SiB2 outputs. In addition, the energy and CO_2 exchanges simulated by SiB2 have tended to focus on forests and grasslands, with relatively less concern for agroecosystems, especially plains regions and different species of wheat [1,4,12,14,44]. In China, the only simulations of agroecosystems have focused on North and Northeast China [45]. Considerable uncertainty remains as to the current performance of simulations in East China.

East China is one of the country's main grain-producing areas, with wheat and rice being the primary crops. The wheat-field ecosystem, as a key component of the broader terrestrial ecosystem, is important for investigating global-scale ecology, energy balances, and regional climate [46]. At the same time, rapid urbanization and economic development are prominent features in East China, both of which can modify nearby surface energy exchanges and the boundary layer structure [47,48]. Consequently, accurate simulation of the fluxes of surface energy and CO_2 for the wheat-field ecosystem in East China will help to better understand energy and carbon budgets, more accurately assess the influence of climate change, and increase crop production [31]. The objectives of the present work were to: (1) quantify the seasonal and diurnal variations in radiation, H, LE, and CO_2 fluxes; (2) compare the radiation, turbulence, soil heat, and CO_2 fluxes modeled by SiB2 against direct measurements; and (3) correct the outputs of the SiB2 model using RF machine learning algorithms.

2. Materials and Methods

2.1. Site Description

The experiment was conducted at a 300 m × 300 m wheat field in Dongtai County, Jiangsu Province, China (32.76° N, 120.47° E; 2 m above sea level; Figure 1) from January to May 2015–2017. The site was relatively flat and homogeneous, with a clay soil texture. The climate is classed as "subtropical monsoon", with an annual mean (1951–2015) air temperature of approximately 14.8 °C and rainfall of 1063 mm [49]. Meanwhile, the average annual sunshine duration and frost-free period are 2213 h and 220 days, respectively [50]. During the observation period, the "Yangmai 16" variety of winter wheat was planted around the EC tower. The wheat grew in good drainage and no silt soil conditions. Nitrogen fertilizer (urea) was applied at 180 kg ha^{-1}. The wheat growing season was divided into the vegetative stage (15 December–28 February), reproductive stage (1 March–15 April), and ripening stage (16 April–31 May) [10].

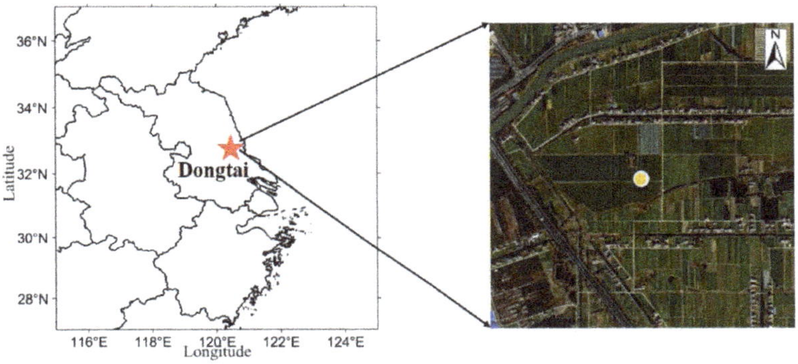

Figure 1. Map showing the location of Dongtai station (red star) and a satellite image showing the location of the study site (yellow point).

2.2. Instruments and Data Processing

The *H*, *LE*, and CO_2 fluxes were collected from an EC tower at 10-m above ground level (AGL), which consisted of a three-dimensional sonic anemometer (Campbell Scientific, Inc., UT, USA, CSAT3) and a CO_2/H_2O open path gas analyzer (LI-COR Biosciences, Inc., NE, USA, LI-7500). Downward shortwave/longwave and upward shortwave/longwave radiation measurements were obtained from a four-component net radiometer (Kipp and Zonen, Inc., CNR-4) at 3-m AGL. The air temperature, humidity (Vaisala, Inc., Helsinki, HMP45A), and wind speed (Met One, Inc., 034B) were measured at 3, 5, 8, and 10-m AGL. The soil heat flux (Hukseflux Thermal Sensors, Netherlands, HFP01), soil temperature (Campbell Scientific PT100), and soil water content (Campbell Scientific CS616) observations were collected at depths of 0.05, 0.1, 0.2, and 0.4-m. The data were averaged over 30-min intervals. In addition, the surface atmospheric pressure (Vaisala, Helsinki, PTB110) and precipitation (Campbell Scientific TE525MM) were also measured. More details about the instruments can be found in Duan et al. [10] and Li et al. [51].

Firstly, the Campbell Scientific LoggerNet 4.2.1 software was used to transform the raw 10-Hz EC data into the 30-min binaries. Then, the LI-COR EddyPro 5.2.1 software was applied to process the EC 30-min binaries, with the main steps involving averaging and statistical tests [52], time delay compensation, double rotation, spectral corrections [53], and compensation of density fluctuations [54]. The quality flags in EddyPro consist of "excellent" (flag 0), "moderate" (flag 1), and "exclude" (flag 2). The EC data on rainy or foggy days were discarded [10,47].

The 16-day Normalized Difference Vegetation Index (NDVI) data for the period 2015–2017, available from the 250-m resolution MODIS MOD13Q1 product (https://ladsweb.modaps.eosdis.nasa.gov/search/), were employed (accessed: 13 September 2022). The leaf area index (LAI), a fraction of photosynthetically active radiation (FPAR), green leaf fraction (*N*), and vegetation cover fraction (*V*) data were calculated based on the NDVI data following Sellers et al. [20] and Zhang et al. [11].

2.3. Methods

2.3.1. The SiB2 Model

The SiB2 model is a widely used and biophysics-based land surface model developed by Sellers et al. [20,21]. SiB2 contains a set of physics-based equations that couple the water balance, energy balance, and vegetation biochemical processes to simulate the exchanges of water, carbon, momentum, and energy fluxes among the atmosphere, a single canopy layer, and three soil layers (surface soil layer, root zone layer, and deep soil layer) [2,12,21,55]. As a parameterization scheme describing the processes of exchange between land and atmosphere, it simulates a more realistic vegetation physiological process owing to the incorporation of a canopy photosynthesis conductance submodel [3,30].

The SiB2 model requires soil, land-surface properties, initial conditions, and meteorological forcing data as inputs. The soil in the wheat field studied here consisted of clay, and thus parameter type 5 (Clay→clay loam) was selected from Table 4 in Sellers et al. [20]. In addition, the land surface category was defined as "agriculture or C3 grassland" (biome type 9) (Table 2 in Sellers et al. [20]). The parameter settings in the model for Dongtai are listed in Table A1. Six meteorological forcing variables—downward shortwave radiation, downward longwave radiation, vapor pressure, air temperature, wind speed, and precipitation—are shown in Figure 2. From January to May, the daily maximum radiation, vapor pressure, and air temperature increased substantially, with values of 1020 W m^{-2}, 442 W m^{-2}, 6 hPa, and 301 K, respectively. The daily average wind speed fluctuated between 1 and 6 m s^{-1} during the observation period. The seasonal cumulative precipitation was 247 mm, with a maximum daily value of 35 mm on 17 March 2015.

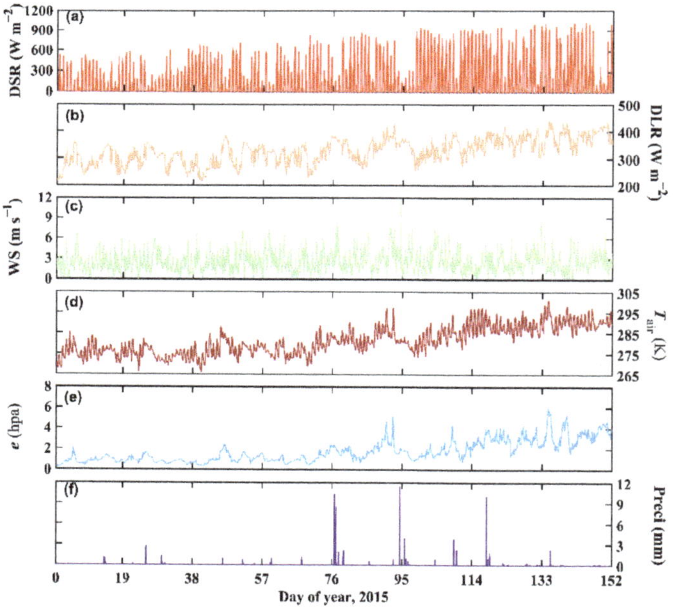

Figure 2. Time series of the 30-min meteorological forcing data collected during the observation period from 1 January to 31 May 2015: (**a**) downward shortwave radiation (DSR); (**b**) downward longwave radiation (DLR); (**c**) wind speed (WS); (**d**) air temperature (T_{air}); (**e**) vapor pressure; (**f**) precipitation (Preci).

2.3.2. The RF Model

The RF algorithm is an extensively used machine learning method that excels at classification and regression owing to its efficiency and flexibility [56]. Multidimensional and multicollinear data can be dealt with satisfactorily, and there is less sensitivity to overfitting [57]. The method's feature selection tool can be used to pass judgement on how significant a predictor is, with the definition of feature importance being the weight of each of the model's input factors, and significant variables having a stronger influence on the outcomes of the model evaluation [58].

In this study, we applied the RF framework to correct heat and carbon fluxes simulated by the SiB2 model (Figure 3). First, a set of explanatory variables (Table A2) were selected based on previous research [43,59–62] and currently available in situ measurements. Second, 90% of the outputs of the SiB2 model and explanatory variables (Table A2) from January to May 2016–2017 were used to train the RF model, with the remaining 10% of them and 100% of the data in 2015 used to validate estimation performance of the model. A 10-fold cross-validation method was applied in the RF model to find the best hyperparameters and avoid the issue of overfitting. Two statistical metrics, the coefficient of determination (R^2) and root-mean-square error (RMSE), were used to evaluate the performance of the 10-fold cross-validation results.

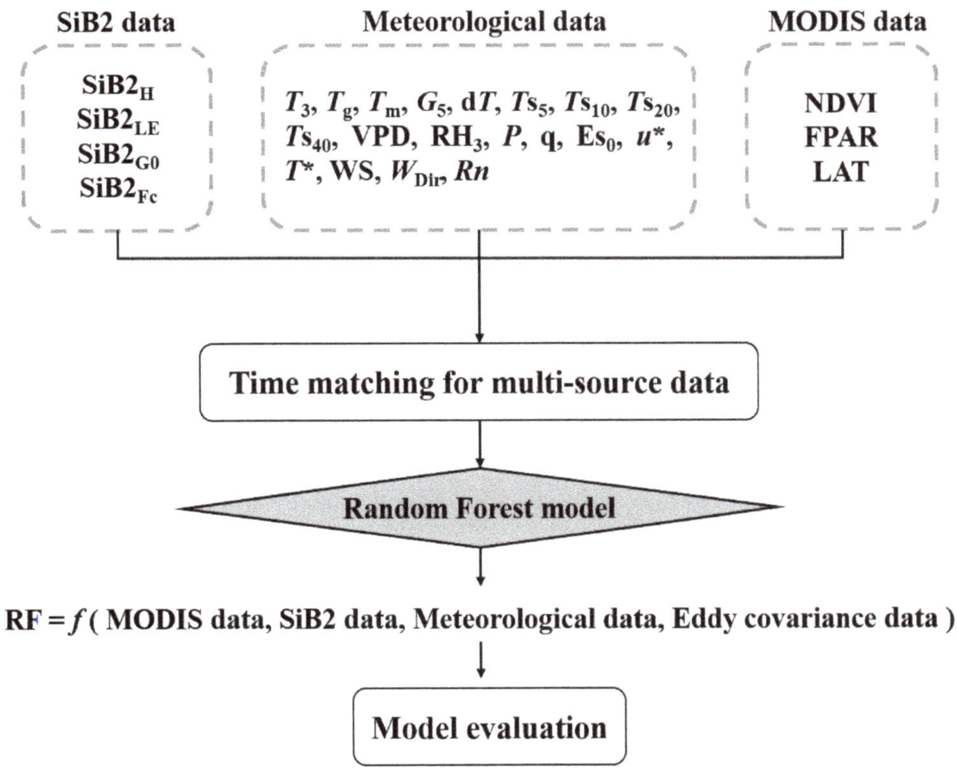

Figure 3. Flowchart of the RF model's three-stage correction of the SiB2 model outputs.

2.3.3. Radiation and Surface Energy Fluxes

Solar radiation is the key driver of surface energy, momentum, carbon, and water fluxes [4]. The Rn consists of DSR, USR, DLR, and ULR [7]:

$$Rn = \text{DSR} + \text{DLR} - \text{USR} - \text{ULR}. \quad (1)$$

where Rn is the net radiation, DSR is the downward shortwave radiation, DLR is the downward longwave radiation, USR is the upward shortwave radiation, and ULR is the upward longwave radiation. The surface energy balance can be estimated by [10]

$$Rn = H + LE + G_0 + \varepsilon, \quad (2)$$

where H is the sensible heat flux, LE is the latent heat flux, G_0 is the soil surface heat flux, and ε is the residual energy term, such as canopy heat storage or the energy consumption of photosynthesis and respiration. The soil surface heat flux (G_0) can be estimated according to the formula given by Liu et al. [63].

2.3.4. Statistical Analysis

In this study, the traditional statistical analysis indexes (the standard deviation, R^2, and RMSE) were used to evaluate the accuracy of SiB2 and RF models. The comparison statistics were calculated as follows:

The standard deviation (S), R^2, and RMSE:

$$S = \sqrt{\frac{\sum_{i-1}^{n}(x_i - \overline{x})^2}{n-1}}, \quad (3)$$

where x_i is the value of the ith point in the data set, $\overline{x}$ is the mean value of the data set, and n is the number of the data points in the data set. The standard deviation is the average amount of variability in the data set.

$$R^2 = 1 - \frac{\sum_{i=1}^{n}(M_i - O_i)^2}{\sum_{i=1}^{n}(O_i - \overline{O})^2}, \quad (4)$$

where M_i are the values modeled by the SiB2/RF model, O_i are the observed values, and $\overline{O}$ is the mean value of the observation. When the R^2 is high, the simulations of the SiB2/RF model are close to the observations.

$$RMSE = \sqrt{\frac{\sum_{i-1}^{n}(M_i - O_i)^2}{n}}. \quad (5)$$

The discrete situation between the simulation and observation is indicated by RMSE.

3. Results

3.1. Radiation, Turbulence, and CO_2 Fluxes

Figure 4 shows remarkable diurnal variations in the median Rn, H, LE, G_0, and F_c in the vegetative, reproductive, and ripening stages. Rn, H, LE, and G_0 began to increase after sunrise (around 05:00–07:00 LST), reached their highest values of 280–615, 59–77, 98–406, and 34–107 W m^{-2}, respectively, in the middle of the day (11:00–14:00 LST), and then gradually decreased to stable values at around 17:00–19:00 LST. F_c had an opposite trend of variation to the surface energy fluxes. The positive nocturnal values of 0.4–3.9 µmol m^{-2} s^{-1} would have mainly been associated with the respiration of wheat [47], a lower boundary layer height [64], and poor atmospheric mixing [65], whilst the negative daytime (07:30–17:30 LST) values of approximately −18.5 to −0.4 µmol m^{-2} would have been closely related to the strong photosynthesis of wheat and favorable dispersion conditions [66].

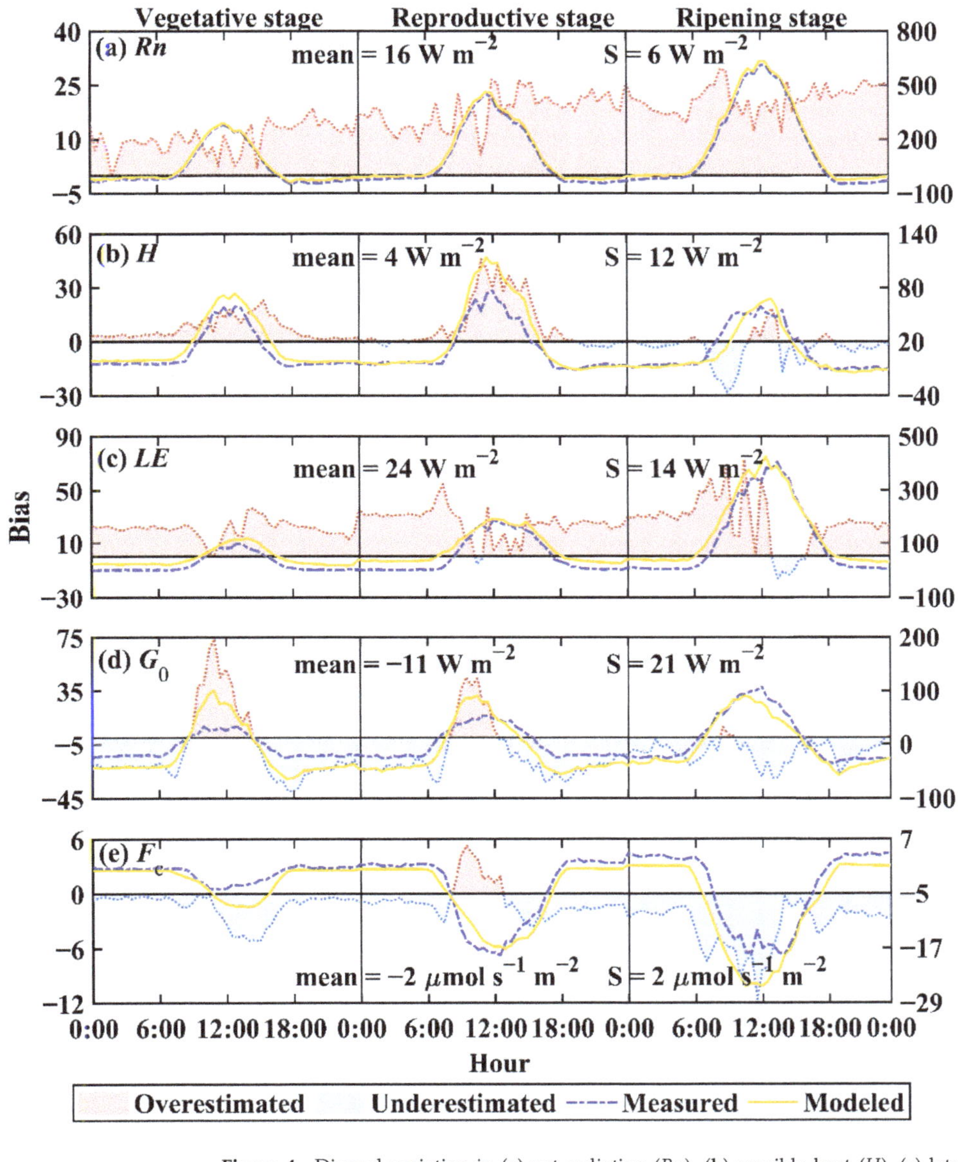

Figure 4. Diurnal variation in (**a**) net radiation (*Rn*), (**b**) sensible heat (*H*), (**c**) latent heat (*LE*), (**d**) surface soil heat (G_0), and (**e**) CO_2 fluxes (F_c) modeled using SiB2 (yellow lines) against direct measurements (blue lines) in the vegetative, reproductive, and ripening stage. The filters are the biases between the SiB2-modeled and directly measured results (former minus the latter). S is the standard deviation of the biases between the SiB2-modeled and directly-measured results and mean is the mean value of that.

In addition, the *Rn*, *H*, *LE*, G_0, and F_c showed significant seasonal variations in the vegetative, reproductive, and ripening stages. The middle-of-the-day maximum *Rn*, *LE*, and G_0 increased from the vegetative stage (280, 98, and 34 W m^{-2}) to the ripening stage (615, 406, and 107 W m^{-2}); the *H* in the ripening stage (59 W m^{-2}) was slightly lower than that in the vegetative and reproductive stage (77 and 60 W m^{-2}; Figure 4a); and the F_c

varied with the wheat phenology. The wheat field served as a carbon sink in the vegetative stage, with a mean value of −0.4 μmol m^{-2} despite the low photosynthetic rate. The wheat field then became a CO_2 sink in the reproductive and ripening stages, with mean values of −4.2 and −3.7 μmol m^{-2} s^{-1}, indicating stronger biological activities of the wheat (i.e., photosynthetic rate) during these stages.

3.2. SiB2 Evaluation

The diurnal variations in *Rn*, *H*, *LE*, G_0, and F_c modeled by SiB2 were consistent with the results obtained by direct measurements (Figure 4). The simulation of *Rn* depended largely on the simulation of USR and ULR, since DSR and DLR were given as inputs. As can be seen in Figure 4a, *Rn* was simulated very well, with both its peak together and diurnal variation being closely captured. In addition, the R^2 and RMSE of the overall growth period were 1.00 and 18.73 W m^{-2}, respectively. However, the biases indicated that the SiB2 model overestimated *Rn*, and this was slightly more apparent at nighttime, similar to the findings of Jing et al. [2], which may have been caused by underestimated values of the surface effective radiative temperature at night in the simulation. In addition, the median bias values in the different growth stages, in ascending order, were 11 W m^{-2} for the ripening stage, 18 W m^{-2} for the reproductive stage, and 21 W m^{-2} for the vegetative stage; and the median (mean) values of observed and simulated *Rn* were −13 (82) and 3 (98) W m^{-2}, respectively. The mean bias and standard deviation for bias of observed and simulated *Rn* were 16 W m^{-2} and 6 W m^{-2}. Overall, the SiB2 model overestimated *Rn* by 13%.

Figure 4b compares the *H* between the SiB2 simulation and observation. The diurnal variation that was again captured by SiB2 reason well, especially in the vegetative stage, however, its pattern was less regular than that of *Rn*. In addition, the R^2 and RMSE of the overall growth period were 0.59 and 32.92 W m^{-2}, respectively. As we can see, the simulation of *H* at night was better than during the daytime, especially around the middle of the day in the reproductive and ripening stages, when SiB2 overestimated the *H*. Further, a small part of the biases was attributable to underestimation in the reproductive and ripening stages. The median (average) values of the observed and simulated *H* were −4 (13) and −1 (20) W m^{-2}, respectively. The mean bias and standard deviation for bias of observed and simulated *H* were 4 W m^{-2} and 12 W m^{-2}. In general, the SiB2 model overestimated *H* by 25%.

It can be seen from Figure 4c that the simulation of *LE* was better than that of *H* and closer to that of *Rn*, with R^2 and RMSE values that reached 0.75 and 72.87 W m^{-2}, respectively. The simulation of *LE* in the vegetative stage was much better than in the later stages. SiB2 basically overestimated *LE*, with only a slight underestimation in the daytime during the reproductive and ripening stages. In addition, the median (average) values of the observed and simulated *LE* were 11 (75) and 39 (92) W m^{-2}, respectively. The mean bias and standard deviation for bias of observed and simulated *LE* were 24 W m^{-2} and 14 W m^{-2}. Overall, the SiB2 model overestimated *LE* by 36%.

The observed and simulated G_0 are compared in Figure 4d, which reveals less consistent temporal changes than for *Rn*. The R^2 and RMSE of the overall growth period for G_0 were 0.60 and 34.33 W m^{-2}, respectively. It is apparent that SiB2 overestimated the soil heat flux after sunrise in the vegetative and reproductive stages and mainly underestimated it at nighttime. The simulation of G_0 in the ripening stage was relatively better. In addition, the median (mean) values of the observed and simulated G_0 were −19 (0.4) and −34 (−12) W m^{-2}, respectively. The mean bias and standard deviation for bias of observed and simulated G_0 were −11 W m^{-2} and 21 W m^{-2}. Overall, the SiB2 model underestimated G_0 by 37%. The negative mean value may have resulted from underestimation, especially at nighttime, in the ripening stage. This would be regarded as a simulation error when less than −100 W m^{-2}. The reason why SiB2 overestimated G_0 in the daytime of the vegetative stage may be the lower vegetation cover fraction set in the parameters resulting in higher absorption of radiation by the ground surface.

Significant diurnal variation is a well-known characteristic of CO_2 flux in cropland, in which the crop absorbs CO_2 by photosynthesis and emits it into the atmosphere. As shown in Figure 4e (in which positive values of biases between the simulation and observation represent emission and negative values indicate absorption), SiB2 estimated the CO_2 fluxes well, capturing the temporal variations accurately. The R^2 and RMSE were 0.62 and 6.82 μmol m^{-2} s^{-1}, respectively; and the median (mean) values of CO_2 in the observation and simulation were 0.7 (−2) and 0.2 (−5) μmol m^{-2} s^{-1}, respectively. The mean bias and standard deviation for bias of observed and simulated F_c were −2 μmol m^{-2} s^{-1} and 2 μmol m^{-2} s^{-1}. In general, the SiB2 model underestimated F_c by 40%. In addition, the simulation in the vegetative stage was the best among the three stages, and the ripening stage was the worst. The simulation of F_c was predominantly underestimated, with only a slight overestimation in the reproductive stage. There were some daytimes when SiB2 overestimated F_c, which may have been caused a weaker photosynthesis resulting from less DSR and lower temperatures before and after rainfall. Additionally, the simulation of photosynthesis increased and decreased as the winter wheat grew and died following higher and lower LAI, FPAR, and *V*.

3.3. Driving Factors of Turbulence and CO_2 Fluxes

As demonstrated in Section 3.2, SiB2 captured the diurnal variation in *H*, *LE*, G_0, and F_c well, albeit with the simulation results still showing certain errors. Given that the R^2 of *Rn* reached 1.00, there was less possibility to improve its simulation accuracy. Therefore, we constructed the RF model to correct the SiB2 model outputs and improve the simulation accuracies of *H*, *LE*, G_0, and F_c. The RF model examined the potential drivers and assessed their relative contributions to *H*, *LE*, G_0, and F_c (Figure 5). Correlations among the turbulence and CO_2 fluxes and input variables were calculated with all the training data from the Dongtai site (Figure 6).

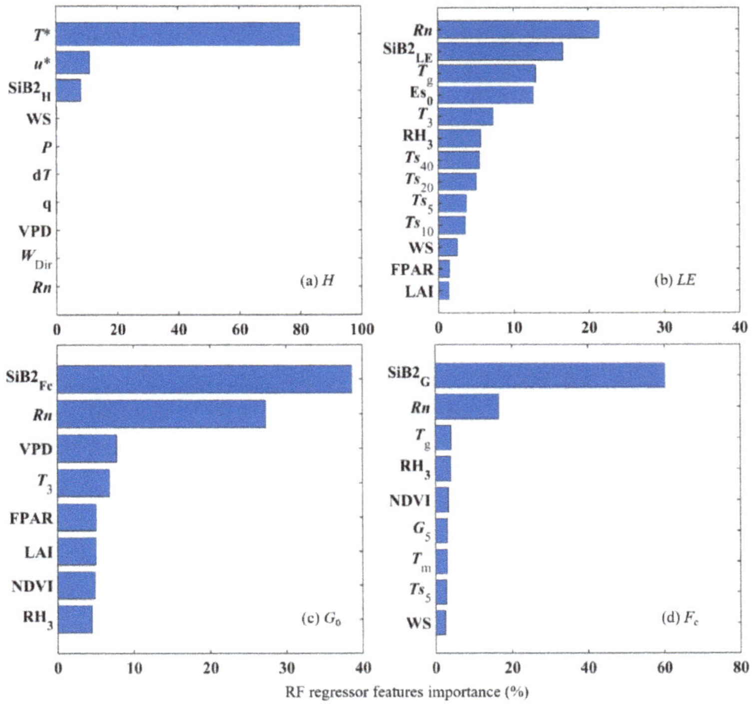

Figure 5. Feature importance for the RF model in Dongtai County.

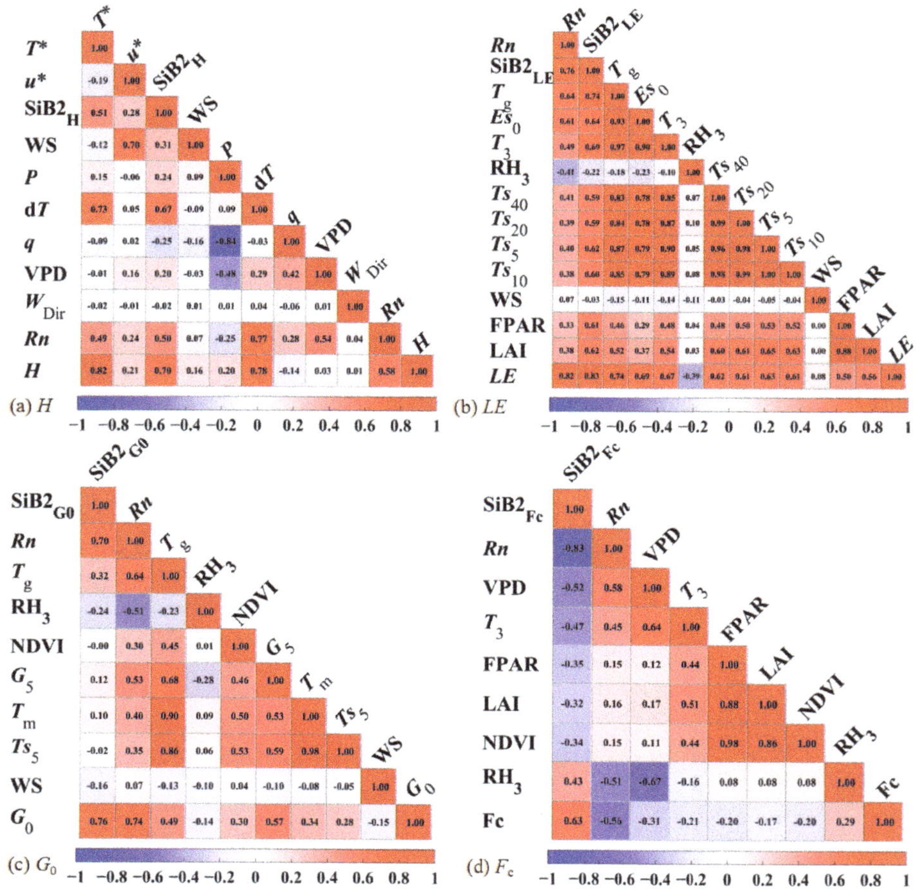

Figure 6. Correlations among the fluxes of turbulence and CO_2 and input variables. The correlations were calculated with all the training data from the Dongtai site.

As we can see from Figure 5a, T^* showed the greatest importance in the modulation of *H*, with a weight of 81% of all variables. Moreover, Figure 6a shows that *H* correlated most strongly with T^*, consistent with the variable weighting value in Figure 5a, whose Pearson correlation coefficient (*r*) was 0.82. Apart from T^*, there were two other critical variables, u^* and $SiB2_H$, with weights of 11% and 8%, respectively. T^* and u^* have been mentioned previously as significant variables in the calculation of *H* in EC observations [10]. Comparatively speaking, the influence of the remaining variables was negligible, with importance values of less than 1%. Moreover, u^* was microcorrelated with *H* despite its high importance in the RF model, while $SiB2_H$ correlated closely with *H* even though its importance was lower than that of u^*.

Figure 5b shows that *Rn* was the most essential variable in modulating *LE*, accounting for 21% of the importance of all variables. Although the *r* of *Rn* reached 0.82, the strongest positive correlation with *LE* was not *Rn* but $SiB2_{LE}$ (Figure 6b), with the highest *r* of 0.83. Besides *Rn*, three other variables, $SiB2_{LE}$, T_g, and Es_0, impacted strongly on *LE*, with importance values of 17%, 13%, and 13%, respectively, since Es_0 and T_g play important roles in calculating *LE* in EC measurements. In contrast, the remaining variables were less important in modulating *LE*, with importance values ranging from 7% down to values approaching 0.

As illustrated in Figure 5c, SiB2$_{G0}$ had the strongest influence in modulating G_0, accounting for 60% of the total variable importance. SiB2$_{G0}$ had the greatest positive correlation with G_0, with the highest r of 0.76 in Figure 6c. Additionally, Rn was the second most important variable, with a weighting of 17%, and also had the second highest correlation with G_0 (r = 0.74). Both SiB2$_{G0}$ and Rn had close consistency in their variable weighting and correlation. Aside from SiB2$_{G0}$ and Rn, the remaining variables showed weaker influences in modulating G_0, with importance values ranging from 4% down to values approaching 0.

Figure 5d shows that SiB2$_{Fc}$ played the most critical role in modulating F_c, with an importance weighting of 39% of all variables. SiB2$_{Fc}$ had the greatest positive correlation with F_c, with the highest r of 0.63 in Figure 6d. In addition, as the second most significant variable, the weighting of Rn was 27% and its r was −0.56, which was the second highest negative correlation. These results were similar to those of SiB2$_{G0}$ and Rn in their modulation of G_0, being consistent in both variable weighting and correlation. In contrast, there was no apparent influence of VPD, T_3, FPAR, LAI, NDVI, and RH$_3$ on F_c, showing lower relative weights, with importance values of 8%, 7%, 5%, 5%, 5%, and 4%, respectively. Note that Rn, VPD, T_3, LAI, FPAR, and NDVI all correlated negatively with F_c.

3.4. RF Model Evaluation

It is clear that the RF model performed excellently in correcting the H (Figure 7a) in all three stages. The degree of overestimation in the daytime was slightly reduced, especially in the reproductive stage. It also resolved the problem of underestimation in the ripening stage. Accordingly, the estimation with the RF model was consistent with the observation, basically catching both the peak and diurnal variation. The mean bias and standard deviation for bias of observed and simulated H were 5 W m^{-2} and 5 W m^{-2}, respectively. The R^2 of the RF model for H was 0.99, larger than that of 0.59 from SiB2 alone. Likewise, the RMSE was 4.73 W m^{-2}, which was smaller than that of 32.92 W m^{-2} with SiB2.

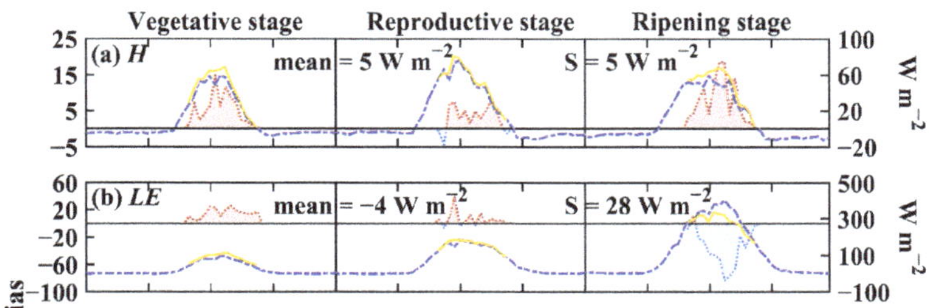

Figure 7. Daily variations of the median RF-modeled (yellow lines) and measured fluxes (blue lines), and their biases (red shading, overestimated; blue shading, underestimated) for (**a**) sensible heat (H), (**b**) latent heat (LE), (**c**) soil heat (G_0), and (**d**) CO_2 (F_c), in the vegetative, reproductive, and ripening stages. S is the standard deviation of the biases between the RF-modeled and directly measured results and mean is the mean value of that.

The simulation of *LE* with the RF model (Figure 7b) was relatively less consistent with the observaton and worse than it was for *H*. It was able to capture the diurnal variation but failed to catch the peak, especially in the ripening stage. The mean bias and standard deviation for bias of observed and simulated *LE* were -4 W m^{-2} and 28 W m^{-2}, respectively. The correction by the RF model did alleviate the degree of overestimation, especially in the ripening stage. However, there was a relatively greater underestimation than the SiB2 model in the ripening stage. Nonetheless, the results indicated that the correction still worked, with the R^2 reaching 0.85 and the RMSE reduced to 54.92 W m^{-2}.

It can be seen from Figure 7c that the simulation of G_0 after correction with the RF model was more consistent with the observation. Although the bias of the simulation for G_0 after the correction was still dominated by underestimation, more peaks were captured. The mean bias and standard deviation for bias of observed and simulated G_0 were -14 W m^{-2} and 15 W m^{-2}, respectively. The biases, in terms of both overestimation and underestimation, became much smaller in the vegetative and reproductive stages, resulting in the simulation by SiB2 in these two stages improving. The R^2 reached 0.78 and the RMSE reduced to 25.53 W m^{-2}. Compared with the results of SiB2 alone, an improvement was still apparent after correction with the RF model.

Figure 7d shows a better simulation of CO_2 following correction with the RF model. It also displays a better consistency with the observation, especially at nighttime, with the overall biases at night becoming smaller. The mean bias and standard deviation for bias of observed and simulated F_c were -4 μmol m^{-2} s^{-1} and 28 μmol m^{-2} s^{-1}. It is apparent that the simulation values became larger after correction, which were close to 0 originally, making the simulation more reasonable. Meanwhile, the biases of F_c in the vegetative and ripening stage decreased and the biases (overestimation) in the reproductive stage increased. In other words, the correction of the biases in the vegetative and ripening stages was better. Moreover, the R^2 increased from 0.62 to 0.71 and the RMSE decreased from 6.82 μmol m^{-2} s^{-1} to 4.70 μmol m^{-2} s^{-1}.

3.5. Comparison of SiB2 and RF

Figure 8a shows the R^2 values for *H*, *LE*, G_0, and F_c in the different growth stages for the SiB2 and RF-corrected outputs (hereafter referred to simply as RF). During the vegetative stage, for the simulation of SiB2 and RF, *H* had the largest R^2, followed by G_0, *LE*, and then F_c. The *H* improved significantly in the vegetative stage, reaching 41%, followed by F_c (33%), G_0 (23%), and *LE* (13%). In the reproductive stage, for the simulation of SiB2, the largest R^2 was that of *LE*, followed by *H*, F_c, and then G_0; while for RF, the largest R^2 was that of *H*, followed by *LE*, G_0, and F_c. In addition, the degree of improvement for G_0 was the highest, reaching 63%, followed by 46% for *H*, 17% for *LE*, and then 9% for F_c. During the ripening stage, for the simulation of SiB2, F_c had the largest R^2, followed by G_0, *LE*, and then *H*; while for RF, *H* also had the largest R^2, followed by *LE*, G_0, and then F_c. Additionally, the degree of improvement for *H* was the highest, reaching 111%, followed by G_0 (33%), *LE* (23%), and then F_c (23%). For the simulation of SiB2, for *H*, the R^2 in the vegetative stage performed better; while for RF, the best performing R^2 was in the vegetative stage along with the reproductive stage. The effect of correction was better in the ripening stage. For the R^2 of the simulated *LE*, both SiB2 and RF performed best in the reproductive stage. For the simulation of G_0, the R^2 of SiB2 and RF were also best in the same stage, namely the ripening stage. However, the effect of correction in the reproductive stage was the best. For the F_c simulated by SiB2, the best performance in terms of R^2 was in the reproductive stage; for RF, the R^2 in the ripening stage performed better. Overall, the mean R^2 of SiB2 (RF) for the reproductive (ripening) stage was the largest, and the effect of correction in the ripening stage was better. During the whole growth period, the R^2 for *H*, *LE*, G_0, and F_c all improved greatly (68%, 13%, 30%, and 15%) (Table 1).

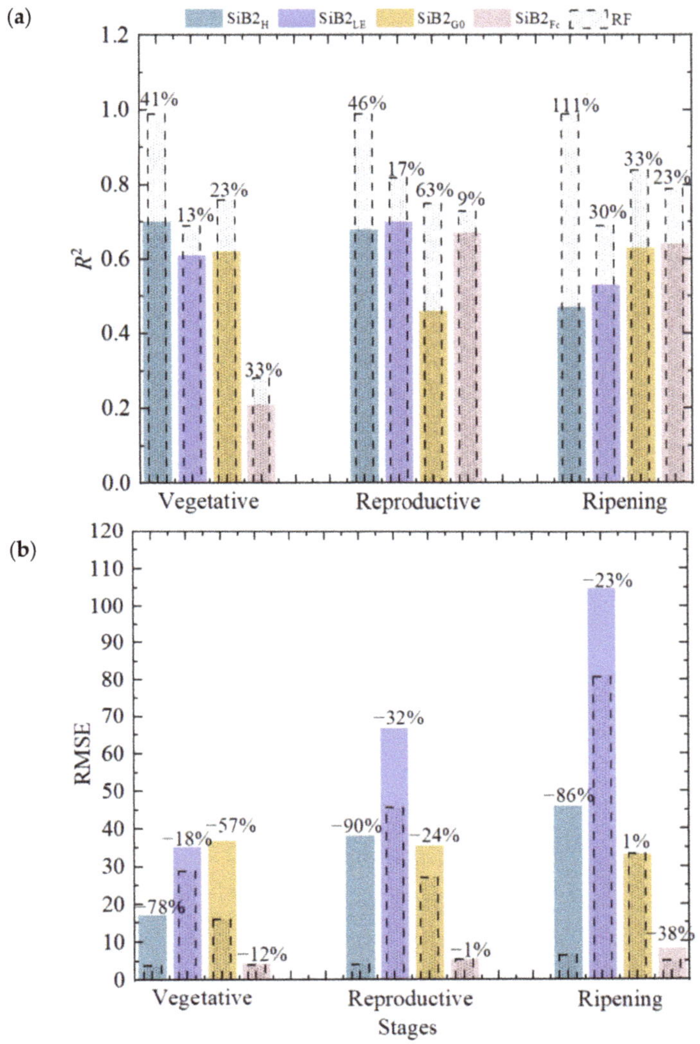

Figure 8. Comparison of the (**a**) R^2 and (**b**) RMSE values for the fluxes of sensible heat (H), latent heat (LE), soil heat (G_0), and CO_2 (F_c) during the different growth stages of wheat in the SiB2 and RF-corrected model outputs. The degree of improvement is indicated by the percentages over each bar.

Table 1. The values of R^2 and RMSE for the overall growth period in the SiB2 and RF-corrected model outputs.

Flux	SiB2		RF	
	R^2	RMSE	R^2	RMSE
H	0.59	32.92	0.99	4.73
LE	0.75	72.87	0.85	54.92
G_0	0.60	34.33	0.78	25.53
F_c	0.62	6.82	0.71	4.70

Note: H, sensible heat flux; LE, latent heat flux; G_0, soil heat flux; F_c, CO_2 flux.

Figure 8b shows the RMSE for H, LE, G_0, and F_c in the different growth stages as simulated by SiB2 and RF. For the simulation of SiB2, the RMSE of H, LE, and F_c became increasingly larger as the wheat grew. In other words, the order of RMSE values for H, LE, and F_c was vegetative stage < reproductive stage < ripening stage, which was similar to the RMSE of RF for H, LE, and G_0. Meanwhile, the RMSE of SiB2 for G_0 was lowest in the ripening stage, while that of RF for F_c was lowest in the vegetative stage. For the vegetative stage, the best degree of correction was for H, reaching −78%, and the worst was for F_c, at only −12%. For the reproductive stage, the correction for H also performed best, while for F_c it was worst. For the ripening stage, the correction with RF also performed best for H, and for G_0 it was negative. Furthermore, the best effect of the correction for H, LE, G_0, and F_c was in the reproductive, reproductive, vegetative, and ripening stages, respectively. Overall, the best mean correction effect was in the vegetative stage. During the whole growth period, all the RMSEs for H, LE, G_0, and F_c were reduced (Table 1).

4. Discussion

We have evaluated the turbulence and CO_2 fluxes with the in situ observations and remote sensing data. H, LE, G_0, and F_c simulated by SiB2 have standard deviations of 12 W m^{-2}, 14 W m^{-2}, 21 W m^{-2} and 2 μmol m^{-2} s^{-1}, respectively. H estimation has a 25% positive bias, which was similar to those reported by Chu et al. [14], Gao et al. [7], Lei et al. [31], Li et al. [3], and Xue et al. [29]. The positive H bias can be attributed to the higher initial input canopy temperature, which was set to the canopy air space temperature. The estimated LE has a 36% positive bias, which was consistent with the results of Gao et al. [7], Yan et al. [30], and Yuan et al. [67]. The soil moisture sensors were not mounted directly in the wheat field, resulting a in higher soil wetness fraction and measured LE. G_0 was underestimated by 37%, which was similar to the results reported by Zhang et al. [5]. The SiB2 model underestimated F_c by 40% and most simulated values were concentrated near to the value of 0, which was also found in Chu et al. [14], and Yuan et al. [4,67]. The soil respiration (R_{soil}) was set to 0 by default (R_{soil} = 0) in SiB2, leading to a relatively lower simulation of respiration in the wheat field, which can be regarded as a kind of model error. In other words, SiB2 has a tendency to underestimate ecosystem respiration at night during the growth period [26].

Given the bias of simulation by SiB2, the RF model was used to correct the results modeled by SiB2 and made great improvements to H, LE, G_0, and F_c with values of 68%, 13%, 30%, and 15%, respectively, during the whole growth period. There has also been some research carried out into revising SiB2 to address biases that arise from the model's complexity and diversity of study area and vegetation. Lei et al. [31] adjusted the physical equation of soil respiration and calibrated the Ball-Berry stomatal conductance model. Li et al. [3] adjusted optimum growth and inhibition temperature parameters. Jing et al. [2] revised the SiB2 model by adding an irrigation module and adjusting parameters. The previous research mentioned above made improvements to the output of the SiB2 model, nevertheless their corrections, based on the observation, only applied to specific study regions. It is important to take into consideration the interactions among parameters and the physical implications of the parameters. Otherwise, there would be great overall uncertainty for the results of the SiB2 model.

This study only investigated the wheat field in Dongtai County for one growth season. In the next study, we will make an effort to extend the simulation of the combined SiB2 model and RF model from Dongtai County to the whole of East China, extend the species studied from wheat to a rotation of summer rice and winter wheat, and extend the study period from one growing season to several crop years. Additionally, investigating land surface processes and improving the accuracy of simulation are of great importance in future work.

5. Conclusions

In this study, radiation, turbulence, and CO_2 fluxes were observed with an EC system in a winter wheat field in eastern China from 1 January to 31 May 2015–2017. The *Rn*, *H*, *LE*, G_0, and F_c modeled by SiB2 showed obvious diurnal and seasonal variations during the whole winter wheat growing season, with R^2 values of 1.00, 0.59, 0.81, 0.60, and 0.62 against the direct observations, respectively. The SiB2 model overestimated the *Rn*, *H*, and *LE* (13%, 25%, and 36%) and underestimated the G_0 (−37%) and F_c (−40%). Thus, an RF model was designed to correct the results modeled by SiB2. The RF-corrected model showed that *T**, *Rn*, $SiB2_{G0}$, and $SiB2_{Fc}$ were the key driving factors in the modulation of *H*, *LE*, G_0, and F_c. Compared with the results modeled by SiB2, the RF model performed well and made great improvements to *H*, *LE*, G_0, and F_c with values of 68%, 13%, 30%, and 15% during the whole growth period.

The input parameters in SiB2 were dynamically updated every week to reflect the process of vegetation phenology, making the simulated turbulence and CO_2 fluxes more reasonable and realistic.

Author Contributions: S.Z. (Shiqi Zhang): Conceptualization, Methodology, Software, Validation, Formal analysis, Investigation, Writing—original draft preparation, Writing—review and editing. Z.D.: Methodology, Software, Validation, Formal analysis, Data curation, Writing—review and editing. S.Z. (Shaohui Zhou): Methodology, Software, Validation, Formal Analysis. Z.G.: Conceptualization, Validation, Formal analysis, Data curation, Writing—review and editing, Funding acquisition. All authors have read and agreed to the published version of the manuscript.

Funding: This research was funded by the National Natural Science Foundation of China (grant number 41875013) and the Second Tibetan Plateau Scientific Expedition and Research Program (Grants 2019QZKK0102).

Institutional Review Board Statement: Not applicable.

Informed Consent Statement: Not applicable.

Data Availability Statement: MODIS products used in this paper can be downloaded from https://ladsweb.modaps.eosdis.nasa.gov/search/, accessed on 13 September 2022. The data of automatic stations and eddy covariance flux data are not available to the public. Please direct any inquiries regarding the data to the first author (20201203032@nuist.edu.cn).

Acknowledgments: We sincerely appreciate the reviewer's thoroughness and insightful criticism, which helped us significantly improve the manuscript.

Conflicts of Interest: The authors declare no conflict of interest.

Appendix A

Table A1. Parameter settings in the SiB2 model for Dongtai.

Parameter		Value	Parameter		Value
Z_2	Canopy-top height (m)	0.15, 0.58, 0.86	S_6	Half-inhibition high temperature, respiration (K)	328
Z_1	Canopy-base height (m)	0.1	T_{opt}	Optimum temperature for vegetation growth (K)	298
χ_L	Leaf-angle distribution factor	−0.02	S_3	Low temperature stress factor, photosynthesis (K^{-1})	0.2
D_r	Root depth (m)	0.1, 0.14, 0.21	S_4	Half-inhibition low temperature, photosynthesis (K)	281
ψ_c	One-half inhibition water potential	−200	S_1	High temperature stress factor, photosynthesis (K^{-1})	0.3
$\delta_{V,l}$	Leaf transmittance, visible, live	0.07	S_2	Half-inhibition high temperature, photosynthesis (K)	308

Table A1. *Cont.*

Parameter		Value	Parameter		Value
$\delta_{V,d}$	Leaf transmittance, visible, dead	0.25	D_T	Total soil depth (m)	0.4
$\delta_{N,l}$	Leaf transmittance, near IR, live	0.22	α_{sV}	Soil reflectance, visible	0.1
$\delta_{N,d}$	Leaf transmittance, near IR, dead	0.38	α_{sN}	Soil reflectance, near IR	0.15
$\alpha_{v,l}$	Leaf reflectance, visible, live	0.105	B	Soil wetness exponent	8.52
$\alpha_{v,d}$	Leaf reflectance, visible, dead	0.58	ψ_s	Soil tension at saturation (m)	−0.36
$\alpha_{N,l}$	Leaf reflectance, near IR, live	0.36, 0.18	K_s	Hydraulic conductivity at saturation (m s^{-1})	2.5×10^{-6}
$\alpha_{N,d}$	Leaf reflectance, near IR, dead	0.58, 0.4	θ_s	Soil porosity (volume fraction)	0.48
ε	Intrinsic quantum efficiency (mol mol^{-1})	0.08	$\varnothing_s$	Mean topographic slope (radians)	0.176
M	Stomatal slope factor	13.0	V_{max0}	Maximum rubisco capacity, top leaf (mol m^{-2} s^{-1})	1.5×10^{-4}
b	Minimum stomatal conductance (mol m^{-2} s^{-1})	0.01	$G(\mu)/\mu$	Time-mean leaf projection	1.0
f_d	Leaf respiration factor	0.015	G_1	Augmentation factor for momentum transfer coefficient	1.449
βce	Photosynthesis coupling coefficient	0.98	G_4	Transition height factor for momentum transfer coefficient	11.785
βps	Photosynthesis coupling coefficient	0.95	z_{wind}	Wind observation height (m)	10.0
S_5	High temperature stress factor, respiration (K^{-1})	1.3	z_{met}	Air temperature and humidity observation height (m)	10.0

Table A2. Variables selected to train the RF model.

Variable	Unit	Description	Variable	Unit	Description
NDVI	–	Normalized difference vegetation index	RH_3	%	Relative humidity at 3 m
LAI	–	Leaf area index	P	hPa	Pressure
FPAR	–	Fraction of photosynthetically active radiation	q	g g^{-1}	Specific humidity at 3 m
T_3	K	Air temperature observed at 3 m	VPD	hPa	Vapor pressure deficit at 3 m
T_g	K	Temperature of land surface	u^*	m s^{-1}	Friction velocity
T_m	K	Average temperature of air at 3 m and ground	T^*	K	Disturbances in temperature
G_5	W m^{-2}	Soil heat flux at the depth of 5 cm	WS	m s^{-1}	Wind speed at 3 m
dT	K	Bias of temperature for canopy air space and observation height	W_{Dir}	degrees from north	Wind direction at 3 m
Ts_5	K	Temperature of soil at the depth of 5 cm	Rn	W m^{-2}	Net radiation
Ts_{10}	K	Temperature of soil at the depth of 10 cm	$SiB2_H$	W m^{-2}	The H modeled by SiB2
Ts_{20}	K	Temperature of soil at the depth of 20 cm	$SiB2_{LE}$	W m^{-2}	The LE modeled by SiB2
Ts_{40}	K	Temperature of soil at the depth of 40 cm	$SiB2_{G0}$	W m^{-2}	The G_0 modeled by SiB2
Es_0	hPa	Saturated vapor pressure of land surface	$SiB2_{Fc}$	μmol m^{-2} s^{-1}	The F_c modeled by SiB2

References

1. Chen, C.; Li, D.; Gao, Z.; Tang, J.; Guo, X.; Wang, L.; Wan, B. Seasonal and Interannual Variations of Carbon Exchange over a Rice-Wheat Rotation System on the North China Plain. *Adv. Atmos. Sci.* **2015**, *32*, 1365–1380. [CrossRef]
2. Jing, Z.; Jing, Y.; Zhang, F.; Qiu, R.; Wido, H. Application of the Simple Biosphere Model 2 (SiB2) with Irrigation Module to a Typical Low-Hilly Red Soil Farmland and the Sensitivity Analysis of Modeled Energy Fluxes in Southern China. *Water* **2019**, *11*, 1128. [CrossRef]
3. Li, Y.; Sun, R.; Liu, S. Vegetation Physiological Parameter Setting in the Simple Biosphere Model 2 (SiB2) for Alpine Meadows in the Upper Reaches of Heihe River. *Sci. China Earth Sci.* **2015**, *58*, 755–769. [CrossRef]
4. Yuan, Z.; Shen, Y.; Chu, Y.; Qi, Y. Characteristics and Simulation of Heat and CO_2 Fluxes over a Typical Cropland During the Winter Wheat Growing in the North China Plain. *Envrion. Sci.* **2010**, *31*, 41–48. [CrossRef]
5. Zhang, G.; Lu, L.; Jiang, L.; Jiang, L.; Baker, I. Study of the surface energy flux at the three different sites over China based on SiB2 and SiB3. *Acta Meteorol. Sin.* **2013**, *71*, 692–708. [CrossRef]
6. Dickinson, R.E.; Henderson-Sellers, A.; Rosenzweig, C.; Sellers, P.J. Evapotranspiration Models with Canopy Resistance for Use in Climate Models, a Review. *Agric. For. Meteorol.* **1991**, *54*, 373–388. [CrossRef]
7. Gao, Z.; Chae, N.; Kim, J.; Hong, J.; Choi, T.; Lee, H. Modeling of Surface Energy Partitioning, Surface Temperature, and Soil Wetness in the Tibetan Prairie Using the Simple Biosphere Model 2 (SiB2). *J. Geophys. Res. Atmos.* **2004**, *109*, D06102. [CrossRef]

8. Rowntree, P.R. Atmospheric Parameterization Schemes for Evaporation over Land: Basic Concepts and Climate Modeling Aspects. In *Land Surface Evaporation*; Schmugge, T.J., André, J.-C., Eds.; Springer: New York, USA, 1991; pp. 5–29. ISBN 978-0-387-97359-3.
9. Baldocchi, D. Assessing the Eddy Covariance Technique for Evaluating Carbon Dioxide Exchange Rates of Ecosystems: Past, Present and Future. *Glob. Chang. Biol.* **2003**, *9*, 479–492. [CrossRef]
10. Duan, Z.; Grimmond, C.S.B.; Gao, C.Y.; Sun, T.; Liu, C.; Wang, L.; Li, Y.; Gao, Z. Seasonal and Interannual Variations in the Surface Energy Fluxes of a Rice–Wheat Rotation in Eastern China. *J. Appl. Meteorol. Climatol.* **2021**, *60*, 877–891. [CrossRef]
11. Zhang, G.; Lu, L.; Jiang, L.; Jiang, L.; Baker, I. Modeling of CO_2 Fluxes at Cropland by Using SiB3 Model. *Environ. Sci.* **2013**, *34*, 4000–4008. [CrossRef]
12. Lei, H.; Yang, D.; Cai, J.; Wang, F. Long-Term Variability of the Carbon Balance in a Large Irrigated Area along the Lower Yellow River from 1984 to 2006. *Sci. China Earth Sci.* **2013**, *56*, 671–683. [CrossRef]
13. Lokupitiya, E.; Denning, S.; Paustian, K.; Baker, I.; Schaefer, K.; Verma, S.; Meyers, T.; Bernacchi, C.; Suyker, A.; Fischer, M. Incorporation of Crop Phenology in Simple Biosphere Model (SiBcrop) to Improve Land-Atmosphere Carbon Exchanges from Croplands. *Biogeosciences* **2009**, *6*, 969–986. [CrossRef]
14. Chu, Y.; Yuan, Z.; Liu, C.; Shi, J.; Zhang, J. An Application of SiB2 in the Yellow River Irrigation Region. *Res. Soil Water Convervation* **2011**, *18*, 159–163.
15. Jiang, L.; Lu, L.; Xing, W.; Zhang, L.; Baker, I.; Zhang, G.; Zuo, J. Evaluating surface energy budgets simulated by SiB3 at three different climate-ecosystem tower sites. *J. Meteorol. Sci.* **2011**, *31*, 493–500. [CrossRef]
16. Liu, X. Parameter Estimation of Terrestrial Ecosystem Process Model and Its Application in Carbon and Water Fluxes Simulation—The CEVSA Model as an Example. Master's Thesis, Lanzhou University, Lanzhou, China, 2020.
17. Manabe, S. Climate and the Ocean Circulation, I, The Atmospheric Circulation and the Hydrology of the Earth's Surface. *Mon. Weather Rev.* **1969**, *97*, 739–774. [CrossRef]
18. Dickinson, R.E. Land Surface Processes and Climate—Surface Albedos and Energy Balance. In *Advances in Geophysics*; Saltzman, B., Ed.; Theory of Climate; Elsevier: Amsterdam, The Netherlands, 1983; Volume 25, pp. 305–353.
19. Sellers, P.J.; Mintz, Y.; Sud, Y.C.; Dalcher, A. A Simple Biosphere Model (SIB) for Use within General Circulation Models. *J. Atmos. Sci.* **1986**, *43*, 505–531. [CrossRef]
20. Sellers, P.J.; Tucker, C.J.; Collatz, G.J.; Los, S.O.; Justice, C.O.; Dazlich, D.A.; Randall, D.A. A Revised Land Surface Parameterization (SiB2) for Atmospheric GCMS. Part II: The Generation of Global Fields of Terrestrial Biophysical Parameters from Satellite Data. *J. Clim.* **1996**, *9*, 706–737. [CrossRef]
21. Sellers, P.J.; Randall, D.A.; Collatz, G.J.; Berry, J.A.; Field, C.B.; Dazlich, D.A.; Zhang, C.; Collelo, G.D.; Bounoua, L. A Revised Land Surface Parameterization (SiB2) for Atmospheric GCMS. Part I: Model Formulation. *J. Clim.* **1996**, *9*, 676–705. [CrossRef]
22. Xie, X.; Li, A.; Jin, H. The simulation models of the forest carbon cycle on a large scale: A review. *Acta Ecol. Sin.* **2018**, *38*, 41–54. [CrossRef]
23. Tian, H.; Liu, M.; Zhang, C.; Ren, W.; Xu, X.; Chen, G.; Lv, C.; Tao, B. The Dynamic Land Ecosystem Model (DLEM) for Simulating Terrestrial Processes and Interactions in the Context of Multifactor Global Change. *Acta Geogr. Sin.* **2010**, *65*, 1027–1047. [CrossRef]
24. Li, G.; Hao, Y.; Yang, T.; Xiao, W.; Pan, M.; Huo, S.; Lyu, T. Enhancing Bioenergy Production from the Raw and Defatted Microalgal Biomass Using Wastewater as the Cultivation Medium. *Bioengineering* **2022**, *9*, 637. [CrossRef] [PubMed]
25. Li, G.; Hu, R.; Wang, N.; Yang, T.; Xu, F.; Li, J.; Wu, J.; Huang, Z.; Pan, M.; Lyu, T. Cultivation of Microalgae in Adjusted Wastewater to Enhance Biofuel Production and Reduce Environmental Impact: Pyrolysis Performances and Life Cycle Assessment. *J. Clean. Prod.* **2022**, *355*, 131768. [CrossRef]
26. Hanan, N.P.; Berry, J.A.; Verma, S.B.; Walter-Shea, E.A.; Suyker, A.E.; Burba, G.G.; Denning, A.S. Testing a Model of CO_2, Water and Energy Exchange in Great Plains Tallgrass Prairie and Wheat Ecosystems. *Agric. For. Meteorol.* **2005**, *131*, 162–179. [CrossRef]
27. Liu, F.; Tao, F.; Xiao, D.; Zhang, S.; Wang, M.; Zhang, H.; Bai, H. The contributions of leaf area index and precipitation to surface energy balance in the process of land cover change. *Geogr. Res.* **2014**, *33*, 1264–1274. [CrossRef]
28. Wang, Y. The Observation and Simulation of Turbulence Fluxes Over Rice Paddy and Dry Farming Land. Master's Thesis, Chinese Academy of Meteorological Sciences, Beijing, China, 2003.
29. Xue, Y.; Gao, Z.; Sha, W. Modeling of Surface Energy Budget in the Tibetan Plateau Using Simple Biosphere Model SiB2. *Chin. J. Atmoshperic Sci.* **2010**, *34*, 131–142. [CrossRef]
30. Yan, X.; Li, H.; Liu, F.; Gao, Z.; Liu, H. Modeling of Surface Flux in Tongyu Using the Simple Biosphere Model 2 (SiB2). *J. For. Res.* **2010**, *21*, 183–188. [CrossRef]
31. Lei, H.; Yang, D.; Shen, Y.; Liu, Y.; Zhang, Y. Simulation of Evapotranspiration and Carbon Dioxide Flux in the Wheat-Maize Rotation Croplands of the North China Plain Using the Simple Biosphere Model. *Hydrol. Process.* **2011**, *25*, 3107–3120. [CrossRef]
32. Wang, H.; Tan, Z.; Yang, M.; Zhang, Q.; Jiang, L. Research on air temperature product examination of three numerical forecast and a method of error correction. *J. Meteorol. Environ.* **2018**, *34*, 22–29. [CrossRef]
33. Chen, Y.; Ning, Y.; Tang, R.; Xie, X. Tropical Temperature Correction for Numerical Forecast in Hainan Ba on Spatiotemporal Independence Random Forest Model. *Nat. Sci. J. Hainan Univ.* **2020**, *38*, 356–364. [CrossRef]
34. Sun, Q.; Jiao, R.; Xia, J.; Yan, Z.; Li, H.; Sun, J.; Wang, L.; Liang, Z. Adjusting Wind Speed Prediction of Numerical Weather Forecast Model Based on Machine Learning Methods. *Meteorol. Mon.* **2019**, *45*, 426–436. [CrossRef]
35. Wang, X.; Shi, G.; Zhou, Z.; Zhen, Y. Revision of solar radiation product ERA5 based on random forest algorithm. *Remote Sens. Nat. Resour.* **2022**, *34*, 105–111. [CrossRef]

36. Zeng, J. Temperature Correction Model Based on Machine Learning and Multi-Meteorological Factor Model. Master's Thesis, Zhejiang Normal University, Jinhua, China, 2020.
37. Xu, L.; Wang, A.; Wang, C.; Chen, Y.; Chen, Y.; Zhou, Z.; Chen, X.; Xing, J.; Liu, K.; Huang, X. Research on correction method of marine environment prediction based on machine learning. *Mar. Sci. Bull.* **2020**, *39*, 695–704. [CrossRef]
38. Wang, T.; Brender, P.; Ciais, P.; Piao, S.; Mahecha, M.D.; Chevallier, F.; Reichstein, M.; Ottlé, C.; Maignan, F.; Arain, A.; et al. State-Dependent Errors in a Land Surface Model across Biomes Inferred from Eddy Covariance Observations on Multiple Timescales. *Ecol. Model.* **2012**, *246*, 11–25. [CrossRef]
39. Abramowitz, G.; Pitman, A.; Gupta, H.; Kowalczyk, E.; Wang, Y. Systematic Bias in Land Surface Models. *J. Hydrometeorol.* **2007**, *8*, 989–1001. [CrossRef]
40. Esteban, J.; McRoberts, R.; Fernández-Landa, A.; Tomé, J.; Næsset, E. Estimating Forest Volume and Biomass and Their Changes Using Random Forests and Remotely Sensed Data. *Remote Sens.* **2019**, *11*, 1944. [CrossRef]
41. Tramontana, G.; Jung, M.; Schwalm, C.R.; Ichii, K.; Camps-Valls, G.; Ráduly, B.; Reichstein, M.; Arain, M.A.; Cescatti, A.; Kiely, G.; et al. Predicting Carbon Dioxide and Energy Fluxes across Global FLUXNET Sites with Regression Algorithms. *Biogeosciences* **2016**, *13*, 4291–4313. [CrossRef]
42. Chen, L.; Zhou, G.; Du, H.; Liu, Y.; Mao, F.; Xu, X.; Li, X.; Cui, L.; Li, Y.; Zhu, D. Simulation of CO_2 Flux and Controlling Factors in Moso Bamboo Forest Using Random Forest Algorithm. *Sci. Silvae Sin.* **2018**, *54*, 1–12. [CrossRef]
43. Duan, Z.; Yang, Y.; Zhou, S.; Gao, Z.; Zong, L.; Fan, S.; Yin, J. Estimating Gross Primary Productivity (GPP) over Rice–Wheat-Rotation Croplands by Using the Random Forest Model and Eddy Covariance Measurements: Upscaling and Comparison with the MODIS Product. *Remote Sens.* **2021**, *13*, 4229. [CrossRef]
44. Lei, H.; Yang, D. Seasonal and Interannual Variations in Carbon Dioxide Exchange over a Cropland in the North China Plain: CARBON DIOXIDE EXCHANGE OVER A CROPLAND. *Glob. Chang.Biol.* **2009**, *16*, 2944–2957. [CrossRef]
45. Li, G.; Zhang, J.; Chen, C.; Tian, H.; Zhao, L. Research progress on carbon storage and flux in different terrestrial ecosystem in China under global climate change. *Ecol. Environ. Sci.* **2013**, *22*, 873–878. [CrossRef]
46. Peng, J.; Hu, J.; Liu, R.; Ma, Q.; Yang, G. Characteristics Analysis of CO_2 and Heat Flux in Winter Wheat Fields. *J. Henan Agric. Sci.* **2012**, *41*, 15–19. [CrossRef]
47. Duan, Z.; Gao, Z.; Xu, Q.; Zhou, S.; Qin, K.; Yang, Y. A Benchmark Dataset of Diurnal- and Seasonal-Scale Radiation, Heat, and CO_2 Fluxes in a Typical East Asian Monsoon Region. *Earth Syst. Sci. Data* **2022**, *14*, 4153–4169. [CrossRef]
48. Yang, Y.; Zhang, M.; Li, Q.; Chen, B.; Gao, Z.; Ning, G.; Liu, C.; Li, Y.; Luo, M. Modulations of Surface Thermal Environment and Agricultural Activity on Intraseasonal Variations of Summer Diurnal Temperature Range in the Yangtze River Delta of China. *Sci. Total Environ.* **2020**, *11*, 139445. [CrossRef] [PubMed]
49. Wang, X.; Dong, Z.; Han, R. Characteristics of Annual Mean Temperature and Precipitation of Jiangsu Dongtai in Recent 65 Years. *Water Resour. Power* **2017**, *35*, 6–9+21.
50. Wang, X. Analysis on the trend of rainfall in recent years in the coastal area of Dongtai City. *Jiangsu Water Resour.* **2019**, *9*, 37–40. [CrossRef]
51. Li, X.; Gao, Z.; Li, Y.; Tong, B. Comparison of Sensible Heat Fluxes Measured by a Large Aperture Scintillometer and Eddy Covariance System over a Heterogeneous Farmland in East China. *Atmosphere* **2017**, *8*, 101. [CrossRef]
52. Lee, X.; Massman, W.J.; Law, B.E. (Eds.) *Handbook of Micrometeorology: A Guide for Surface Flux Measurement and Analysis*; Atmospheric and Oceanographic Sciences Library; Kluwer Academic: Dordrecht, The Netherlands; London, UK, 2004; ISBN 978-1-4020-2264-7.
53. Moncrieff, J.; Clement, R.; Finnigan, J.; Meyers, T. Averaging, Detrending, and Filtering of Eddy Covariance Time Series. In *Handbook of Micrometeorology: A Guide for Surface Flux Measurement and Analysis*; Lee, X., Massman, W., Law, B., Eds.; Atmospheric and Oceanographic Sciences Library; Springer Netherlands: Dordrecht, The Netherlands, 2005; pp. 7–31. ISBN 978-1-4020-2265-4.
54. Webb, E.K.; Pearman, G.I.; Leuning, R. Correction of Flux Measurements for Density Effects Due to Heat and Water Vapour Transfer. *Q. J. R. Meteorol. Soc.* **1980**, *106*, 85–100. [CrossRef]
55. Zhang, P.; Bounoua, L.; Thome, K.; Wolfe, R. Modeling Impact of Urbanization in Us Cities Using Simple Biosphere Model SiB2. In Proceedings of the 2016 IEEE International Geoscience and Remote Sensing Symposium (IGARSS), Beijing, China, 10–15 July 2016; pp. 6758–6761.
56. Breiman, L. Random Forests. *Mach. Learn.* **2001**, *45*, 5–32. [CrossRef]
57. Belgiu, M.; Drăguţ, L. Random Forest in Remote Sensing: A Review of Applications and Future Directions. *ISPRS J. Photogramm. Remote Sens.* **2016**, *114*, 24–31. [CrossRef]
58. Liu, J.; Yunjiang, Z.; Nannan, W.; Yuan, F.; Zhu, X.; Zhang, L.; Zhang, J.; Sun, Y.; Guo, Z.; Guo, Y.; et al. Comparative Analysis of Two Machine Learning Algorithms in Predicting Site-Level Net Ecosystem Exchange in Major Biomes. *Remote Sens.* **2021**, *13*, 2242. [CrossRef]
59. Chen, X.; Yu, Y.; Chen, J.; Zhang, T.; Li, Z. Study of Estimation of Soil Heat Flux at a Wheat Field in Semi-Arid Area Loess Plateau. *Plateau Meteorol.* **2014**, *33*, 1514–1525. [CrossRef]
60. Sun, Y. Regional Turbulent Water and Heat Fluxes from Airborne Eddy Covariance Measurements. Ph.D. Thesis, University of Chinese Academy of Sciences, Beijing, China, 2018.
61. Xing, L.; Cui, N.; Dong, J. Study on LSTM deep learning model-based prediction of reference crop evapotranspiration in North China. *Water Resour. Hydropower Eng.* **2019**, *50*, 64–72. [CrossRef]

62. Zhu, X.; Liu, Y.; Wu, G. An assessment of summer sensible heat flux on the Tibetan Plateau from eight data sets. *Sci. China Earth Sci.* **2012**, *42*, 1104–1112. [CrossRef]
63. Liu, C.; Gao, Z.; Li, Y.; Gao, C.Y.; Su, Z.; Zhang, X. Surface Energy Budget Observed for Winter Wheat in the North China Plain During a Fog–Haze Event. *Bound.-Layer Meteorol* **2019**, *170*, 489–505. [CrossRef]
64. Yang, Y.; Zhou, M.; Wang, T.; Yao, B.; Han, P.; Ji, D.; Zhou, W.; Sun, Y.; Wang, G.; Wang, P. Spatial and Temporal Variations of CO_2 Mole Fractions Observed at Beijing, Xianghe, and Xinglong in North China. *Atmos. Chem. Phys.* **2021**, *21*, 11741–11757. [CrossRef]
65. Matthews, B.; Schume, H. Tall Tower Eddy Covariance Measurements of CO_2 Fluxes in Vienna, Austria. *Atmos. Environ.* **2022**, *274*, 118941. [CrossRef]
66. Duan, Z.; Yang, Y.; Wang, L.; Liu, C.; Fan, S.; Chen, C.; Tong, Y.; Lin, X.; Gao, Z. Temporal Characteristics of Carbon Dioxide and Ozone over a Rural-Cropland Area in the Yangtze River Delta of Eastern China. *Sci. Total Environ.* **2021**, *757*, 143750. [CrossRef]
67. Yuan, Z.; Shen, Y.; Wang, L.; Liu, C.; Cao, S.; Lin, Z. Simulation of Energy and Carbon Fluxes over a Typical Cropland during the Summer Maize Growing in the Yellow River Irrigation Region by Use of SIB2. In Proceedings of the 2010 International Conference on Machine Learning and Cybernetics, Qingdao, China, 11–14 July 2010; pp. 2309–2312.

atmosphere

MDPI

Article

A Simple Parameterization to Enhance the Computational Time in the Three Layer Dry Deposition Model for Smooth Surfaces

Omar M. M. Nofal [1], Omar Al-Jaghbeer [2], Zaid Bakri [3] and Tareq Hussein [1,2,*]

1 Environmental and Atmospheric Research Laboratory (EARL), Department of Physics, School of Science, The University of Jordan, Amman 11942, Jordan; amr9170120@ju.edu.jo
2 Institute for Atmospheric and Earth System Research (INAR/Physics), University of Helsinki, FI-00014 Helsinki, Finland; amr8190216@ju.edu.jo
3 Physics Department and Atmospheric Science Program, Michigan Technological University, Houghton, MI 49931, USA; zbakri@mtu.edu
* Correspondence: tareq.hussein@helsinki.fi

Abstract: Optimization of dry deposition velocity calculation has been of great interest. Every time, determining the value of the concentration boundary layer (CBL) thickness led to a waste of numerical calculation time, which appears as a huge time in large-scale climate models. The goal of this study is to optimize the numerical calculation time in the three-layer deposition model for smooth surfaces through the development of a MATLAB code that can parameterize the appropriate concentration boundary layer height (y^{+}_{cbl}) and internal integral calculation intervals for each particle diameter D_p (0.01–100 μm) and friction velocity u^* (0.01–100 m/s). The particle concentration, as a solution to the particle flux equation, is obtained and modeled numerically by performing the left Riemann sum using MATLAB software. On the other hand, the number of subdivisions N of the Riemann sum was also parameterized for each D_p and u^* in order to lessen the numerical calculation time. From a numerical point of view, the new parameterizations were tested by several computers; about 78% on the average of the computation time was saved when compared with the original algorithm. In other words, on average, about 1.2 s/calculation was gained, which is valuable in climate models simulations when millions of dry deposition calculations are needed.

Keywords: three-layer deposition model; dry deposition velocity; Brownian diffusion; Eddy diffusion; gravitational settling; concentration boundary layer thickness; parameterization; numerical calculation time; global models; friction velocity

Citation: Nofal, O.M.M.; Al-Jaghbeer, O.; Bakri, Z.; Hussein, T. A Simple Parameterization to Enhance the Computational Time in the Three Layer Dry Deposition Model for Smooth Surfaces. *Atmosphere* **2022**, *13*, 1190. https://doi.org/10.3390/atmos13081190

Academic Editors: Yubin Li and Jie Tang

Received: 21 June 2022
Accepted: 22 July 2022
Published: 27 July 2022

1. Introduction

Particle dry deposition is important in many aspects of applications, such as environmental (atmospheric climate, rooms clean, deposition on vegetation, etc.), industrial (paper industry, copying on papers, painting, ventilation ducts, insecticides, etc.), medical (deposition in pulmonary airway replicas, respiratory tract deposition, aerosol contaminant deposition on human skin, etc.), pharmaceutical (treatment by deposition of pharmaceutical aerosol, etc.), etc. [1–13]. Therefore, accurate estimation of the number of particles being transferred from the airborne state into the deposit state is a vital aspect. The development in the computational part is as important as that in the theoretical or the experimental part in this matter. For example, the computational time required to estimate (or calculate) the deposition velocity of the particle flux towards a surface becomes tremendously long and expensive when the dry deposition is taken into account in global aerosol models. In that sense, parameterizations are introduced to make the computational resources less expensive, i.e., save computational time. In addition to that, parameterization is a valuable method in situations where the details are not apparent or cannot be measured experimentally, as in medical, pharmaceutical, and climate modeling, and is an affordable method in situations in which the wrong results are expensive, as in industrial applications, so we can use modeling and, as a sequence, parameterizations to test a hypothesis [14].

In general, dry deposition onto a surface is assumed to occur through two major processes. First, advection and turbulent mixing transport the airborne particles to the region adjacent to the deposition surface, reaching the so-called concentration boundary layer (CBL), which is, from a fluid mechanics point of view, a thin layer formed by fluid in the immediate neighborhood of surrounding surface. Second, transport mechanisms transport the particles from the air and deposit them onto the surface. It is believed that the second process controls the particle flux towards the surface when the fluid adjacent to the surface is turbulently well mixed [15]. The major particle transport mechanisms within the CBL include Fickian diffusion (Brownian and Eddy), gravitational settling, and turbophoresis, and the minor mechanisms can be thermophoresis, electrophoresis, magnetophoresis, etc. [16].

The dry deposition velocity (V_d), which is a measure of the effectiveness of the deposition mechanism, is derived from the particle flux equation. It is governed by certain boundary conditions [16–19]. For example, V_d for particles being transported under gravitational settling and under the effect of turbophoresis is dominant for particles with large relaxation time τ_p (i.e., particles with larger mass) [20]. Another example is the influence of the friction velocity (u^*); increasing u^* results in enhancement in V_d for particles with small τ_p (i.e., small particles). In the situations of inhomogeneous air turbulence, increasing u^* can also enhance the deposition of particles with large τ_p because of enhanced turbophoresis.

From a modeling point of view, we are interested in developing more robust models either by enhancing the theory or by introducing new algorithms. With respect to theoretical development, several approaches have been introduced, trying to formulate the best mathematical expressions that can construct the proper model that fits well with the measured data [18,19,21]. However, the lack of understanding of the nature of the CBL parameters (y_0 and y_{cbl}, which are presented in Figure 1) is considered a major challenge. For example, Kallio and Reeks [22] developed a power law expression to achieve a mathematical representation based on a Direct Numerical Simulation (DNS) after Kim et al. [23]. After that, it was assumed that CBL height could be set to 30 (dimensionless units; $y^+{}_{cbl}$) above a smooth surface [18,19]. This height was set to 100 for large particles [15] and 200 for rough surfaces [21]. However, the chosen value for the CBL height (i.e., y_{cbl}) and starting level (i.e., y_0) above the surface area ought to be defined based on the particle concentration profile within the boundary layer. This depends on several factors, including u^*, particle diameter (D_p), and the deposition mechanism.

Here, we introduced a simple parameterization for the y_{cbl} as a function of u^* and D_p. This parameterization was developed for a wide particle size range (D_p 0.01–100 μm) and u^* within the range 0.01–100 m/s. This reflects typical conditions for dry deposition in most physical systems and applications. The parameterization was then implemented in the well-known three-layer dry deposition model in order to improve the computational time required to calculate the dry deposition velocity V_d.

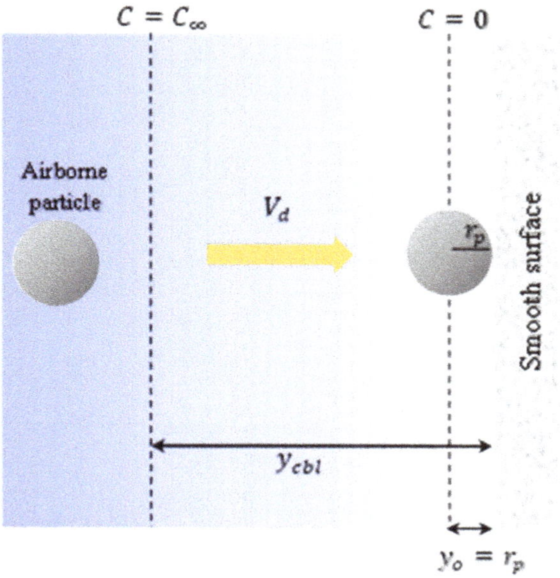

Figure 1. Schematic diagram for dry deposition on a smooth surface of airborne particle with radius r_p across the concentration boundary layer. The concentration of particle number across the boundary layer grows from lowest value ($C = 0$) at height y_0 (= r_p) to its highest value ($C = C_\infty$) at height y_{cbl}. V_d is the dry deposition velocity through the concentration boundary layer.

2. Materials and Methods

2.1. Three-Layer Dry Deposition Model

The mathematical formulation of the well-known three-layer dry deposition model is based on Eulerian approach for the particle flux across the CBL towards the surface [4,15,16,18,19,21,24–31]

$$J = J_{Fickian} + J_{Gravitational} + \sum_n J_n, \tag{1}$$

where J is the total particle flux, $J_{Fickian}$ is the particle flux due to Fickian diffusion (Brownian and Eddy), $J_{Gravitational}$ is the particle flux due to gravitational settling, and the sum represents the particle fluxes due to other mechanisms not included in this part of research. For this equation to be valid, a number of assumptions must be confirmed: steady-state particle flux perpendicular to the surface, the particle concentration gradient exists only very close to the deposition surface (i.e., within the CBL), there are no sources or sinks of particles within the boundary layer, and the surface is a perfect sink for particles.

As illustrated in Figure 1, the particle is considered deposited when its center is at a distance y_0 from the surface (i.e., $y_0 = r_p$); thus, the particle concentration is zero in the fluid right above the surface. The CBL has an upper limit above which particle concentration becomes homogeneous (i.e., $dC/dy = 0$) [27]. This implies that the top of the CBL layer (i.e., y_{cbl}) is set at the maximum concentration (i.e., C_∞). In that sense, the key parameters in accurate estimation for the dry deposition velocity are to have the right value for the height of the concentration boundary layer (i.e., y_{cbl}) and to have a well-behaved profile for the particle concentrations (i.e., C) within the boundary layer.

The particle flux due to Fickian diffusion and gravitational settling is given by

$$J = -(\varepsilon_p + D)\frac{dC}{dy} - iV_s C \tag{2}$$

where ε_p [m^2s^{-1}] is the turbulent (Eddy) diffusivity coefficient, C [m^{-3}] is the particle concentration, D [m^2s^{-1}] is the Brownian diffusivity, y [m] is the height from the surface, $i = 0, 1, -1$ according to the surface orientations, vertical, horizontal facing up (floor) and horizontal facing down (ceiling), respectively, and V_s [ms^{-1}] is the particle velocity of gravitational settling, which is the constant velocity (terminal) towards the surface [28,29]:

$$V_s = \left[\frac{4}{3}\frac{gD_p(\rho_p - \rho)C_c}{\rho C_D}\right]^{1/2} \quad (3)$$

where g [m s^{-2}] is the gravitational acceleration, D_p [m] is the particle diameter, ρ_p [kg m^{-3}] is the particle density, ρ [kg/m^{-3}] is the gas density, C_D [unitless] is the drag coefficient and C_c [unitless] is the Cunningham slip correction coefficient. The deposition velocity V_d is calculated as

$$V_d = \frac{|J(y=0)|}{C_\infty}, \quad (4)$$

For convenience, the model is described in the dimensionless formulation (Abbreviations)

$$\frac{V_d^+}{D^+} = \frac{dC^+}{dy^+} + \frac{iV_s^+}{D^+}C^+ = \frac{dC^+}{dy^+} + p(y^+)C^+, \quad (5)$$

which has the general solution for the dimensionless particle concentration C^+ as a function of the dimensionless height y^+ [32]

$$C^+ = \frac{1}{F(y^+)}\int_{y_o^+}^{y^+}\frac{V_d^+}{D^+}F(x).dx, \quad (6)$$

$$\frac{1}{V_d^+} = \frac{1}{F(y_{cbl}^+)}\int_{y_o^+}^{y_{cbl}^+}\frac{1}{D^+}F(x).dx, \quad (7)$$

$$F(x) = exp(\int_{y_o^+}^{x}p(y^+).dy^+), \quad (8)$$

$$C^+|_{y_0^+ \to r_p^+} = 0 \quad C^+|_{y^+ \to y_{cbl}^+} = 1, \quad (9)$$

It is evident that accurate calculation of ${y^+}_o$ and ${y^+}_{cbl}$ leads to accurate calculation for V_d. The determination of the appropriate value of ${y^+}_{cbl}$ for each D_p at a certain u^* will be discussed hereafter.

2.2. Parametrization for ${y^+}_{cbl}$

In the numerical investigation, we treated ${y^+}_{cbl}$ as an unknown quantity to be determined for each particle size and friction velocity. In other words, the upper limit of the integral in Equation (7) was set to ${y^+}_{max}$, which satisfies the second boundary condition in the numerical investigation; when C^+ reaches 1. The method of left Reimann sum was adopted to evaluate the integral using MATLAB™ software (including Simulink toolboxes). It was evaluated by limiting the height from the surface with the parameter ${y^+}_{max}$ to a certain value aiming for the determination of ${y^+}_{cbl}$. The distance from the surface to ${y^+}_{max}$ was divided equally into a proper number of intervals (N) that lead to an accurate solution for V_d.

The numerical investigations were performed for the particle diameter range 0.01–100 μm, friction velocity range 0.01–100 m/s, ${y^+}_{max}$ up to 1000, and N up to 1000 subdivisions for y^+. The ${y^+}_{cbl}$ and N for each parameter varied to obtain convergent solutions for V_d. The variation of ${y^+}_{cbl}$ (0–1000) was performed continuously and repeatedly using a MATLAB™ code that can repeat the solution of Equation (7) until the value of V_d convergent to a certain value; at that end, the code gives us the values of V_d and ${y^+}_{cbl}$ as outputs at a certain friction velocity where N = 1000 at this stage. After we parameterize ${y^+}_{cbl}$ and V_d get known and convergent to non-parameterized value we proceeding the parameterization for N in order to diminish the calculation time further by using y^+_{cbl} as input and N as variable in this stage until the same V_d achieved.

In the literature, as mentioned above in the introduction, each study case takes a certain value of y^+_{cbl} according to certain assumptions that meet the conditions considered in the model needs. For example, it was assumed to have y^+_{cbl} = 30 for a smooth surface and fine particles [18,19,33]. For micron particles, it was suggested that y^+_{cbl} = 100 [15]. for a rough surface it was suggested that y^+_{cbl} = 200 [21]. Figure 2 illustrates the variation of y^+_{cbl} with D^+_p by using a certain u^* (= 100 m/s) and N (= 1000 subdivisions)

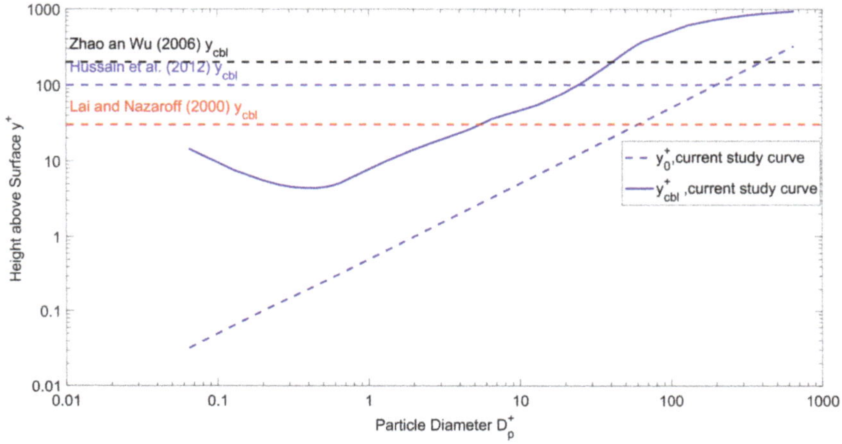

Figure 2. Illustration of the Variation of the concentration boundary layer height above the surface in dimensionless unit (y^+) with the dimensionless particle diameter (D^+_p) for the current study. The horizontal dashed lines resemble the limit value for y^+_{cbl} used in the calculation by Zhao and Wu [19], Hussain et al. [15], and Lai and Nazaroff [18]. Friction velocity u^* = 100 m/s, and number of subdivisions N = 1000 subdivisions.

As will be shown later in the results and discussion and also made evident in Figure 2, one can conclude that if the D^+_p is about 50 and above, none of the previous assumptions for y^+_{cbl} is satisfactory. Furthermore, for D^+_p about 0.3, it is a waste of computational time for calculation to take the y^+_{cbl} larger than 5, so our new parameterizations, as will be shown in the next section, determine a proper y^+_{cbl} and N for each D^+_p at a certain u^*.

We developed two parameterizations. The first one for y^+_{cbl} as a function of D_p and u^*. The second one for N as a function of D_p and u^*. These parameterizations were utilized in the three-layer deposition model so that the most suitable y^+_{cbl} and N are used as pre-set input parameters in the V^+_d calculation. The enhancement in the computational time was then compared between the original algorithm and the new one with these parameterizations.

3. Results and Discussion

3.1. A Parameterization for Fickian Diffusion

In the beginning, the effect of Fickian diffusion (Eddy and Brownian diffusion) was parameterized. In order to investigate the concentration boundary layer height, the y^+_{cbl} parameter, the integral in Equation (7), was evaluated using MATLAB code by taking the height from the surface $y^+_o = r^+_p$ to y^+_{max} = 1000, aiming to determine y^+_{cbl}. The distance from the surface to y^+_{max} was divided equally into a number of subdivisions N = 1000, which leads to an accurate solution for V^+_d. That was calculated for each particle diameter and friction velocity determined in the range mentioned above. Figure 3a shows the y^+_{cbl} profile; from the figure, one can notice the dependence of y^+_{cbl} on the dimensionless particle relaxation time τ^+_p (or particle size D_p) and u^*. Figure 3b shows the 3-D matrix we obtained for y^+_{cbl} as a function D_p and u^*.

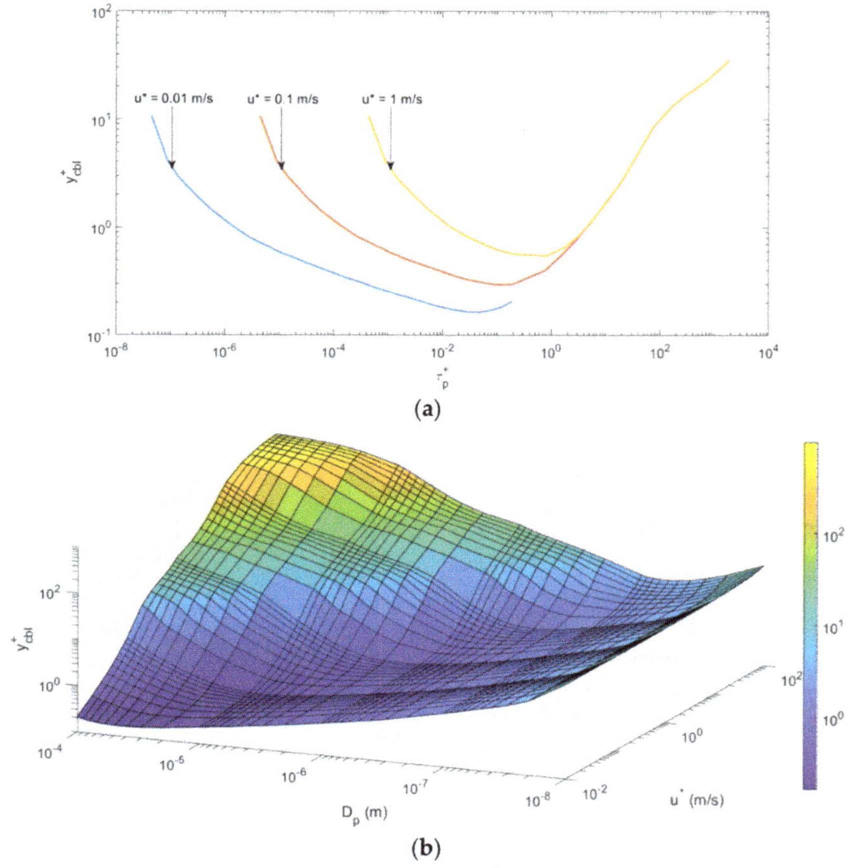

Figure 3. Dependence of y^+_{cbl} on particle size (τ^+_p or D_p) and friction velocity (u^*) for Fickian diffusion: (**a**) profile of y^+_{cbl} as a function of τ^+_p for three friction velocities. (**b**) y^+_{cbl} as a function of D_p (10 nm–100 µm) and u^* (0.01–100 m/s).

From Figure 3b, it seems that the y^+_{cbl} decreases as the D_p increases for small $u^* < 0.02$ m/s, where for 0.02 m/s < u^* < 0.3 m/s the y^+_{cbl} decreases for D_p (<8 μm) and then increases smoothly as D_p increases. For 0.3 m/s < u^* < 10 m/s, the behavior is almost the same, but the y^+_{cbl} decreases for small D_p (<0.3 μm) and then increases steeper than the previous range as D_p increases. Finally, for u^* > 10 m/s, the y^+_{cbl} decreases for small D_p (<0.3 μm) and then increases steeply as D_p increases.

In order to get the number of subdivisions N at which the conversion is achieved for the dry deposition velocity (V^+_d) for each particle diameter at a certain friction velocity, we varied N in the code until we obtained the accurate value of V^+_d for each D_p and u^*. The 3-D matrix we obtained for N as a function D_p and u^* is illustrated in Figure 4. The largest N (= 100) is for the smallest D_p and u^* and then decreases (the smallest value is 50) in general as D_p or u^* increases (Figure 4). This behavior has varied for many particle diameters, where N fluctuated for large values of u^* and D_p.

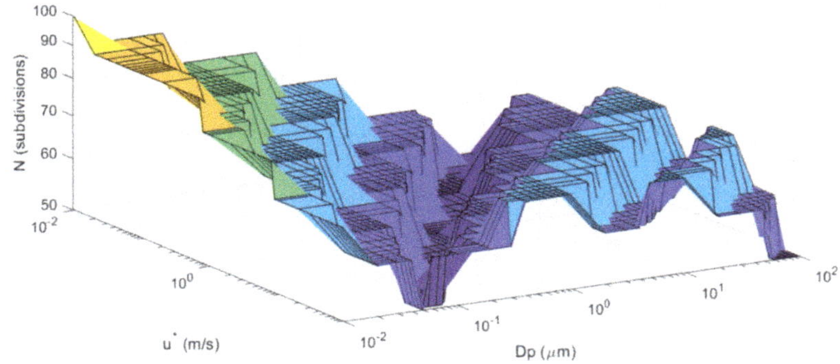

Figure 4. Appropriate number of subdivisions N for each pair of D_p (10 nm–100 μm) and u^* (0.01–100 m/s).

3.2. *The Inclusion of Gravitational Settling*

In this case, the effect of gravitational settling in addition to the effect of Fickian diffusion was studied. The $y^+{}_{cbl}$ parameter was investigated by solving the integral in equation (6) was evaluated using MATLAB code and taking the height from the surface with the parameter $y^+{}_{max}$ = 1000 again. The area under the curve was calculated by partitioning the area to N = 1000 subdivisions, and then the integral was calculated for each D_p and u^* and in the range determined. The profile of the $y^+{}_{cbl}$ and the 3-D matrices we obtained for $y^+{}_{cbl}$ as a function D_p and u^* is shown in Figure 5a,b; respectively.

Due to the mixing of two mechanisms, Fickian diffusion, and gravitational settling, the surface obtained in Figure 5b is not as smooth as that of Fickian diffusion alone. Despite the fact that there are some small fluctuations, the general behavior is the same. We notice that the conversion is achieved for $V^+{}_d$ at the same N for the same D_p and u^* in this case is identical with that of the Fickian diffusion effect alone.

The parameterization was finished after we carried out the 4-D matrix for $y^+{}_{cbl}$ and N as a function D_p and u^*. At this stage, we transformed to another phase, which is the goal of our study, which was the optimization of calculation time by updating the code to select the appropriate $y^+{}_{cbl}$ and N based on the input u^* and D_p.

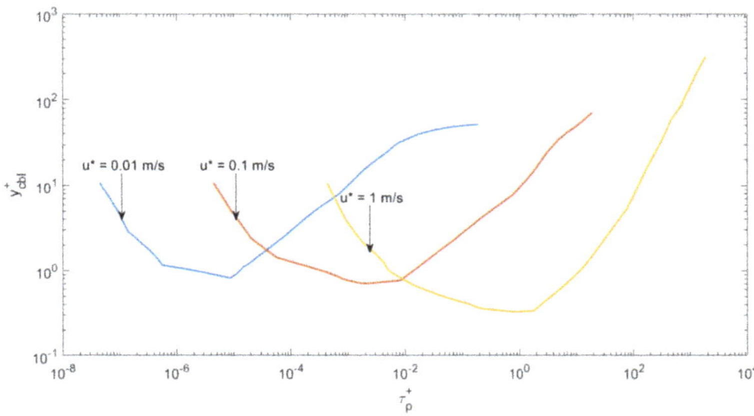

(a)

Figure 5. *Cont.*

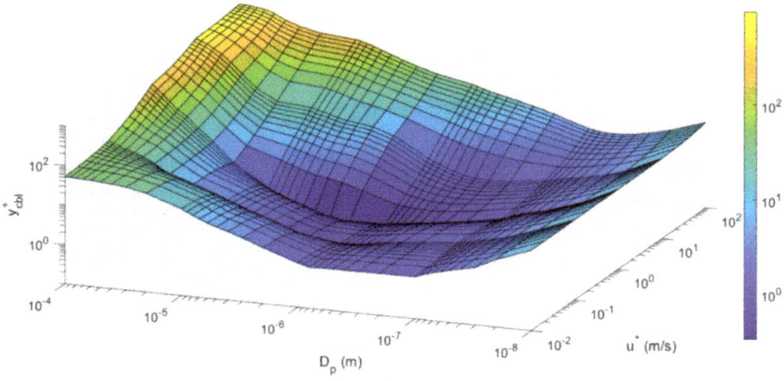

(b)

Figure 5. Dependence of y^+_{cbl} on particle size (τ^+_p or D_p) and friction velocity (u^*) when gravitational settling is added to Fickian diffusion: (**a**) profile of y^+_{cbl} as a function of the dimensionless relaxation time τ^+_p for three friction velocities. (**b**) y^+_{cbl} as a function of D_p (10 nm–100 µm) and u^* (0.01–100 m/s).

3.3. Computation Advantage by the Parameterization

To verify the optimization, in the case of Fickian diffusion alone and in the case of Fickian diffusion and gravitational settling together, we utilized the optimized code and the original one in order to make a comparison between them to find V^+_d. The calculations were performed by four computers for the two codes for selected particle diameters for three different friction velocities (u^* = 0.01, 0.1, and 1 m/s). Specifications of computers used for testing running time are summarized in Table 1.

Table 1. Computers specifications.

Computer	Processor (CPU)	Memory (RAM)	Storage
PC-1	Core i7 10th, generation	8 GB	256 SSD
PC-2	Core i5 2nd, generation	4 GB	256 SSD
PC-3	AMD RYZON 3, 3rd generation	4 GB	256 SSD
PC-4	Core i7 3rd, generation	8 GB	250 SSD

Here we compare the time elapsed for each code to obtain the accurate value of V^+_d. To add a flavor to our calculations, we found the time gained as percent time gained (% time) by the percent error method and as the time difference. The results are summarized in Tables 2–4. Notice that Table 2 is for friction velocity u^* = 0.01 m/s, Table 3 is for friction velocity u^* = 0.1 m/s, and Table 4 is for friction velocity u^* = 1 m/s.

Several conclusions can arrive from a deep look at Tables 2–4 details. First, the percent of the average time gained in each part of Table 2, Table 3, and Table 4 and/or each friction velocity (u^* = 0.01, 0.1, and 1 m/s) are 77%, 78%, and 79% respectively. Second, the percent of the average time gained for PC-1 is 79%, for PC-2 is 74%, for PC-3 is 80%, and for PC-4 is 80%. Or in general, the percent of the overall average time gained for all PCs over the range of friction velocities is 78%, which means that we need only 22% of the running time of the original code to run the optimized code. From a computer specification point of view, PC-1 and PC-4 have the same processor, memory (RAM), and storage, but the generation is different, despite that the percent of the average time gained is 79% and 80%, respectively. PC-2 and PC-3 have the same memory (RAM) and storage, but the generation and processor are different, which caused the percent of the average time gained to be 74% and 80%, respectively. From this point of view, we can deduce that processors significantly affect the percent of the average time gained rather than the other computer specifications.

From another point of view, the average time difference for the selected friction velocities for each computer (PC-1, PC-2, PC-3, and PC-4) is about 0.47 s, 1.0 s, 2.0 s, and 1.1, respectively, with a total average for all computers, is about 1.2 s. On the other hand, the average time difference for all computers for each friction velocity (i.e., for each table) is roughly the same, which is about 1.2 s. That means if we need to make just a million calculations by optimized code and saving 1.2 s/calculation, we will save about 280 h out of 356 h (or 11.6 days out of 14.8 days); this simple example opens the imagination to the effectiveness of the optimization in global models' calculation by regular computers instead of super computers. For example, in climate modeling as a global model, if we divide the Earth's surface area (510,072,000 km^2) [34] into a grid whose spatial resolution is 50 km × 50 km or at most 100 km × 100 km, which is relatively high resolution, the computer time on the fastest computers to simulate an experiment along one century may spend several weeks typically due to the large number of calculations required [14]. It seems affordable to use parameterizations that can save about 78% of the computational time (i.e., increasing the computing power), which stimulates us to include more factors that affect the dry deposition and/or use a grid with higher resolution.

Table 2. A comparison between computation time for unoptimized code (without parameterization) and optimized one (with parameterization). for four computers for a selected particle diameters (D_p) for friction velocity u^* = 0.01 m/s.

PC	D_p (μm)	Calculation Time (s) Without Parameterization	With Parameterization	% Time	Time Difference (s)
PC-1	0.01	0.60	0.15	75%	0.45
	0.1	0.67	0.17	75%	0.50
	1	0.63	0.15	76%	0.48
	10	0.61	0.13	78%	0.48
	100	0.56	0.14	74%	0.42
PC-2	0.01	1.6	0.38	76%	1.2
	0.1	1.5	0.40	73%	1.1
	1	1.5	0.38	74%	1.1
	10	1.4	0.37	74%	1.0
	100	1.3	0.36	72%	0.94
PC-3	0.01	2.5	0.52	79%	2.0
	0.1	2.8	0.52	81%	2.3
	1	2.3	0.51	78%	1.8
	10	2.4	0.48	80%	1.9
	100	2.7	0.46	83%	2.2
PC-4	0.01	1.4	0.31	78%	1.1
	0.1	1.4	0.28	80%	1.1
	1	1.3	0.30	77%	1.0
	10	1.3	0.26	80%	1.0
	100	1.3	0.25	80%	1.0

Table 3. A comparison between calculation time for unoptimized code (without parameterization) and optimized one (with parameterization) for four computers for a selected particle diameters (D_p) for friction velocity u^* = 0.1 m/s.

PC	D_p (μm)	Calculation Time (s) Without Parameterization	With Parameterization	% Time	Time Difference (s)
PC-1	0.01	0.70	0.10	86%	0.60
	0.1	0.66	0.15	77%	0.51
	1	0.63	0.13	79%	0.50
	10	0.60	0.13	78%	0.47
	100	0.58	0.13	77%	0.45
PC-2	0.01	1.6	0.40	76%	1.2
	0.1	1.5	0.37	75%	1.1
	1	1.4	0.36	75%	1.0
	10	1.4	0.40	70%	1.0
	100	1.2	0.37	68%	0.8
PC-3	0.01	2.7	0.49	82%	2.2
	0.1	2.2	0.48	78%	1.7
	1	2.4	0.46	81%	1.9
	10	2.2	0.50	78%	1.7
	100	2.7	0.55	79%	2.2
PC-4	0.01	1.5	0.27	81%	1.2
	0.1	1.3	0.25	81%	1.0
	1	1.8	0.24	87%	1.6
	10	1.2	0.30	74%	0.9
	100	1.4	0.26	82%	1.1

Table 4. A comparison between calculation time for unoptimized code (without parameterization) and optimized one (with parameterization) for four computers for a selected particle diameters (D_p) for friction velocity u^* = 1 m/s.

PC	D_p (μm)	Calculation Time (s) Without Parameterization	With Parameterization	% Time	Time Difference (s)
PC-1	0.01	0.68	0.13	81%	0.55
	0.1	0.65	0.13	80%	0.52
	1	0.75	0.13	83%	0.62
	10	0.62	0.13	79%	0.49
	100	0.59	0.13	78%	0.46
PC-2	0.01	1.6	0.41	74%	1.2
	0.1	1.6	0.38	76%	1.2
	1	1.5	0.39	74%	1.1
	10	1.2	0.38	69%	0.82
	100	1.1	0.26	77%	0.84

Table 4. *Cont.*

PC	D_p (μm)	Calculation Time (s) Without Parameterization	With Parameterization	% Time	Time Difference (s)
PC-3	0.01	2.7	0.46	83%	2.2
	0.1	2.4	0.50	79%	1.9
	1	2.6	0.53	80%	2.1
	10	2.4	0.45	81%	2.0
	100	2.3	0.44	81%	1.9
PC-4	0.01	1.2	0.26	78%	0.94
	0.1	1.2	0.24	80%	1.0
	1	1.5	0.26	83%	1.2
	10	1.3	0.30	77%	1.0
	100	1.7	0.26	85%	1.4

Since $y^+{}_{cbl}$ and N parameters are the factors that affect the running time; the time gained in the case of gravitational settling and Fickian diffusion together is identical to that we obtained for Fickian diffusion alone because N does not change in the two cases, and we use in the optimized code the $y^+{}_{cbl}$ matrix that shown in Figure 5b since it includes $y^+{}_{cbl}$ matrix for Fickian diffusion (i.e., we have used just one matrix) in order to shorten the time elapsed for codes to obtain the accurate value of $V^+{}_d$.

3.4. The Effect of Parameterization on $V^+{}_d$ Calculations

Here we compare the values of $V^+{}_d$ itself before and after optimization. The results we obtained after optimization, which appear as dashed curves, are almost identical to the results we have before optimization, which appear as solid curves, as shown in Figure 6a for Fickian diffusion. This figure implicitly tells us that we gain time by using the optimized code without accuracy loss. We verify that by quantitative comparison between parameterized versus non-parameterized dry deposition velocities shown in Figure 6b, notice that the residuals of the order of magnitude of 10^{-5} and the residual points are distributed around zero, which indicates that the parameterization is valuable from the point of calculation of view since the parameterization will not affect the accuracy of the calculation in any application including climate modeling.

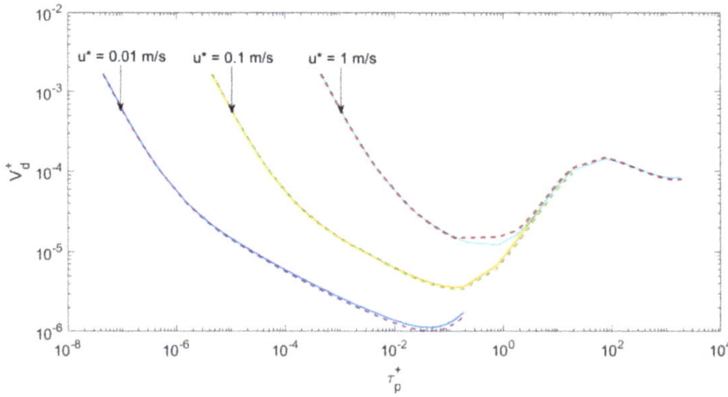

(a)

Figure 6. *Cont.*

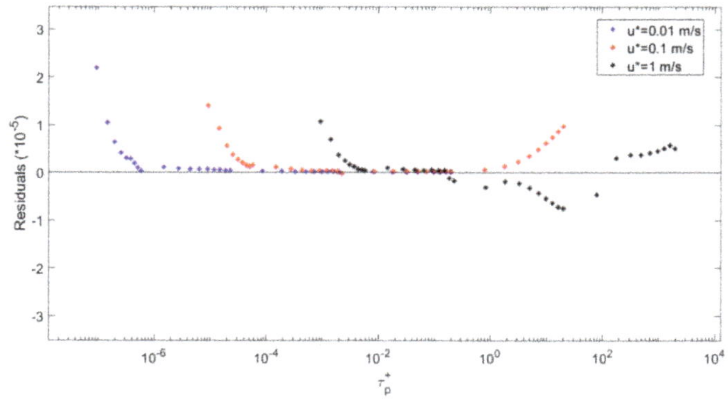

(**b**)

Figure 6. (**a**) A comparison, for Fickian diffusion only, between the optimized code (dashed curves) and the unoptimized code (solid curves) for dry deposition velocity $V^+{}_d$. (**b**) the residuals between both codes.

The comparison in the case of gravitational settling and Fickian diffusion together is shown in Figure 7a. It seems that the accuracy is high between the optimized code (dashed curves) and the code we have used before optimization (solid curves). Figure 7b shows the residuals in this case, where it seems that the residuals diverge when the dimensionless particle relaxation time ($\tau^+{}_p > 0.07$) for small friction velocity ($u^* = 0.01$ m/s) when we take all the residuals into account, and the other residuals point almost lie on residual = 0. In fact, the behavior of residuals, in this case, is selective due to the gravitational settling mechanism.

By magnifying Figure 7b we found that for friction velocity ($u^* = 0.01$ m/s) the divergence appears when $\tau^+{}_p > 6.46 \times 10^{-6}$ (i.e., $D_p > 0.4$ μm), where the residuals diverge when $\tau^+{}_p > 8.45 \times 10^{-5}$ (i.e., $D_p > 1$ μm) for friction velocity ($u^* = 0.1$ m/s), and finally the divergence accrue after $\tau^+{}_p = 7.24 \times 10^{-4}$ (i.e., $D_p > 5$ μm) for friction velocity ($u^* = 1$ m/s), as shown in Figure 7c. These details led us to conclude that our parameterizations, in the case of gravitational settling mechanism is included, are excellent in the case of sub-micron (i.e., fine and ultrafine) particles, whereas in the case of super-micron particles (1 μm < D_p < 10 μm) the parameterizations are excellent for $u^* \geq 1$ m/s since the residuals of an order of magnitude 10^{-5} and the residual points surrounding the residual is zero.

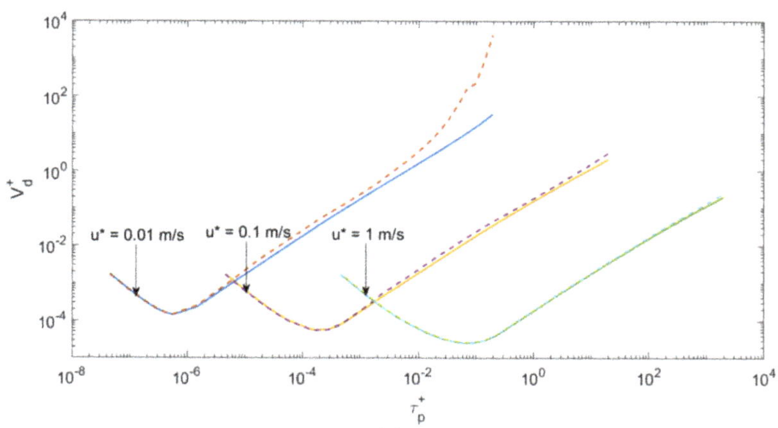

(**a**)

Figure 7. *Cont.*

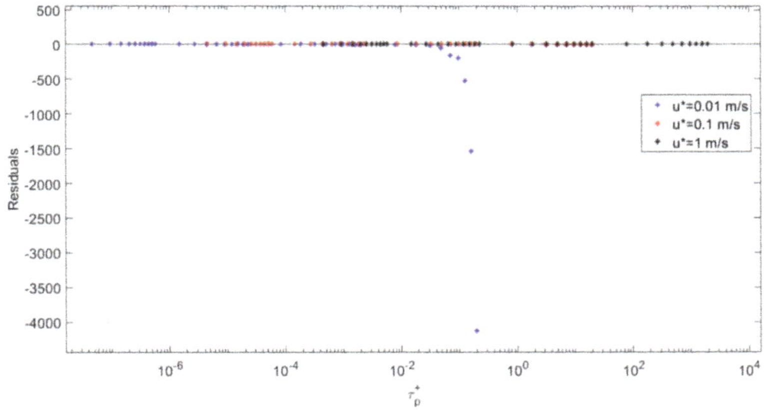

(b)

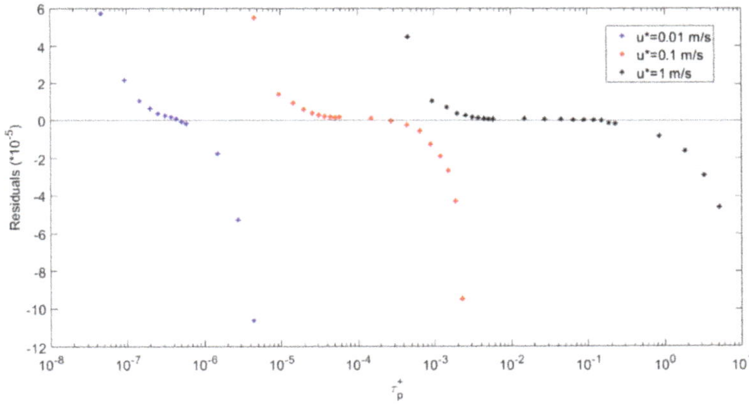

(c)

Figure 7. (**a**) A comparison, for gravitational settling and Fickian diffusion together, between the optimized code (dashed curves) and the unoptimized code (solid curves) for dry deposition velocity $V^+{}_d$. (**b–c**) the residuals between both codes.

Depending on Figures 6 and 7, the differences between non-parameterized and parameterized $V^+{}_d$ will not affect the accuracy of any calculations, including $V^+{}_d$ in the climate models and other applications.

4. Conclusions

The present article presented a parameterization to optimize the calculation time for the dry deposition velocity in the three-layer deposition model. That was based on determining a proper concentration boundary layer thickness ($y^+{}_{cbl}$) that gives accurate dry deposition velocity ($V^+{}_d$) above a smooth surface in a shorter calculation time according to the particle's diameter (D_p) and the friction velocity (u^*).

The formulation of the three-layer deposition model was used in the manipulation of Fickian (Eddy and Brownian) diffusion and gravitational settling mechanisms since they are the dominant mechanisms for small and large particles, respectively.

The solution to the particle flux (J) equation, which is the particle concentration within the boundary layer, was performed by the Riemann sum, so we found a suitable number of subdivisions (N) at which the conversion was achieved for each case along D_p range (10 nm–100 µm) and u^* range (0.01–100 m/s) in addition to the $y^+{}_{cbl}$.

After we obtained the optimization parameterization matrix, we plugged it into a MATLAB code of the three-layer deposition model that can pick up or interpolate a convenient value of y^{+}_{cbl} and N depending on D_p and u^* as the first step of the calculation. This procedure saved up to 78% (average) for each calculation when compared to the time taken for the same code without parameterization, where the comparison was by four computers that have different specifications. In other words, on average, 1.2 s/calculation can be saved, which means that our parameterization can lessen the accumulated time in the case of big data calculations in large-scale climate models.

The verification of our development was performed by comparing the V^{+}_{d} we obtained without parameterization with that we obtained with parameterization. The results confirm that the accuracy did not affect the results for the dry deposition calculation value, but the calculation time gain is valuable.

The inclusion of the minor mechanisms, such as turbophoresis, thermophoresis, electrophoresis, magnetophoresis, etc., in addition to the effect of other factors, such as temperature and pressure, on the calculation time is our future work.

Author Contributions: Conceptualization, T.H., O.M.M.N., O.A.-J. and Z.B.; methodology, T.H., O.M.M.N., O.A.-J. and Z.B.; software, O.M.M.N., O.A.-J. and Z.B.; validation, O.M.M.N.; formal analysis, O.M.M.N. and O.A.-J.; investigation, T.H., O.A.-J. and O.M.M.N.; resources, T.H.; data curation, O.M.M.N. and O.A.-J.; writing—original draft preparation, T.H., O.M.M.N., O.A.-J. and Z.B.; writing—review and editing, T.H., O.M.M.N., O.A.-J. and Z.B.; visualization, O.M.M.N. and O.A.-J.; supervision, T.H.; project administration, T.H.; funding acquisition, T.H. All authors have read and agreed to the published version of the manuscript.

Funding: This research was supported by the Deanship of Scientific Research (DSR) at the University of Jordan.

Institutional Review Board Statement: Not applicable.

Informed Consent Statement: Not applicable.

Data Availability Statement: Not applicable.

Acknowledgments: The authors acknowledge the support of the Deanship of Scientific Research (DSR) at the University of Jordan. T.H. acknowledges the support of the Atmosphere and Climate Competence Center (ACCC) Flagship funded by the Academy of Finland (grant# 337549); and the Eastern Mediterranean and Middle East Climate and Atmosphere Research (EMME-CARE) project, which received funding from the European Union's Horizon 2020 Research and Innovation Programme (Grant Agreement Number 856612) and the Government of Cyprus. The sole responsibility of this publication lies with the authors. Open access funding was provided by the University of Helsinki.

Conflicts of Interest: The authors declare no conflict of interest.

Abbreviations

Symbol	Unit	Description
C	m^{-3}	Particle concentration within the boundary layer. In dimensionless form $C^{+} = C/C_{\infty}$ C_{∞} is the particle concentration above the boundary layer or far away from the surface
C_c	-	Cunningham slip correction coefficient
D	$m^2\ s^{-1}$	Brownian diffusivity of the particle, $D = k_B\ T\ Cc/3\pi\mu\ D_p$ in dimensionless form $D^{+} = (\varepsilon_p + D)/\nu$
D_p	m	Particle diameter, in dimensionless form $D^{+}_{p} = D_p\ u^*/\nu$
J	$m^{-2}\ s^{-1}$	Total particle flux across the concentration boundary layer towards the surface. $J_{Fickian}$ is particle flux due to Brownian and Eddy diffusions. J_n is the particle flux across the concentration boundary layer due to other mechanisms to be included in the model in the future

k_B	Joule/K	Boltzmann constant
m_p	kg	Particle mass
r_p	m	Particle radius, in dimensionless form $r^+_p = r_p\, u^*/\nu$
T	K	Absolute temperature
u^*	m s^{-1}	Friction velocity
V_d	m s^{-1}	Deposition velocity onto a surface, in dimensionless form $V^+_d = V_d/u^*$
$\left\langle V'^2_y \right\rangle$	m^2 s^{-2}	Air wall normal fluctuating velocity intensity, in dimensionless [16,22]: $\left\langle V'^2_y \right\rangle^+ = \frac{\left\langle V'^2_y \right\rangle}{(u*)^2} = \left[\frac{0.005(y^+)^2}{1+0.002923(y^+)^{2.128}}\right]^2$
$\left\langle V'^2_{py} \right\rangle$	m^2 s^{-2}	Particle wall normal fluctuating velocity intensity [31]: $\left\langle V'^2_{py} \right\rangle = \left\langle V'^2_y \right\rangle \left[1 + \frac{\tau_p}{\tau_L}\right]^{-1}$ $\left\langle V'^2_{py} \right\rangle^+ = \left\langle V'^2_{py} \right\rangle / (u*)^2$
y	m	Vertical distance from the surface, in dimensionless form $y^+ = y\, u^*/\nu$
y_0	m	Distance from the surface at which the particle with a radius r_p is deposited, in dimensionless form $y^+_o = y_0\, u^*/\nu$
y_{cbl}	m	Depth of the concentration boundary layer above which $dC/dy = 0$ in dimensionless form $y^+_{cbl} = y_{cbl}\, u^*/\nu$
μ	kg m^{-1} s^{-1}	Dynamic viscosity of the fluid
ρ	kg m^{-3}	Fluid density
τ_L	s	Lagrangian time-scale of the fluid [31]: $\tau_L = \nu_t / \left\langle V'^2_y \right\rangle$ $\tau^+_L = \tau_L (u*)^2/\nu$
τ_p	s	Particle relaxation time $\tau_p = m_p C_c / 3\pi\mu D_p$ $\tau^+_p = \tau_p (u*)^2/\nu$
ε_p	m^2 s^{-1}	Eddy diffusivity of the particle. For relatively small particles and homogeneous isotropic turbulence [18] $\varepsilon_p = \nu_t$ For any particle size [21,29] $\varepsilon_p = \left[1 + \frac{\tau_p}{\tau_L}\right]^{-1} \nu_t$
ν	m^2 s^{-1}	Kinematic viscosity of the fluid, $\nu = \mu/\rho$
$\nu\tau$	m^2 s^{-1}	Air turbulent viscosity. For smooth surfaces it is [17] $\frac{\nu_t}{\nu} = \begin{cases} 7.67 \times 10^{-4} (y^+)^3, & 0 \le y^+ \le 4.3 \\ 10^{-3} (y^+)^{2.8214}, & 4.3 \le y^+ \le 12.5 \\ 1.07 \times 10^{-2} (y^+)^{1.8895}, & 12.5 \le y^+ \le 30 \end{cases}$ and for rough surfaces it is [16] $\frac{\nu_t}{\nu} = \begin{cases} \left(\frac{y^+}{11.15}\right)^3, & 0 \le y^+ \le 3 \\ \left(\frac{y^+}{11.4}\right)^3 - 0.049774, & 3 \le y^+ \le 52.108 \\ 0.4y^+, & 52.108 \le y^+ \end{cases}$

References

1. Bozlaker, A.; Muezzinoglu, A.; Odabasi, M. Atmospheric concentrations, dry deposition and air-soil exchange of polycyclic aromatic hydrocarbons (PAHs) in an industrial region in Turkey. *J. Hazard Mater.* **2008**, *153*, 1093–1102. [CrossRef] [PubMed]
2. El-Batsh, H. Modelling Particle Deposition on Compressor and Turbine Blades. Ph.D. Thesis, Vienna University of Technology, Vienna, Austria, 2001.
3. Liu, B.Y.H.; Ahn, K.H. Particle Deposition on Semiconductor Wafers. *Aerosol Sci. Technol.* **1987**, *6*, 215–224. [CrossRef]
4. Lobo, P.; Durdina, L.; Brem, B.T.; Crayford, A.P.; Johnson, M.P.; Smallwood, G.J.; Siegerist, F.; Williams, P.I.; Black, E.A.; Llamedo, A.; et al. Comparison of Standardized Sampling and Measurement Reference Systems for Aircraft Engine Non-Volatile Particulate Matter Emissions. *J. Aerosol Sci.* **2020**, *145*, 105557. [CrossRef]
5. Pui, D.Y.H.; Ye, Y.; Liu, B.Y.H. Experimental Study of Particle Deposition on Semiconductor Wafers. *Aerosol Sci. Technol.* **1990**, *12*, 795–804. [CrossRef]
6. Song, L.; Elimelech, M. Particle Deposition onto a Permeable Surface in Laminar Flow. *J. Colloid Interface Sci.* **1995**, *173*, 165–180. [CrossRef]
7. Tong, X.; Dong, J.; Shang, Y.; Inthavong, K.; Tu, J. Effects of Nasal Drug Delivery Device and Its Orientation on Sprayed Particle Deposition in a Realistic Human Nasal Cavity. *Comput. Biol. Med.* **2016**, *77*, 40–48. [CrossRef] [PubMed]

8. Tong, Z.X.; Li, M.J.; He, Y.L.; Tan, H.Z. Simulation of Real Time Particle Deposition and Removal Processes on Tubes by Coupled Numerical Method. *Appl. Energy* **2017**, *185*, 2181–2193. [CrossRef]
9. Andersson, K.G.; Roed, J.; Byrne, M.A.; Hession, H. Deposition of contaminant aerosol on human skin. *J. Environ. Radioact.* **2006**, *85*, 182–195. [CrossRef]
10. Casal, J.; Planas-Cuchi, E.; Moreso, J.M.; Casal, J. Forecasting virus atmospherical dispersion. Studies with foot-and-mouth disease. *J. Hazard. Mater.* **1995**, *43*, 229–244. [CrossRef]
11. Karlsson, E.; Huber, U. Influence of desorption on the indoor concentration of toxic gases. *J. Hazard. Mater.* **1996**, *49*, 15–27. [CrossRef]
12. Kleinstreuer, C.; Zhang, Z.; Li, Z. Modeling airflow and particle transport/deposition in pulmonary airways. *Respir. Physiol. Neurobiol.* **2008**, *163*, 128–138. [CrossRef] [PubMed]
13. Londah, J.; Pagels, J.; Swietlicki, E.; Zhou, J.; Ketzel, M.; Massling, A.; Bohgard, M. A set-up for field studies of respiratory tract deposition of fine and ultrafine particles in humans. *J. Aerosol Sci.* **2006**, *37*, 1152–1163. [CrossRef]
14. Goosse, H.; Barriat, P.Y.; Lefebvre, W.; Loutre, M.F.; Zunz, V. Introduction to Climate Dynamics and Climate Modeling. Online Textbook. Available online: http://www.climate.be/textbook (accessed on 17 July 2022).
15. Hussein, T.; Smolik, J.; Kerminen, V.M.; Kulmala, M. Modeling Dry Deposition of Aerosol Particles onto Rough Surfaces. *Aerosol Sci. Technol.* **2012**, *46*, 44–59. [CrossRef]
16. Guha, A. A Unified Eulerian Theory of Turbulent Deposition to Smooth and Rough Surfaces. *J. Aerosol Sci.* **1997**, *28*, 1517–1537. [CrossRef]
17. Corner, J.; Pendlebury, E.D. The Coagulation and Deposition of a Stirred Aerosol. *Proc. Phys. Soc. Sect. B* **1951**, *64*, 645–654. [CrossRef]
18. Lai, A.C.K.; Nazaroff, W.W. Modeling Indoor Particle Deposition from Turbulent Flow onto Smooth Surfaces. *J. Aerosol Sci.* **2000**, *31*, 463–476. [CrossRef]
19. Zhao, B.; Wu, J. Modeling Particle Deposition from Fully Developed Turbulent Flow in Ventilation Duct. *Atmos. Environ.* **2006**, *40*, 457–466. [CrossRef]
20. Hussein, T.; Ibrahim, S.; Malek, S. Basic Concepts and Development of Dry Deposition Modelling. *Jordan J. Phys.* **2019**, *12*, 113–132.
21. Zhao, B.; Wu, J. Modeling Particle Deposition onto Rough Walls in Ventilation Duct. *Atmos. Environ.* **2006**, *40*, 6918–6927. [CrossRef]
22. Kallio, G.A.; Reeks, M.W. A Numerical Simulation of Particle Deposition in Turbulent Boundary Layers. *Int. J. Multiph. Flow* **1989**, *15*, 433–446. [CrossRef]
23. Kim, J.; Moin, P.; Moser, R. Turbulence Statistics in Fully Developed Channel Flow at Low Reynolds Number. *J. Fluid Mech.* **1987**, *177*, 133–166. [CrossRef]
24. Crump, J.G.; Flagan, R.C.; Seinfeld, J.H. Particle Wall Loss Rates in Vessels. *Aerosol Sci. Technol.* **1982**, *2*, 303–309. [CrossRef]
25. McMurry, P.H.; Rader, D.J. Aerosol Wall Losses in Electrically Charged Chambers. *Aerosol Sci. Technol.* **1985**, *4*, 249–268. [CrossRef]
26. Nazaroff, W.W.; Cass, G.R. Mathematical Modeling of Indoor Aerosol Dynamics. *Environ. Sci. Technol.* **1984**, *23*, 157–166. [CrossRef]
27. Bejan, A. *Convection Heat Transfer*, 2nd ed.; John Wiley & Sons: New York, NY, USA, 1995.
28. Seinfeld, J.H.; Pandis, S.N. *Atmospheric Chemistry and Physics: From Air Pollution to Climate Change*, 3rd ed.; John Wiley & Sons: New York, NY, USA, 1998.
29. Hinze, J.O. *Turbulence*, 2nd ed.; McGraw-Hill: New York, NY, USA, 1975.
30. Hinds, W.C. *Aerosol Technology*, 2nd ed.; John Wiley & Sons: New York, NY, USA, 1999.
31. Johansen, S.T. The Deposition of Particles on Vertical Walls. *Int. J. Multiph. Flow* **1991**, *17*, 355–376. [CrossRef]
32. Boyce, W.E.; DiPrima, R.C. *Elementary Differential Equations and Boundary Value Problems*, 10th ed.; John Wiley & Sons: New York, NY, USA, 2012.
33. Piskunov, V.N. Parameterization of Aerosol Dry Deposition Velocities onto Smooth and Rough Surfaces. *J. Aerosol Sci.* **2009**, *40*, 664–679. [CrossRef]
34. Pidwirny, M. *Surface Area of Our Planet Covered by Oceans and Continents*; University of British Columbia: Kelowna, BC, Canada, 2006.

atmosphere

MDPI

Article

An Evaluation of Algebraic Turbulence Length Scale Formulations

Stephanie Reilly [1], Ivan Bašták Ďurán [1,2,*], Anurose Theethai Jacob [3] and Juerg Schmidli [1,2]

[1] Faculty of Geosciences and Geography, Institute for Atmospheric and Environmental Sciences, Goethe University, 60438 Frankfurt am Main, Germany; reilly@iau.uni-frankfurt.de (S.R.); schmidli@iau.uni-frankfurt.de (J.S.)

[2] Hans Ertel Centre for Weather Research, 63067 Offenbach am Main, Germany

[3] National Center for Medium Range Weather Forecasting, Ministry of Earth Sciences, Noida 201309, India; anurose.tj@gov.in

* Correspondence: bastakduran@iau.uni-frankfurt.de; Tel.: +49-69-7980-40242

Abstract: Turbulence kinetic energy (TKE) schemes are routinely used for turbulence parameterization in numerical weather prediction models. A key component of these schemes is the so-called turbulence length scale. Novel scale-aware, budget-based diagnostics that account for the cross-scale transfer of variances are used to evaluate the performance of selected turbulence length scale formulations in the gray zone of turbulence. The diagnostics are computed using the coarse-graining method on high resolution large eddy simulation data for selected idealized cases. The vertical profiles and the temporal evolution of the turbulence length scales are analyzed. Additionally, the local normalized root mean square error and a non-local three-component technique tailored specifically to the turbulence length scale profiles are used for the evaluation. Based on our analyses, we recommend using turbulence length-scale formulations that depend not only on the boundary layer height, but also on the TKE and stratification. Such formulations are able to perform satisfactorily in different flow regimes, but their scale-awareness is still limited. Only the Honnert et al. formulation shows a stronger scale-awareness thanks to its cut-off relationship in the gray zone. However, in contrast to the turbulence length scale diagnostics, its resolution dependence does not change with height.

Keywords: turbulence length scale; turbulence parameterization; evaluation; large eddy simulation

Citation: Reilly, S.; Bašták Ďurán, I.; Theethai Jacob, A.; Schmidli, J. An Evaluation of Algebraic Turbulence Length Scale Formulations. *Atmosphere* **2022**, *13*, 605. https://doi.org/10.3390/atmos13040605

Academic Editors: Yubin Li and Jie Tang

Received: 26 February 2022
Accepted: 5 April 2022
Published: 9 April 2022

1. Introduction

Turbulent motions in the atmosphere occur on a range of different scales, from millimeter to kilometer scales [1,2]. As some turbulent motions occur on scales that are always smaller than the resolution of an atmospheric model, basically all atmospheric models require a turbulence parameterization to account for the influence of sub-grid-scale turbulence. The level to which the turbulent flow is parameterized is thus dependent on the resolution of the model. In numerical weather prediction (NWP) and global circulation (GC) models, all turbulent motions, including both large and small eddies, are parameterized. However, in large-eddy simulation (LES) models, only the isotropic turbulent eddies (the smaller eddies) need to be parameterized, because the energy-containing anisotropic turbulent eddies (the larger eddies) are resolved in LES [1,3].

The parameterization of turbulence in models with a resolution between the classical NWP/GC and the LES models is naturally also required, but it is more difficult. This is because the anisotropic turbulent eddies are already partly resolved, and only the unresolved part of the large eddies needs to be parameterized. In this gray zone of turbulence, some classical NWP assumptions need to be revisited [4–6]. We focus in this paper particularly on the fact that the net sub-grid cross-scale transfer of turbulence kinetic energy (TKE) is not zero anymore in the gray zone [1].

In order to account for the influence of the cross-scale transfer of TKE, Reference [3] proposed a modification of the turbulence length scale in TKE turbulence schemes. The turbulence length scale is one of the key components in TKE turbulence schemes. Primarily, the turbulence length scale is used to parameterize the viscous dissipation in the prognostic TKE equation. Consequently, the primary output of a turbulence scheme, the vertical turbulent fluxes, are directly affected by the turbulence length scale (see [3,7] for more details). The modifications of the turbulence length scale should lead to a scale-aware formulation of the turbulence length scale, which would enable a better representation of turbulence in the gray zone.

The reference for such a new formulation is a scale-aware diagnostic, where the turbulence length scale is defined through the viscous dissipation of TKE. The TKE dissipation term is calculated from the TKE budget using high-resolution LES data. To obtain the length scale at different resolutions, a coarse-graining method is used on the LES data (averaging over sub-domains of different sizes) [3].

While this TKE-based turbulence length scale diagnostic is accurate in most situations, the use of a TKE budget faces a potential problem when gravity waves are present in and above the atmospheric boundary layer (ABL). Specifically, the velocity fluctuations in the LES data cannot be easily separated between turbulence and wave kinetic energy [2]. In the presence of gravity waves, the TKE can be over-estimated, which yields unrealistic values of the turbulence length scale [3]. This problem can be partly avoided if the length scale is diagnosed from the budgets of the following scalar variances: liquid water potential temperature variance and total specific water content variance. Therefore, we also extend the budget-based diagnostic to scalar variances in this paper.

All three diagnostics are then used for the evaluation of several existing algebraic formulations: the height-dependent Blackadar (1962) [8] and Bastak Duran et al. (2018) [9] formulations, the TKE-dependent Bougeault and Lacarrere (1989) [10] and Honnert et al. (2021) [11] formulations, and the combined Nakanishi and Niino (2009) [12,13] formulation. This particular set represents the most frequently used algebraic turbulence length scale formulations in NWP models. The evaluation is performed for several idealized cases, covering different atmospheric boundary layer conditions.

The paper is organized as follows. Section 2 gives a detailed description of the different budget-based turbulence length scale diagnostics and algebraic formulations, along with a description of the LES model used to simulate a variety of different boundary layer cases. The main results of the study are shown in Section 3, looking in particular at the scale dependence of the budget-based diagnostics, the evaluation of the algebraic turbulence length scale formulations, and the temporal development of the turbulence length scale over the simulated time period. Finally, a summary of the study's findings along with the main conclusions are found in Section 4.

2. Formulations and Methods

2.1. Turbulence Length Scale Diagnostics

In this study, we evaluate existing algebraic turbulence length scale formulations using turbulence length scale diagnostics computed from coarse-grained LES data (see [3]). Two sets of diagnostics are used for this purpose.

First, the turbulence length scale is diagnosed without consideration of the transfer of TKE or scalar variances across scales. The turbulence length scales are derived from the definition of the viscous dissipation term in the prognostic equations for the TKE and from the definitions of the molecular dissipation terms for the variance of the liquid water potential temperature, θ_l, or the total specific water content, q_t (see [3,9] for more details):

$$L_{\epsilon,e_k} = C_\epsilon \frac{\langle e_k(l_c)\rangle^{\frac{3}{2}}}{\epsilon}, \tag{1}$$

$$L_{\epsilon,\theta_l} = \frac{C_\epsilon}{C_p} \langle e_k(l_c)\rangle^{\frac{1}{2}} \frac{\left\langle \left[\theta_l'^2\right]^{l_c} \right\rangle}{\epsilon_{\theta_l}}, \tag{2}$$

$$L_{\epsilon,q_t} = \frac{C_\epsilon}{C_p}\langle e_k(l_c)\rangle^{\frac{1}{2}}\frac{\left\langle \left[q_t''^2\right]^{l_c}\right\rangle}{\epsilon_{q_t}}, \tag{3}$$

$$e_k(l_c) \equiv \frac{1}{2}\left(\left[u''^2\right]^{l_c} + \left[v''^2\right]^{l_c} + \left[w''^2\right]^{l_c}\right) \tag{4}$$

where the $[]^{l_c}$ operator denotes an average over a horizontal sub-domain of size l_c (equivalent to a horizontal grid box in NWP) and over time (in our case over 1 h; see Section 2.3), and the fluctuations from this average are denoted with $''$:

$$\psi = [\psi]^{l_c} + \psi'', \tag{5}$$

where ψ stands for any relevant quantity, and the average over all sub-domains with size l_c is denoted by $<>$ (see [3] for more details about the LES coarse graining). $e_k(l_c)$, $\left[\theta_l''^2\right]^{l_c}$, and $\left[q_t''^2\right]^{l_c}$ are the sub-grid scale TKE, the variance of θ_l, and the variance of q_t for sub-domains with size l_c (corresponding to a NWP computational mesh with a grid spacing of l_c). L_{ϵ,e_k}, L_{ϵ,θ_l}, and L_{ϵ,q_t} are the respective diagnosed turbulence length scales. $C_\epsilon = 0.871$ and $C_p = 0.3677$ are closure constants [14,15]. ϵ, ϵ_{θ_l}, and ϵ_{q_t} are the viscous/molecular dissipation rates of the TKE, the variance of θ_l, and the variance of q_t in their respective Reynolds-averaged (assuming horizontal homogeneity) prognostic equations (see, e.g., [16]):

$$\frac{\partial e_k}{\partial t} + \overline{u}\frac{\partial e_k}{\partial x} + \overline{v}\frac{\partial e_k}{\partial y} + \overline{w}\frac{\partial e_k}{\partial z} = -\frac{\partial}{\partial z}\left(\overline{u'^2 w'} + \overline{v'^2 w'} + \overline{w'^3} + \frac{1}{\rho_0}\overline{p'w'}\right) - \left(\overline{u'w'}\frac{\partial \overline{u}}{\partial z} + \overline{v'w'}\frac{\partial \overline{v}}{\partial z}\right) + \frac{g}{\theta_{vr}}\overline{w'\theta_v'} - \epsilon, \tag{6}$$

$$\frac{\partial \overline{\theta_l'^2}}{\partial t} + \overline{u}\frac{\partial \overline{\theta_l'^2}}{\partial x} + \overline{v}\frac{\partial \overline{\theta_l'^2}}{\partial y} + \overline{w}\frac{\partial \overline{\theta_l'^2}}{\partial z} = -\frac{\partial}{\partial z}\left(\overline{\theta_l'^2 w'}\right) - 2\,\overline{\theta_l' w'}\frac{\partial \overline{\theta_l}}{\partial z} - \epsilon_{\theta_l}, \tag{7}$$

$$\frac{\partial \overline{q_t'^2}}{\partial t} + \overline{u}\frac{\partial \overline{q_t'^2}}{\partial x} + \overline{v}\frac{\partial \overline{q_t'^2}}{\partial y} + \overline{w}\frac{\partial \overline{q_t'^2}}{\partial z} = -\frac{\partial}{\partial z}\left(\overline{q_t'^2 w'}\right) - 2\,\overline{q_t' w'}\frac{\partial \overline{q_t}}{\partial z} - \epsilon_{q_t}, \tag{8}$$

$$e_k \equiv \frac{1}{2}\left(\overline{u'^2} + \overline{v'^2} + \overline{w'^2}\right), \tag{9}$$

where the $\overline{}$ operator denotes an average over the horizontal domain and in time (in our case over the whole LES domain and 1 h; see Section 2.3), and the fluctuations from this average are denoted with $'$:

$$\psi = \overline{\psi} + \psi'. \tag{10}$$

x, y, and z are horizontal and vertical coordinates; u, v and w are the corresponding components of the wind velocity; t is the time; p is the atmospheric pressure; θ_v is the virtual potential temperature; θ_{vr} is a reference virtual potential temperature; ρ_0 is a reference density; and g is the gravitational acceleration. The first terms on the left-hand side in Equations (6)–(8) are the tendency terms. The remaining terms on the left-hand side are the horizontal and vertical advection terms. Please note that under the assumption of horizontal homogeneity, the contribution from the advection terms is equal to zero (see, e.g., [1]). The first terms on the right-hand side represent the vertical turbulent transport. For TKE, the turbulent transport term also includes the pressure correlation term (the last expression). The second terms in Equations (6)–(8) are the shear production term for TKE and the gradient terms for scalar variances. The third term on the right-hand side in Equation (6) is the buoyancy production/destruction term.

Second, the above formulations are extended to account for the transfer of TKE or scalar variances across scales:

$$L_{C,e_k}(l_c) = C_\epsilon \frac{\langle e_k(l_c)\rangle^{\frac{3}{2}}}{\langle \widetilde{\epsilon}(l_c)\rangle}, \tag{11}$$

$$L_{C,\theta_l}(l_c) = \frac{C_\epsilon}{C_p}\langle e_k(l_c)\rangle^{\frac{1}{2}} \frac{\left\langle [\theta_l''^2]^{l_c}\right\rangle}{\left\langle \widetilde{\epsilon}_{\theta_l}(l_c)\right\rangle}, \tag{12}$$

$$L_{C,q_t}(l_c) = \frac{C_\epsilon}{C_p}\langle e_k(l_c)\rangle^{\frac{1}{2}} \frac{\left\langle [q_t''^2]^{l_c}\right\rangle}{\left\langle \widetilde{\epsilon}_{q_t}(l_c)\right\rangle}, \tag{13}$$

where $\widetilde{\epsilon}$, $\widetilde{\epsilon}_{\theta_l}$, and $\widetilde{\epsilon}_{q_t}$ are the effective viscous/molecular dissipation rates of the TKE, the variance of θ_l, and the variance of q_t, defined as:

$$\tilde{\epsilon}(l_c) \equiv \epsilon - T_r^{e_k}(l_c), \tag{14}$$

$$\tilde{\epsilon}_{\theta_l}(l_c) \equiv \epsilon_{\theta_l} - T_r^{\theta_l}(l_c), \tag{15}$$

$$\tilde{\epsilon}_{q_t}(l_c) \equiv \epsilon_{q_t} - T_r^{q_t}(l_c), \tag{16}$$

where $T_r^{e_k}(l_c)$, $T_r^{\theta_l}(l_c)$, and $T_r^{q_t}(l_c)$ are the cross-scale transfers at the cutoff scale l_c for TKE, the variance of θ_l, and the variance of q_t.

Effective dissipation rates are estimated from the budgets of the corresponding variables for a sub-domain of size l_c (obtained as residuals from all other terms computed from the LES data; see [3]):

$$\begin{aligned}
\frac{\partial e_k(l_c)}{\partial t} &+ [u]^{l_c}\frac{\partial e_k(l_c)}{\partial x} + [v]^{l_c}\frac{\partial e_k(l_c)}{\partial y} + [w]^{l_c}\frac{\partial e_k(l_c)}{\partial z} = \\
&- \frac{\partial}{\partial z}\left(\left[u''^2 w''\right]^{l_c} + \left[v''^2 w''\right]^{l_c} + \left[w''^3\right]^{l_c}\right) \\
&- \frac{\partial}{\partial z}\left(\frac{1}{\rho_0}[p''w'']^{l_c}\right) \\
&- [u''w'']^{l_c}\frac{\partial [u]^{l_c}}{\partial z} - [v''w'']^{l_c}\frac{\partial [v]^{l_c}}{\partial z} \\
&+ \frac{g}{\theta_{vr}}[w''\theta_v'']^{l_c} - \tilde{\epsilon}(l_c),
\end{aligned} \tag{17}$$

$$\begin{aligned}
\frac{\partial [\theta_l''^2]^{l_c}}{\partial t} &+ [u]^{l_c}\frac{\partial [\theta_l''^2]^{l_c}}{\partial x} + [v]^{l_c}\frac{\partial [\theta_l''^2]^{l_c}}{\partial y} + [w]^{l_c}\frac{\partial [\theta_l''^2]^{l_c}}{\partial z} = \\
&- \frac{\partial}{\partial z}\left(\left[\theta_l''^2 w''\right]^{l_c}\right) \\
&- 2\left[\theta_l'' w''\right]^{l_c}\frac{\partial [\theta_l]^{l_c}}{\partial z} - \tilde{\epsilon}_{\theta_l}(l_c),
\end{aligned} \tag{18}$$

$$\begin{aligned}
\frac{\partial [q_t''^2]^{l_c}}{\partial t} &+ [u]^{l_c}\frac{\partial [q_t''^2]^{l_c}}{\partial x} + [v]^{l_c}\frac{\partial [q_t''^2]^{l_c}}{\partial y} + [w]^{l_c}\frac{\partial [q_t''^2]^{l_c}}{\partial z} = \\
&- \frac{\partial}{\partial z}\left(\left[q_t''^2 w''\right]^{l_c}\right) \\
&- 2\left[q_t'' w''\right]^{l_c}\frac{\partial [q_t]^{l_c}}{\partial z} - \tilde{\epsilon}_{q_t}(l_c).
\end{aligned} \tag{19}$$

2.2. Algebraic Turbulence Length Scale Formulation

Five representative algebraic turbulence length scale formulations are evaluated in this study. The length scales are computed for all sub-domain sizes from the corresponding averaged LES data (see Section 2.1). While all the studied algebraic formulations are dependent on different aspects of the boundary layer, all five share a common height dependence, where the turbulence length scale is proportional to height via the von Karman constant, k, in a neutral surface layer.

2.2.1. Blackadar Formulation

The first algebraic formulation used in this study was proposed by Blackadar (1962) [8] and is referred to as L_B for the remainder of this paper. The mixing length l_B is defined as follows:

$$l_B = \frac{k \cdot z}{1 + \frac{k \cdot z}{\lambda}}, \tag{20}$$

where z is the altitude and λ is an asymptotic parameter set to 150 m for unstable layers and to 30 m for stable layers. It should be noted that l_B was derived from similarity law relationship in neutral stratification. The corresponding length scale for a TKE scheme, L_B, is computed from l_B according to [9,17]:

$$L_B = \frac{(C_K C_\epsilon)^{\frac{1}{4}}}{C_K} l_B. \tag{21}$$

$C_K = 0.0882$ is a closure constant [7].

2.2.2. Bastak Duran et al. Formulation

The next formulation used in Bastak Duran et al. [9], L_{BD1}, is an algebraic formulation derived from [18]:

$$L_{BD1} = \frac{(C_K \cdot C_\epsilon)^{\frac{1}{4}}}{C_K} \frac{k \cdot z}{1 + \frac{k \cdot z}{\lambda_m}\left[\frac{1 + exp\left(-a_m \cdot \sqrt{\frac{z}{H_{abl}}} + b_m\right)}{\beta_m + exp\left(-a_m \cdot \sqrt{\frac{z}{H_{abl}}} + b_m\right)}\right]}, \tag{22}$$

where H_{abl} is the ABL height, calculated using the bulk Richardson number. The constants λ_m, a_m, b_m, and β_m are set to be 350 m, 5.5, 3.0, and 0.02, respectively.

2.2.3. Nakanishi and Niino Formulation

The formulation by Nakanishi and Niino (2009) [12,13], L_{NN}, comprises three different turbulence length scale formulations, L_{Surf}, L_T, and L_{Bouy}:

$$\frac{1}{L_{NN}} = \left(\frac{1}{L_{Surf}} + \frac{1}{L_T} + \frac{1}{L_{Bouy}}\right), \tag{23}$$

$$L_{NN,e_k} = \frac{B_1 C_\epsilon}{2^{\frac{5}{2}}} L_{NN}, \tag{24}$$

$$L_{NN,\theta_l} \equiv L_{NN,q_t} = \frac{B_2 C_\epsilon}{C_p 2^{\frac{5}{2}}} L_{NN}, \tag{25}$$

where $B_1 = 24.0$ and $B_2 = 15.0$ are closure constants and L_{NN,e_k}, L_{NN,θ_l}, and L_{NN,q_t} are values used for TKE and scalar variances, respectively. Please note that the introduction of L_{NN,e_k}, L_{NN,q_t}, and L_{NN,θ_l} is only required for our evaluation since the formulations for the dissipation terms in [13] differ from the formulation that was used in our turbulence length scale diagnostic (Equations (1)–(3)).

The three length scales considered for L_{NN} are dependent on different aspects of the boundary layer. The first of the three length scales, L_{Surf}, is the turbulence length scale within the surface layer. Above the surface layer, the value of L_{Surf} increases and therefore becomes less significant in the overall calculation. L_{Surf} is computed as follows (Equation (53) in [13]):

$$L_{Surf} = \begin{cases} k \cdot \frac{z}{3.7} & \zeta \geq 1 \\ k \cdot z \cdot (1 + 2.7 \cdot \zeta)^{-1} & 0 \leq \zeta < 1 \\ k \cdot z \cdot (1 - 100 \cdot \zeta)^{0.2} & \zeta < 0 \\ \zeta = z / L_{MO} \end{cases} \tag{26}$$

where ζ is a stability parameter, $L_{MO} = -\theta_{vr} u_*^3 / (kg(\overline{w'\theta_v'})_s)$ is the Monin–Obukhov length, u_* is the friction velocity, and $(\overline{w'\theta_v'})_s$ is virtual potential temperature flux at the surface.

The second of the three length scales, L_T, is the length scale that is dependent on the boundary layer height [12,13,19]:

$$L_T = 0.23 \cdot \frac{\int_0^h \sqrt{e_k} \cdot z\,dz}{\int_0^h \sqrt{e_k}\,dz}, \tag{27}$$

where h is the vertical limit of the simulated domain.

The final length scale, L_{Bouy} is related to how far a parcel of air, with a given TKE, can travel vertically before the buoyancy force stops the motion:

$$L_{Buoy} = \begin{cases} \frac{\sqrt{2 \cdot e_k}}{N} & \frac{\delta \theta_v}{\delta z} > 0 \ \& \ \zeta \geq 0 \\ \left(1 + 5 \cdot \left(\frac{q_c}{L_T \cdot N}\right)^{0.5}\right) \cdot \frac{\sqrt{2 \cdot e_k}}{N} & \frac{\delta \theta_v}{\delta z} > 0 \ \& \ \zeta < 0 \\ \infty & \frac{\delta \theta_v}{\delta z} \leq 0 \end{cases} \tag{28}$$

where N is the Brunt–Väisälä frequency and q_c is a velocity scale (see [13]). L_{Buoy} is most accurate in stable layers.

2.2.4. Bougeault and Lacarrere Formulation

The Bougeault and Lacarrere [10] length scale, L_{BL}, takes the balance between the TKE and buoyancy into account. This length scale is computed by combining the upward and downward free paths, L_{up} and L_{down}:

$$L_{BL} = \frac{(C_K C_\epsilon)^{\frac{1}{4}}}{C_K} (L_{up} \cdot L_{down})^{\frac{1}{2}}. \tag{29}$$

The value of L_{up} is defined as the movement of a parcel of air with a given TKE in the upward direction until a point where it is stopped by buoyancy effects [10,20]:

$$\int_z^{z+L_{up}} \frac{g}{\theta_{vr}} (\theta_v(z) - \theta_v(z'))\,dz' = e_k(z). \tag{30}$$

The derivation of L_{down} is similar to L_{up}, but it is calculated in the opposite direction:

$$\int_{z-L_{down}}^{z} \frac{g}{\theta_{vr}} (\theta_v(z') - \theta_v(z))\,dz' = e_k(z). \tag{31}$$

It also has an additional requirement in that the parcel cannot go below the surface. The L_{BL} is scaled similarly to L_B (see Equation (21)) to ensure similarity law relationship near the surface.

2.2.5. Honnert et al. Formulation

The final formulation, derived by Honnert et al. (2021) [11], L_{H21}, is a combination of a non-local turbulence length scale and grid box size:

$$L_{H21} = min(L_R, \alpha \cdot L_{LES}), \quad (32)$$

$$L_{LES} = (\Delta x \cdot \Delta y)^{\frac{1}{2}} \quad (33)$$

where Δx and Δy are horizontal sizes of the grid box (in our case sub-domain), $\alpha = 0.5$ is a calibration parameter, and L_R is an extension of L_{BL} by a contribution from shear according to [21]:

$$L_R = \frac{(C_K C_\epsilon)^{\frac{1}{4}}}{C_K} \left(\frac{L_{Rup}^{-\frac{2}{3}} + L_{Rdown}^{-\frac{2}{3}}}{2} \right)^{-\frac{3}{2}}, \quad (34)$$

$$\int_z^{z+L_{Rup}} \frac{g}{\theta_{vr}} \left(\theta_v(z) - \theta_v(z') + C_0 \cdot \sqrt{e_k} \cdot \sigma(z')\right) dz' = e_k(z), \quad (35)$$

$$\int_{z-L_{Rdown}}^{z} \frac{g}{\theta_{vr}} \left(\theta_v(z) - \theta_v(z') + C_0 \cdot \sqrt{e_k} \cdot \sigma(z')\right) dz' = e_k(z), \quad (36)$$

where σ is the vertical wind shear, L_{Rup} and L_{Rdown} are the upward and downward free paths, and $C_0 = 0.5$ is a calibration constant.

2.3. LES Data

A total of five idealized cases with different ABL conditions are used for the evaluation of the turbulence length scale formulations: a continental cumulus case based on data from the Atmospheric Radiation Measurement (ARM) program at the Cloud and Radiation Testbed site in Oklahoma [22], a stratocumulus case with drizzle based on data recorded during the first research flight (RF01) of the second period of the Dynamics and Chemistry of Marine Stratocumulus (DYCOMS-II) campaign [23], a stable boundary layer case based on data from the Global Energy and Water Cycle Experiment (GEWEX) Atmospheric Boundary Layer Study (GABLS1) [24,25], a trade-wind cumulus case from the Barbados Oceanographic and Meteorological Experiment (BOMEX) [26], and a precipitating shallow cumulus case from the Rain in Shallow Cumulus over the Ocean (RICO) campaign [27].

All cases are simulated using the MicroHH LES model [28] with a very high resolution that ensures accuracy of the diagnostic method according to the recommendation in [3]. The LES were first run with a coarse resolution and then with subsequently finer resolutions, until the budget terms converged sufficiently. Sufficient convergence was declared if the values of the second-order moments changed less than by a predefined threshold (5% of the value) with increasing resolution. The finest resolution was then considered as adequate for the turbulence length scale diagnostic. The domain size was chosen to be large enough to be able to include sub-domains with spatial scales above the energy production range of TKE; i.e., the spatial scale of the larger sub-domains correspond to the resolution of typical NWP models. The smallest sub-domains have a spatial scale in the gray zone of turbulence for the particular case.

A detailed description of the model, including its governing equations and parameterizations, can be found in [28]. A summary of the setups for the five LES cases is listed in Table 1 (an adaptive time step was used).

The LES data are sampled every 60 s. The LES statistics and the subsequent turbulence length scale profiles, diagnosed or computed according to the algebraic formulations, are obtained by averaging over one hour in time (moving average).

Table 1. Setup of the five idealized case studies run with the MicroHH LES model.

LES Case	Domain Size (km^3)	Grid Size (m^3)
ARM	12.8 × 12.8 × 4.4	12.5 × 12.5 × 34.4
BOMEX	12.8 × 12.8 × 3.0	12.5 × 12.5 × 23.4
DYCOMS-II	12.8 × 12.8 × 1.5	12.5 × 12.5 × 2.93
RICO	12.8 × 12.8 × 4.4	12.5 × 12.5 × 31.25
GABLS1	0.4 × 0.4 × 0.4	0.78 × 0.78 × 0.78

In order to study the scale-dependence of the turbulence length scale, the diagnostics and the algebraic formulations are computed on coarse-grained data with varying resolution. The horizontal domain of the LES is divided into a series of sub-domains of equal size. The number (N_{sd}) and the size (l_c) of the sub-domains are chosen so that the number of grid points in a sub-domain is large enough to ensure the representativeness of the resulting quantities (see the respective studies of the selected cases). Furthermore, the smallest sub-domains have a spatial scale in the gray zone of turbulence.

A summary of the sizes for all the sub-domains is listed in Table 2. A schematic of the LES domain can be seen in Figure 2 of [3].

Table 2. A summary of the sub-domain sizes and number of grid points in a sub-domain for the cloudy cases (ARM, BOMEX, DYCOMS-II, and RICO) and the GABLS1 case.

No. of	Sub-Domain Size (km^2)		No. of Grid Points in a Sub-Domain	
Sub-Domains	Cloudy Cases	GABLS1	Cloudy Cases	GABLS1
1 × 1	12.8 × 12.8	0.4 × 0.4	1024^2	512^2
2 × 2	6.4 × 6.4	0.2 × 0.2	512^2	256^2
4 × 4	3.2 × 3.2	0.1 × 0.1	256^2	128^2
8 × 8	1.6 × 1.6	0.05 × 0.05	128^2	64^2
16 × 16	0.8 × 0.8	0.025 × 0.025	64^2	32^2
32 × 32	0.4 × 0.4	0.0125 × 0.0125	32^2	16^2

It should be noted that estimation of the dissipation rates and the effective dissipation rates is not always accurate. In such cases, the values of these quantities are either too small or negative.

Typically, these difficulties with accuracy occur above the ABL, where the contribution from gravity waves to the budgets of TKE and scalar variances can be expected. To avoid these regions, all data points that have an effective TKE dissipation rate smaller than a prescribed fraction, $C_{\tilde{\epsilon}} = 5 \times 10^{-5}\,\mathrm{s}^{-1}$, of the estimated surface TKE [29]:

$$e_{k,s} = c_0 u_*^2 + c_1 w_*^2, \tag{37}$$

per second ,are filtered out. Here w_* is the convective velocity scale, and $c_0 = 3.75$ and $c_1 = 0.2$ are empirical parameters [29].

Additionally, values that fall under a prescribed threshold are not used for computation of the turbulence length scale formulations. The prescribed thresholds are $10^{-5}\,\mathrm{m^2\,s^{-3}}$ for dissipation rates of TKE, $10^{-8}\,\mathrm{K^2\,s^{-1}}$ for dissipation rates of the θ_l variance, and $10^{-12}\,\mathrm{kg^2\,kg^{-2}\,s^{-1}}$ for dissipation rates of the q_t variance.

2.4. Evaluation Scores

The turbulence length scale formulations are evaluated using root mean square of a normalized (by H_{abl}) length scale error:

$$\mathrm{RMSE}_n = \sqrt{\frac{1}{n_z\, n_t} \sum_{j=1}^{j=n_t} \sum_{i=1}^{i=n_z} \left(\frac{L_F(z_i, t_j) - L_D(z_i, t_j)}{H_{abl}} \right)^2}, \tag{38}$$

where z_i is the height of the i-th model level, n_z is number of model levels in the ABL, t_j is the time after j time steps, n_t is number of time steps, L_F is length scale computed according to one of the turbulence length scale formulations listed in Section 2.2, and L_D is one of the three length scale diagnostics (L_{C,e_k}, L_{C,θ_l}, and L_{C,q_t}). The RMSE$_n$ score shows the average of the spatially local (independently at each level) accuracy of the length scale formulation.

In order to also evaluate non-local properties of the formulations, such as amplitude and shape of the profile, a three-component score inspired by a three-component technique of [30] is used. In this paper, however, the three-component technique is tailored specifically for the turbulence length scale evaluation in the ABL. The first component is the amplitude component, A, which is defined as the mean (in time) difference in the normalized length scale paths:

$$\mathrm{A} = \frac{1}{n_t}\sum_{j=1}^{j=n_t}\left(\mathrm{L_FP}(t_j) - \mathrm{L_DP}(t_j)\right), \tag{39}$$

$$\mathrm{L_FP} = \frac{1}{H_{abl}^2}\int_{z=0}^{z=H_{abl}} L_F(z)dz, \quad \mathrm{L_DP} = \frac{1}{H_{abl}^2}\int_{z=0}^{z=H_{abl}} L_D(z)dz, \tag{40}$$

where $\mathrm{L_FP}$ and $\mathrm{L_DP}$ are normalized (by H_{abl}) length scale paths in the ABL. A expresses the error in the mixing potential of the whole length scale profile.

When looking at the shape of the length scale profile, the most prominent feature is a peak in the ABL (for details see Section 3.1). Therefore, two shape components are used to specify the mean (in time) error in the normalized (by H_{abl}) peak location, S_l, and peak magnitude, S_m:

$$\mathrm{S_l} = \frac{1}{n_t}\sum_{j=1}^{j=n_t}\frac{\mathrm{z_{F,max}}(t_j) - \mathrm{z_{D,max}}(t_j)}{H_{abl}}, \tag{41}$$

$$\mathrm{S_m} = \frac{1}{n_t}\sum_{j=1}^{j=n_t}\frac{\mathrm{L_{F,max}}(t_j) - \mathrm{L_{D,max}}(t_j)}{H_{abl}}, \tag{42}$$

where $\mathrm{L_{F,max}}$ and $\mathrm{L_{D,max}}$ are the maximum values of the length scales in the ABL, and $\mathrm{z_{F,max}}$ and $\mathrm{z_{D,max}}$ are the heights of these peaks, respectively. In case there are multiple local maxima in the ABL, the lowest maxima is taken. If the maxima are at adjacent model levels, the peak location is set to the average height of these points (typically for L_{H21}).

All three components are then displayed in one scatter plot, where S_l and S_m are the x and y coordinates, and the A value determines the coloring of the point.

Please note that all scores are normalized by H_{abl} to enable a more consistent comparison between different cases and between different points in time. H_{abl} is determined from the vertical profile of $\widetilde{\epsilon}$ for the given sub-domain size. H_{abl} is the height above which $\widetilde{\epsilon}$ drops below a critical value of 10^{-5} m^2 s^{-3}.

3. Results

The turbulence length scale diagnostics based on the budget of TKE and scalar variances for selected idealized cases are presented in this section. Subsequently, these diagnostics are used for the evaluation of the chosen algebraic turbulence length scale formulations.

3.1. Turbulence Length Scale Diagnostics

The new scale-aware turbulence length scales are computed from the effective dissipation rates, which are obtained from the variance budgets (see [3] and Section 2.3 for more details). To visualize the accuracy of the effective dissipation rates diagnostics, the individual terms of the three budgets are presented in Figure 1 for the BOMEX case.

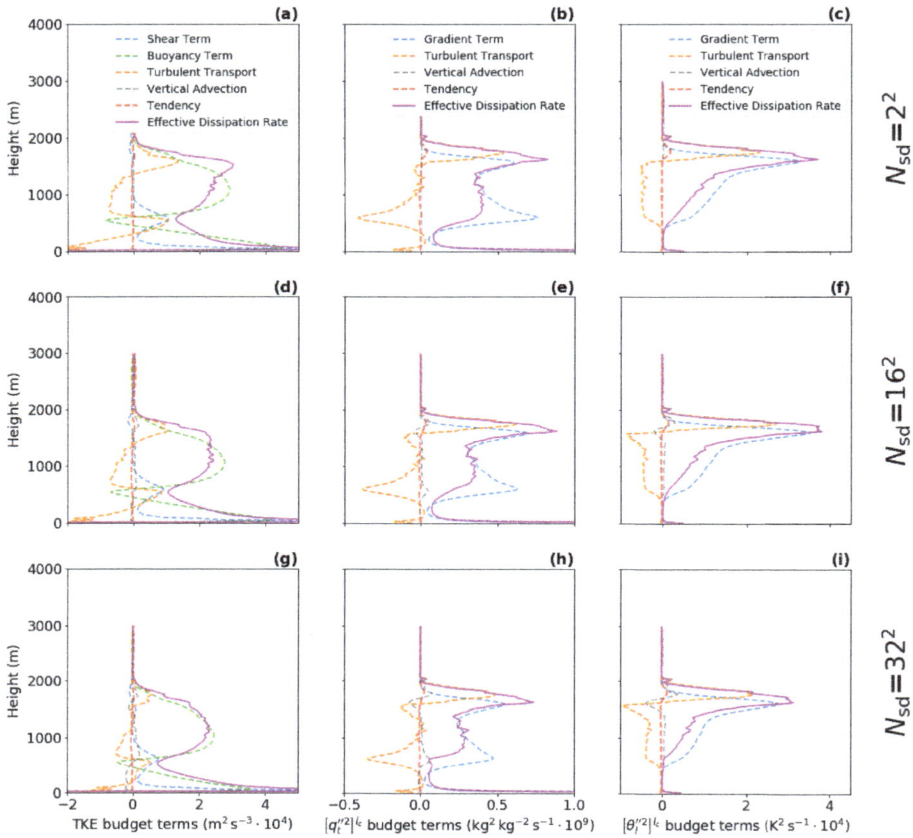

Figure 1. The budget terms for the TKE (**a**,**d**,**g**), $\left[q_t''^2\right]^{l_c}$ (**b**,**e**,**h**), and $\left[\theta_l''^2\right]^{l_c}$ (**c**,**f**,**i**) (see Equations (6)–(8)) for the BOMEX case after 7 h of integration. The budget terms are computed for different sub-domain sizes. N_{sd} indicates the number of sub-domains.

It can be seen that the grid spacing, vertical resolution, and the domain size of the LES are sufficient to obtain smooth profiles of the individual terms for all sub-domain sizes (only the results for the second largest, second smallest, and the smallest sub-domain sizes are shown). Such smooth profiles are necessary for further computation of the effective dissipation rates (violet lines). Similar results are found for all five idealized cases (not showed here). When comparing the diagnostics for different sub-domain sizes, the magnitude of the source terms (buoyancy term—green, gradient terms—blue), the turbulent transport terms (orange) is in general found to be smaller for the smaller sub-domain sizes.

The vertical advection terms (gray) should be equal to zero because horizontal homogeneity is assumed for all sub-domains. However, the vertical advection terms deviate from zero for the smallest sub-domains ($N_{sd} = 32^2$). This implies that the condition of horizontal homogeneity is not fulfilled at this scale and thus the length scale diagnostics for the smallest sub-domain size is less accurate. Therefore, only results for larger sub-domains are used in the evaluation. The accuracy of the method for the smallest sub-domain size can be increased by including horizontal terms (i.e., the buoyancy, shear/gradient, and turbulence transport terms) in the respective budgets.

The turbulence length scale diagnostics resulting from the budgets can be seen in Figure 2 for all cases. As expected, all length scales have a similarly shaped profile. Near the surface, the length scale is proportional to the distance from the surface according to the von Karman theory. At higher levels, the length scale continues to grow, but the growth rate decreases until the length scale reaches a maximum in the ABL. The location and the intensity of this peak depend on the ABL regime, the resolution, and the type of diagnostic (see below). Above the peak, the turbulence length scale decreases towards the top of the ABL, where it reaches either a positive value or it converges to zero. The L_{C,e_k} and L_{ϵ,e_k} diagnostics tend to show a secondary length scale peak near the top of the ABL in the convective cases (ARM, BOMEX, RICO) [3]. Such an increase in the turbulence length scale would imply an increase in the representation of the top entrainment. However, for the L_{C,θ_l}, L_{C,q_t}, L_{ϵ,θ_l}, and L_{ϵ,q_t} diagnostics, the secondary peak is either significantly smaller (ARM, BOMEX, RICO) or non-existent (L_{ϵ,θ_l} and L_{C,θ_l} for RICO). This suggests that the amplitude of the secondary peak in the L_{C,e_k} and L_{ϵ,e_k} diagnostics is over-estimated due to the lower accuracy of the method caused by the presence of gravity waves near the top of the ABL [3].

In the cloudy cases, the location of the ABL peak correlates with the height of the cloud base. In the dry GABLS1 case, the peak is significantly weaker and is located roughly in the middle of the ABL. This means that the presence of clouds and related processes significantly affects the shape and the amplitude of the turbulence length scale profile.

At lower resolutions (larger sub-domains), the six different diagnostics are in general closer to each other for all cases. At higher resolutions (smaller sub-domain sizes), which enter the gray zone of turbulence (starting roughly with 8^2 sub-domains), the differences between the classical (L_{ϵ,e_k}, L_{ϵ,θ_l} and L_{ϵ,q_t}) and new diagnostics (L_{C,e_k}, L_{C,θ_l}, L_{C,q_t}) become more apparent. While the classically diagnosed turbulence length scales monotonously decrease with the sub-domain size due to their clear dependence on the TKE and the scalar variances, the changes in the new diagnostics are height-dependent because of their additional dependence on the effective dissipation rates [3]. Basically, the newly diagnosed length scales decrease more slowly or even increase in the region of the ABL peak and decrease faster in the remaining regions with increasing resolution. Such changes make the shape of the vertical profile of the turbulence length scales resolution-dependent, where the ABL peak is more pronounced at higher resolutions.

Similarly, there are also bigger differences visible between the diagnostic based on TKE (L_{ϵ,e_k}, L_{C,e_k}) and diagnostics based on scalar variances (L_{ϵ,θ_l}, L_{ϵ,q_t}, L_{C,θ_l}, L_{C,q_t}) at higher resolutions. The ABL peak is sharper in the scalar variances diagnostics than in the TKE diagnostic. The scalar variances diagnostics are relatively close to each other with the exception of the RICO case, where the L_{ϵ,θ_l} and L_{C,θ_l} have an additional strong peak in the sub-cloud layer. Differences between the TKE diagnostics and the scalar variance diagnostics in the ABL peak could be attributed to the accuracy of the diagnostic method, because the effective dissipation rates for scalar variances have relatively small values in the sub-cloud layer for higher resolutions (particularly for potential temperature variance). However, despite the potential accuracy issues, which is most evident in the highest resolution, the increase in the ABL peak for the scalar variances in the sub-cloud layer appears consistently in all cloudy cases. The differences between L_{C,e_k} and L_{C,θ_l}, L_{C,q_t} could indicate that the dissipation rates for scalars are not entirely proportional to the dissipation rates of TKE in the gray zone of turbulence as is usually assumed at lower resolutions (see, e.g., [7,14,31]), and that their dependence could be height- or regime-dependent.

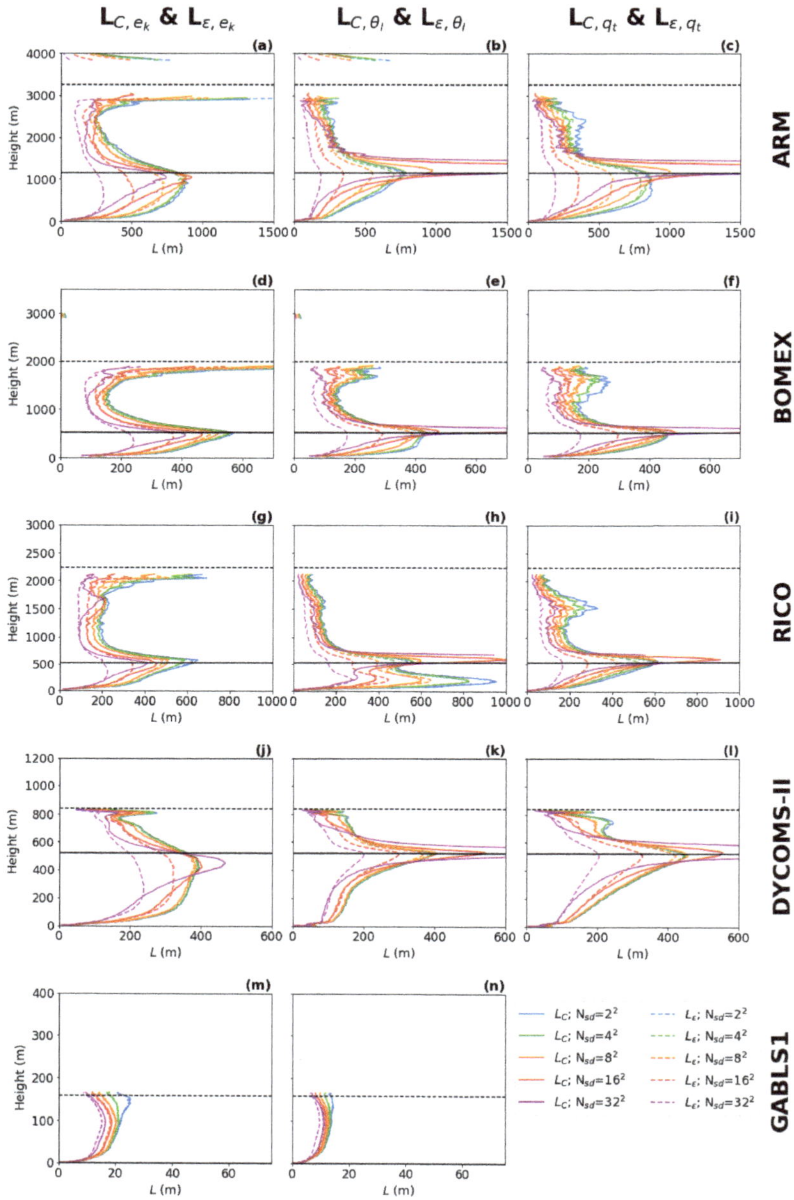

Figure 2. The scale dependence of the turbulence length scale diagnostics based on the TKE budget (1st column; **a**,**d**,**g**,**j**,**m**), the θ_l variance budget (2nd column; **b**,**e**,**h**,**k**,**n**), and the q_t variance budget (3rd column; **c**,**f**,**i**,**l**) for the five boundary layer cases (rows). The different sub-domain sizes are represented by different colors of lines, where the full lines are our new diagnostics (L_C) and dashed lines are the classical diagnostics (L_ϵ). N_{sd} indicates the number of sub-domains. L_{C,q_t} and L_{ϵ,q_t} are not plotted for the dry GABLS1 case. The budgets are plotted for the ARM case after 10 h of integration, for the BOMEX and the RICO case after 7 h of integration and for the DYCOMS-II case and the GABLS1 case after 3 h of integration. The horizontal black dashed lines indicate the ABL height, and the horizontal black solid lines indicate the cloud base height.

3.2. Temporal Evolution of the Turbulence Length Scale Diagnostics

For a better understanding of the behavior of the turbulence length scale, we present the temporal evolution of its diagnostics in Figures 3a–c, 4a–c, 5a–c, 6a–c and 7a–c. Due to potential accuracy issues with the smallest sub-domain sizes (highest resolution), discussed in Section 3.1, the temporal evolution is analyzed for the second smallest sub-domain size.

For all cases, all diagnostics show a similar shape of the vertical profile for the turbulence length scale as described in Section 3.1: a linear increase with height near the surface; a peak in the ABL near the cloud base when clouds are present; and a decrease towards the ABL top, where zero or a positive asymptotic value is reached. As expected, it can be seen that the shapes of the profile follow the changes in the ABL height (dashed black line) and the cloud base height (full black line), which implies that these two heights should play an important role in the parameterization of the turbulence length scale. An exception to the dependence on the ABL height is visible in the initial phase of the GABLS1 case, where the amplitude of the ABL peak changes without any correlation to the ABL height (no clouds are present).

When comparing the three types of diagnostics (TKE, θ_l variance, and q_t variance diagnostics), the results are consistent with findings in Section 3.1 for this (second smallest) sub-domain size, namely that the scalar variance diagnostics tend to have a more pronounced peak (sharper gradients of shading) in the ABL than the TKE diagnostic during the whole integration interval. Furthermore, a stronger secondary peak near the ABL top is present only in the TKE-based diagnostic.

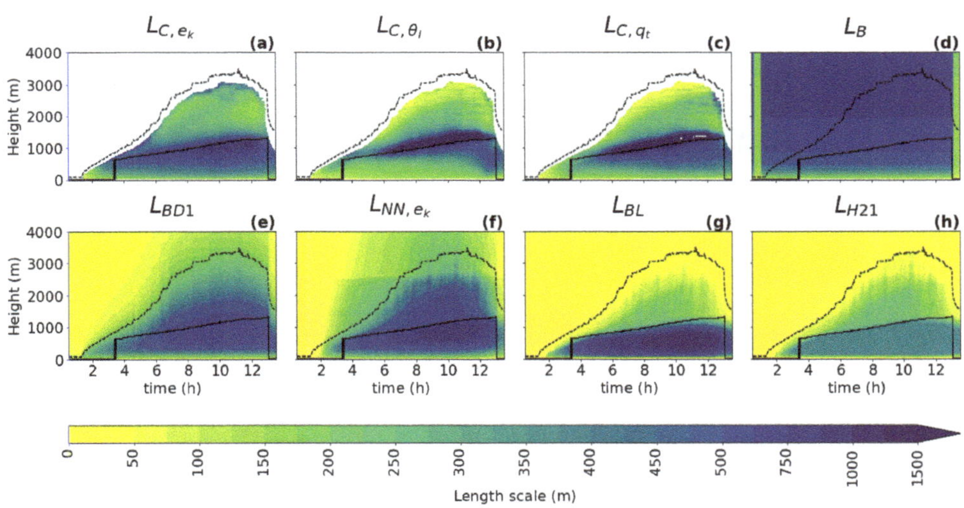

Figure 3. The temporal evolution of the turbulence length scale diagnostics: L_{C,e_k} (**a**), L_{C,q_t} (**b**), and L_{C,Θ_l} (**c**); and the turbulence length scale formulations: L_B (**d**), L_{BD1} (**e**), L_{NN,e_k} (**f**), L_{BL} (**g**), and L_{H21} (**h**); for the ARM case. The length scales are computed for the second smallest sub-domain size ($N_{sd} = 16^2$). The evolution of the cloud base height and the boundary layer height are represented by the black solid and dashed lines, respectively.

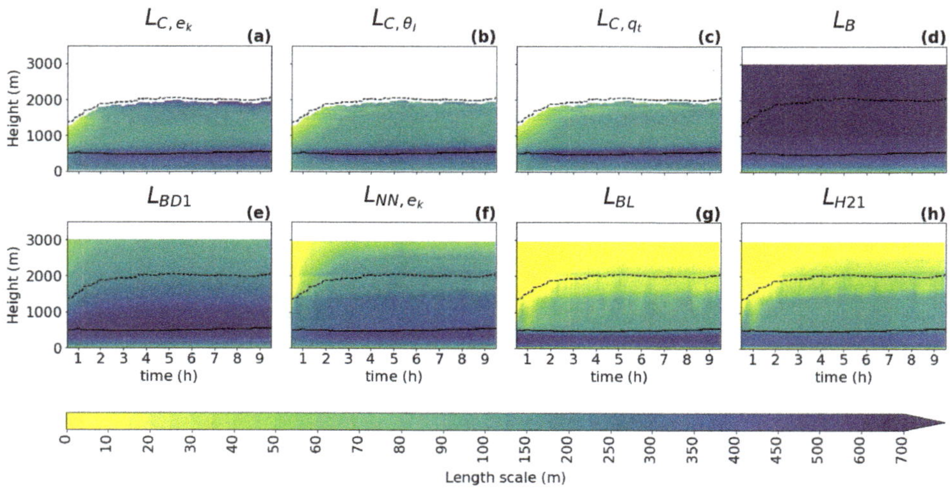

Figure 4. Same content as Figure 3, but for the BOMEX case.

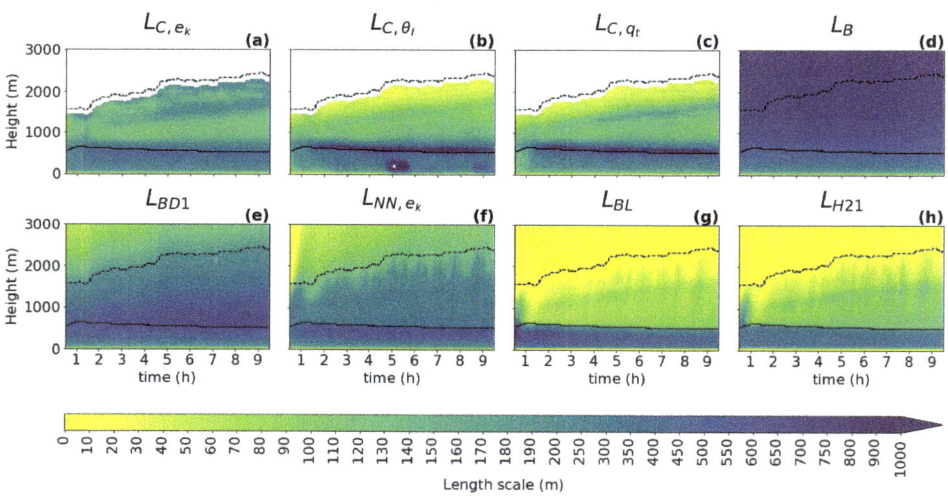

Figure 5. Same content as Figure 3, but for the RICO case.

3.3. Evaluation of the Algebraic Turbulence Length Scale Formulations

The selected algebraic formulations (see Section 2.2) computed from the coarse-grained LES data (see Section 2.3) are compared to the turbulence length scale diagnostics in Figure 8 for all cases.

All formulations exhibit an almost linear growth of the turbulence length scale near the surface as an extension of the von Karman theory, which is valid in the surface layer. A slight over-estimation compared to the diagnostics can be seen in the L_{BL} and L_{H21} formulations, which is caused by the influence of the upward length scale component, L_{up} or L_{Rup}, that can be larger than the distance from the surface. The linear behavior near the surface layer does not significantly change with the resolution.

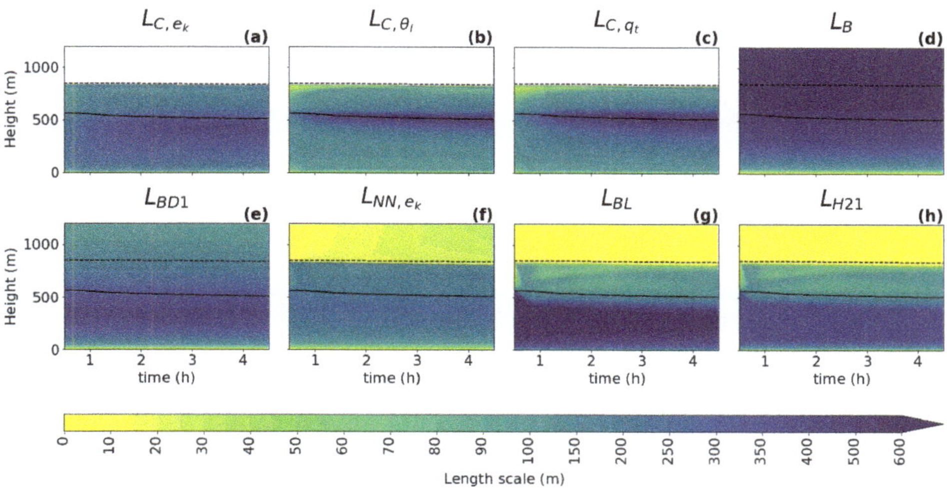

Figure 6. Same content as Figure 3, but for the DYCOMS-II case.

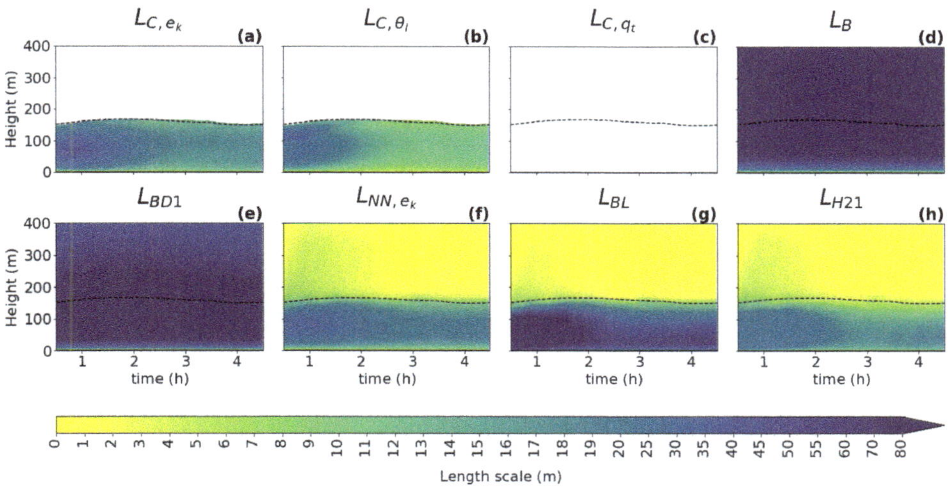

Figure 7. Same content as Figure 3, but for the GABLS1 case. L_{C,q_t} is not plotted for the dry GABLS1 case.

There are differences in the location and amplitude of the ABL peak between the length scale formulations. In order to achieve a more objective assessment of these characteristics, the time-averaged three-component plots (see Section 2.4) for two resolutions are presented in Figures 9 and 10.

L_B does not have an ABL peak, which significantly decreases its accuracy, as can be seen in the local normalized RMSE scores (see Equation (38)) that are presented in Figures 11 and 12 for two selected resolutions. Additionally, L_B is the same for all cases and all resolutions, because it depends only on the height.

L_{BD1} has a built-in ABL peak in its formulation, but its use is more suited for cloudy cases, where the estimation of the height and magnitude of the ABL peak is relatively accurate. L_{BD1} strongly over-estimates the magnitude of the ABL peak in the GABLS1. While L_{BD1} could be improved by calibration for the GABLS1 case, it would decrease its performance in the cloudy cases. Its overall performance across cases is thus limited. The shape of the L_{BD1} profile is closer to the shape of the L_{C,e_k} profile (the less pronounced ABL peak) than the profiles for scalar diagnostics. Therefore, its ABL peak estimation fits the TKE diagnostic better.

Both L_B and L_{BD1} significantly over-estimate the magnitude of the length scale in the GABLS1 case, which can be seen not only in the non-local overall mixing A-component (red color), but also in the local RMSE scores. The L_{BD1} also over-mixes in the cloudy cases, except for the ARM case, but its local RMSE score is relatively good compared to other formulations. L_{BD1} only scales with the ABL height; therefore, its overall performance decreases with increasing resolution.

L_{NN,e_k}, L_{BL}, and L_{H21} have A-component and RMSE scores similar to L_{BD1} for the shallow convection cases (ARM, BOMEX and RICO). For these cases, their ABL peak magnitude closely matches the diagnosed magnitudes. However, their ABL peak is mostly placed below the diagnosed heights.

For the stratocumulus DYCOMS-II case, L_{NN,e_k} under-estimates the ABL peak magnitude, and L_{BL} and L_{H21} over-estimate the ABL peak magnitude and under-estimate the ABL peak height. In the GABLS1 case, L_{BL} over-estimates the amplitude and under-estimates the location of the ABL peak. L_{NN,e_k} shows a better ability in this respect. L_{H21} also shows a good ability due to its explicit dependence on the vertical wind shear (see Equations (35) and (36)). The differences in the ABL peak for DYCOMS-II and GABLS1 cases are also mirrored in the differences in the A-component and the RMSE scores.

Because L_{NN,e_k}, L_{BL}, and L_{H21} scale not only with the ABL height, but also with the TKE and stratification, their case-awareness is better than for L_B and L_{BD1}. Their resolution awareness is also improved due to this dependence. However, only L_{H21} shows significant adjustment to resolution due to its cut-off formulation (see Equation (32)). Therefore, L_{H21} out-performs L_{BL} at higher resolutions, particularly for the DYCOMS-II case, as can be seen in all scores. L_{H21} and L_{BL} are very close to each other at lower resolutions for the cloudy cases. L_{NN,e_k} shows better performance for higher resolutions. This is probably caused by the choice of that particular calibration rather than by the scale-awareness of L_{NN,e_k}, since L_{NN,e_k} changes relatively slowly with resolution.

When all aspects of the evaluation are taken in to account, L_{NN,e_k} shows the best performance among the selected length scale formulations. This is due to its composite formulation that introduces dependence to the ABL height, the TKE, and the stratification. Due to these properties, L_{NN,e_k} adjusts to the specific flow regimes in the selected cases. L_{NN,e_k} also slightly adapts to the changes in the resolution. However, L_{NN,e_k} under-estimates the height of the ABL peak compared to the diagnostics. The scale-awareness of L_{NN,e_k} is rather weak and depends on the specific calibration of the constants. L_{H21} and L_{BL} have similar scores to L_{NN,e_k} for the cloudy cases, but these formulations place the ABL peak even lower than L_{NN,e_k} and also over-estimate the ABL peak magnitude and overall mixing for DYCOMS-II. For the GABLS1 case, L_{H21} is comparable with L_{NN,e_k} and clearly out-performs L_{BL} due to its dependence on the vertical wind shear. While L_{BD1} is well suited for cloudy cases, it strongly over-estimates the turbulence length scale for the GABLS1 case. Additionally, L_{BD1} lacks scale awareness since it only scales with the ABL height. Adjustment to other resolutions and flow regimes can thus be achieved only via recalibration. L_B is the simplest formulation, and thus it is not surprising that it has the worst scores. In particular, it does not have a peak in the ABL, and does not scale with any characteristic of the ABL.

When looking at the temporal evolution of the turbulence length scale formulations (Figures 3–7), the above overall evaluation can be confirmed for all formulations. L_{NN,e_k}, L_{H21}, L_{BL}, and L_{BD1} change according to the changes in the ABL height. L_B does not have this dependence. In addition, the TKE-dependent formulations (L_{NN,e_k}, L_{H21}, L_{BL}) are noisier than L_{BD1} in the shallow convection cases, which could deteriorate their performance when used as an an active component of a numerical model.

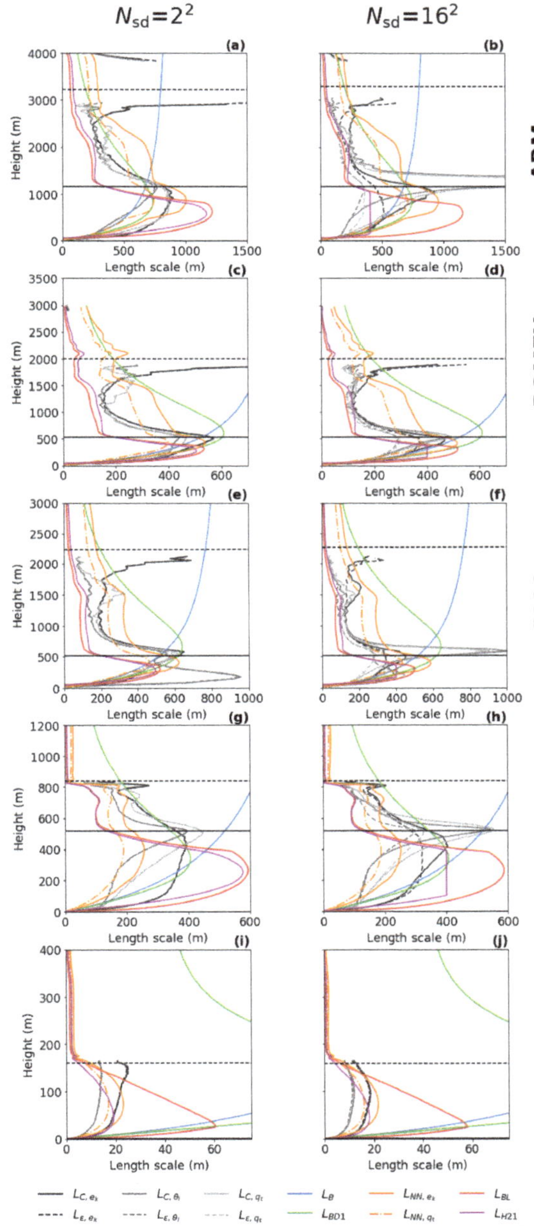

Figure 8. Comparison of the turbulence length scale diagnostics (L_{C,e_k}, L_{C,q_t}, L_{C,θ_l} , L_{ϵ,e_k}, L_{ϵ,q_t}, and L_{ϵ,θ_l}) and the turbulence length scale formulations (L_B, L_{BD1}, L_{NN,e_k}, L_{NN,q_t}, L_{BL}, and L_{H21}). Vertical profiles are plotted for the ARM case after 10 h of integration, for the BOMEX and the RICO case after 7 h of integration, and for the DYCOMS-II case and the GABLS1 case after 3 h of integration. L_{C,q_t} and L_{ϵ,q_t} are not plotted for the dry GABLS1 case. The length scales are computed for the second-largest sub-domain size (1st column; **a**,**c**,**e**,**g**,**i**) and the second-smallest sub-domain size (2nd column; **b**,**d**,**f**,**h**,**j**). N_{sd} indicates the number of sub-domains.

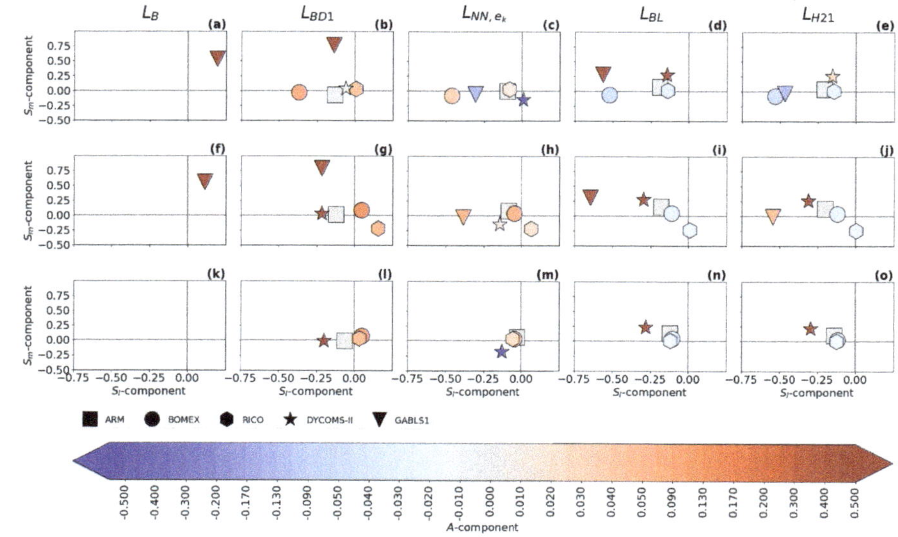

Figure 9. The differences in the time averaged normalized magnitude, S_m, and height/location, S_l, of the ABL peak (see Equations (41) and (42)) between the turbulence length scale formulations: L_B, L_{BD1}, L_{NN,e_k}, L_{NN,q_t}, L_{BL}, and L_{H21} (columns); and the turbulence length scale diagnostics (rows; L_{C,e_k}: **(a–e)**, L_{C,q_t}: **(f–j)**, L_{C,θ_l}: **(k–o)**). The symbols represent the evaluation for each of the boundary layer cases, while the colors represent the time-averaged amplitude component, A (see Equation (39)). The scores are calculated for the second largest sub-domain size ($N_{sd} = 2^2$).

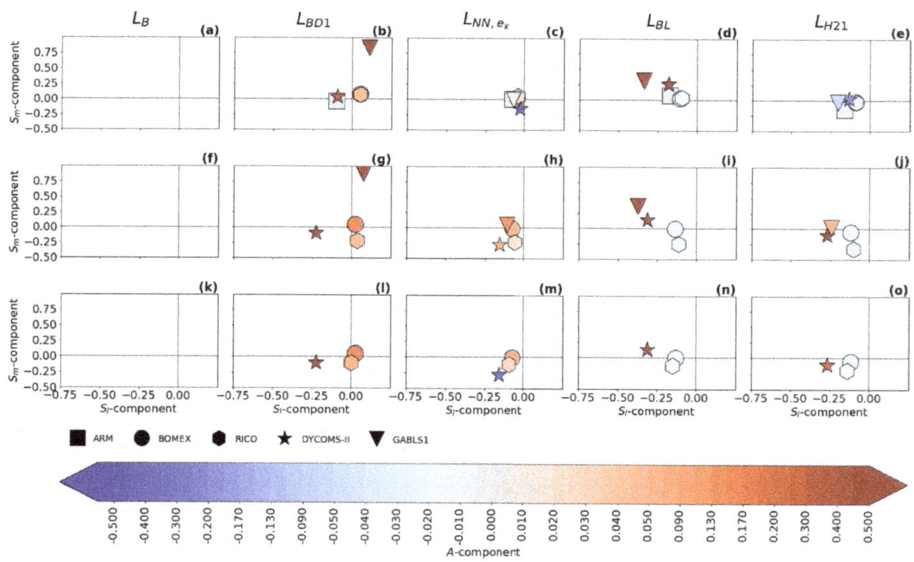

Figure 10. Same content as Figure 9 but for the second smallest sub-domain size ($N_{sd} = 16^2$).

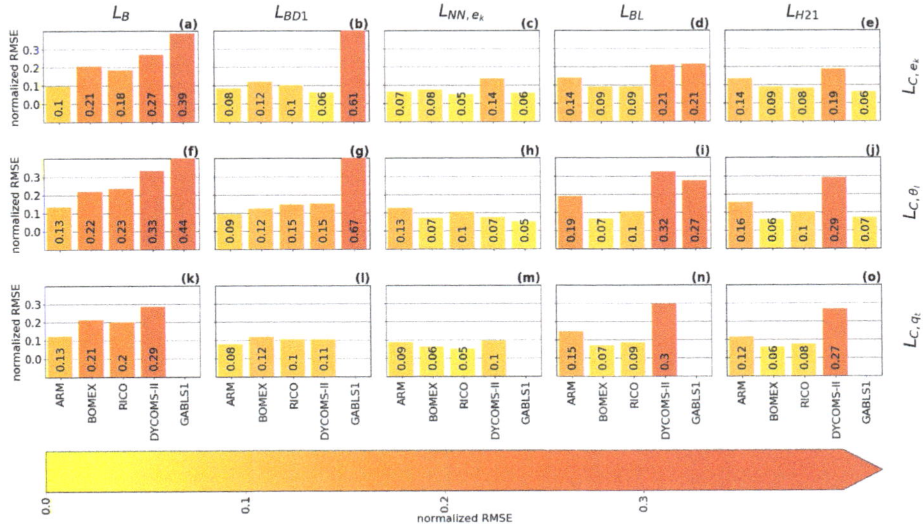

Figure 11. The time-averaged normalized RMSE (see Equation (38)) of the turbulence length scale formulations: L_B, L_{BD1}, L_{NN,e_k}, L_{NN,q_t}, L_{BL}, and L_{H21} (columns); with respect to the three diagnostics (rows; L_{C,e_k}: **(a–e)**, L_{C,q_t}: **(f–j)**, L_{C,θ_l}: **(k–o)**). The normalized RMSE is calculated for the second-largest sub-domain size ($N_{sd} = 2^2$).

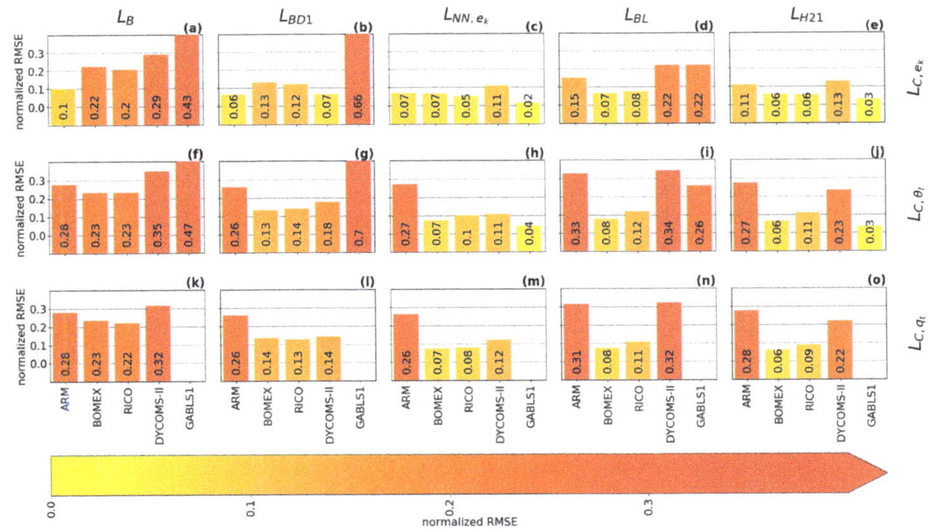

Figure 12. Same content as Figure 11, but for the second smallest sub-domain size ($N_{sd} = 16^2$).

4. Summary, Discussion, and Future Work

We have evaluated selected algebraic turbulence length scale formulations with budget-based turbulence length scale diagnostics that account for the cross-scale transfer of variances [3]. The diagnostics were computed using a coarse-graining method on high resolution LES data for selected idealized ABL cases: ARM (a continental cumulus case), BOMEX (a trade-wind cumulus case), RICO (a precipitating shallow cumulus case), DYCOMS-II (a stratocumulus case with drizzle), and GABLS1 (a weakly stable boundary layer case).

All vertical profiles of the length-scale diagnostics have a typical shape. Near the surface, the length scale is proportional to the distance from surface according to the von Karman theory. At higher levels, the length-scale growth rate decreases until it reaches a maximum in the ABL, whose location and intensity depend on the ABL regime, the resolution, and the type of diagnostic. Above the peak, the turbulence length scale decreases until it reaches an asymptotic value (positive or zero) at the top of the ABL.

In the cloudy cases, the location of the ABL peak correlates with the height of the cloud base. In the dry GABLS1 case, the peak is significantly weaker and is located roughly in the middle of the ABL. The new scale-aware diagnostics change with resolution (sub-domain size), but contrary to the changes in the classical turbulence length scale diagnostics, the resolution changes in the new diagnostics are height-dependent.

Compared to our previous study [3], we used not only the diagnostic based on the TKE budget, but also the diagnostics based on the budgets of the liquid-water potential temperature variance and the total specific-water-content variance. This extension helps to mitigate the accuracy problems of the TKE-based diagnostic in the presence of gravity waves in the convective cases near the top of the ABL, where the TKE diagnostic tends to show a secondary peak. Indeed, the scalar variance diagnostics indicate that the TKE diagnostic probably over-estimates the magnitude of this secondary peak in the ABL. In addition, we can observe that the ABL peak is sharper in the scalar variance diagnostics than in the TKE diagnostic. This could indicate that the dissipation rates for scalars are not entirely proportional to the dissipation rates of TKE in the gray zone of turbulence, as is usually assumed at lower resolutions, and that their dependence could be height- or regime-dependent.

We have used the local normalized RMSE (see Equation (38)) and the non-local three-component technique tailored specifically for the turbulence length scale profiles (see Section 2.4) for the evaluation of the turbulence length scale formulations.

Overall, L_{NN,e_k} shows the best performance among the selected length scale formulations. This is due to its multi-component composition and its dependence on the ABL height, the TKE, and the stratification. Thanks to these properties, L_{NN,e_k} adjusts to the specific flow regimes in the selected cases and also slightly adapts to the changes in the resolution. However, L_{NN,e_k} under-estimates the ABL peak location, and its scale-awareness could be stronger. L_{H21} and L_{BL} have similar scores to L_{NN,e_k} for the cloudy cases but are less accurate in the estimation of the ABL peak location and magnitude, particularly for the DYCOMS-II case. For the GABLS1 case, L_{H21} is comparable with L_{NN,e_k} and clearly out-performs L_{BL} due to its dependence on the vertical wind shear. L_{BD1} is well suited for cloudy cases, but it strongly over-estimates the turbulence length scale for the GABLS1 case. Additionally, L_{BD1} only scales with the ABL height, and thus it has almost no scale awareness. L_B is the simplest formulation, and thus it is not surprising that it has the worst scores. In particular, it lacks the peak in the ABL and does not scale with any characteristic of the ABL.

When looking at the temporal evolution of the turbulence length scale formulations, L_{NN,e_k}, L_{H21}, L_{BL}, and L_{BD1} adequately change according to the changes in the ABL height. In addition, the TKE-dependent formulations (L_{NN,e_k}, L_{H21}, L_{BL}) are noisier than L_{BD1} in the shallow convection cases, which could deteriorate their performance when used as an an active component of a numerical model.

It is clear from the evaluation that a proper scaling with the ABL height, TKE, stratification, and the vertical wind shear can improve the performance of the formulations. Such a scaling behavior makes the formulations more universal in terms of the ABL flow regime. To further improve this behavior, additional scaling variables could be introduced. In particular, all evaluated formulations have difficulties in the determination of the height of the ABL peak. As we have seen in the diagnostics, the ABL peak is usually placed just under the cloud base height. However, none of the formulations have an explicit dependence on the cloud base height, and thus the introduction of such a feature could be beneficial for future turbulence length scale formulations. We plan to investigate the scaling potential of the cloud base height in our future work.

The scale awareness in most of the evaluated formulations is relatively poor. The TKE's dependence on the resolution and the stratification introduces a degree of scale awareness to L_{NN,e_k}, L_{H21}, L_{BL}, but it is not sufficient. Only L_{H21} shows a stronger scale awareness thanks to its cut-off formulation in the gray zone. However, in contrast to the turbulence length scale diagnostics, this dependence on resolution does not change with height. To improve the representation of turbulence in the gray zone of turbulence, we would recommend a turbulence length scale formulation that has a scale dependence that changes with height. We plan to propose such a formulation in our future work.

Better performance of the evaluated formulations can be achieved via recalibration of constants. This option can be used when changing the resolution of the model. Still, such a recalibration is rather time-consuming and does not support seamless changes in model resolution. Furthermore, a recalibration cannot be used to mitigate the lack of flow-regime dependence.

Based on our evaluation, we recommend using the L_{NN,e_k}, L_{H21} or L_{BL} formulation in TKE turbulence schemes. Nevertheless, we would like to point out that the turbulence length scale formulation is only a part of a turbulence scheme, and thus the overall performance of the scheme can decrease if the turbulence length scale formulation is changed. This is because the scheme was previously calibrated with a different length scale formulation. Hence, an introduction of a new turbulence length scale formulation requires recalibration of the whole scheme, and/or recalibration of other physical parameterizations of the model.

Author Contributions: Conceptualization, I.B.Ď.; methodology, I.B.Ď.; software, S.R., I.B.Ď. and A.T.J.; validation, S.R. and I.B.Ď.; formal analysis, S.R., I.B.Ď. and J.S.; investigation, S.R., I.B.Ď. and J.S.; resources, I.B.Ď. and J.S.; data curation, I.B.Ď.; writing—original draft preparation, S.R. and I.B.Ď.; writing—review and editing, S.R., I.B.Ď., J.S. and A.T.J.; visualization, S.R. and I.B.Ď.; supervision, I.B.Ď.; project administration, I.B.Ď.; funding acquisition, I.B.Ď. and J.S. All authors have read and agreed to the published version of the manuscript.

Funding: This research was funded by the Deutsche Forschungsgemenschaft (DFG) grant number 454477837 (GZ: BA 6804/2-1) and by the Hans Ertel Centre for Weather Research of DWD (3rd phase, The Atmospheric Boundary Layer in Numerical Weather Prediction) grant number 4818DWDP4. Furthermore, funded by the Deutsche Forschungsgemeinschaft (DFG, German Research Foundation)—TRR 301—Project-ID 428312742.

Institutional Review Board Statement: Not applicable.

Informed Consent Statement: Not applicable.

Data Availability Statement: The data for this study were generated with the LES model MicroHH and are openly available in Zenodo at https://zenodo.org/record/822842 [32] (accessed on 26 February 2022). The configuration files, outputs, and visualization scripts for the LES are openly available in Zenodo at: https://doi.org/10.5281/zenodo.6372434 [33] (accessed on 26 February 2022).

Acknowledgments: The authors would like to note that this research was made possible using the resources of the Deutsches Kimarechenzentrum (DKRZ) granted by its Scientific Steering Committee (WLA) under project ID bb1096.

Conflicts of Interest: The authors declare no conflict of interest.

Abbreviations

The following abbreviations are used in this manuscript:

ABL	Atmospheric Boundary Layer
ARM	Atmospheric Radiation Measurement
BOMEX	Barbados Oceanographic and Meteorological Experiment
DYCOMS-II	the second Dynamics and Chemistry of Marine Stratocumulus

GABLS1	GEWEX Atmospheric Boundary Layer Study
GEWEX	Global Energy and WAter Cycle Experiment
GCM	Global Circulation Model
LES	Large-Eddy Simulation
NWP	Numerical Weather Prediction
RICO	Rain in Shallow Cumulus Over the Ocean
RMSE	Root Mean Sqaure Error
TKE	Turbulence Kinetic Energy

References

1. Wyngaard, J.C. *Turbulence in the Atmosphere*; Cambridge University Press: Cambridge, UK, 2010. [CrossRef]
2. Kitamura, Y. Estimating Dependence of the Turbulent Length Scales on Model Resolution Based on A Priori Analysis. *J. Atmos. Sci.* **2015**, *72*, 750–762. [CrossRef]
3. Bašták Ďurán, I.; Schmidli, J.; Bhattacharya, R. A Budget-Based Turbulence Length Scale Diagnostic. *Atmosphere* **2020**, *11*, 425. [CrossRef]
4. Wyngaard, J.C. Toward Numerical Modeling in the "Terra Incognita". *J. Atmos. Sci.* **2004**, *61*, 1816–1826. doi: 10.1175/1520-0469(2004)061<1816:tnmitt>2.0.co;2. [CrossRef]
5. Honnert, R.; Masson, V.; Couvreux, F. A Diagnostic for Evaluating the Representation of Turbulence in Atmospheric Models at the Kilometric Scale. *J. Atmos. Sci.* **2011**, *68*, 3112–3131. [CrossRef]
6. Honnert, R. Representation of the grey zone of turbulence in the atmospheric boundary layer. *Adv. Sci. Res.* **2016**, *13*, 63–67. [CrossRef]
7. Bašták Ďurán, I.; Geleyn, J.F.; Váňa, F. A Compact Model for the Stability Dependency of TKE Production–Destruction–Conversion Terms Valid for the Whole Range of Richardson Numbers. *J. Atmos. Sci.* **2014**, *71*, 3004–3026. [CrossRef]
8. Blackadar, A.K. The vertical distribution of wind and turbulent exchange in a neutral atmosphere. *J. Geophys. Res.* **1962**, *67*, 3095–3102. [CrossRef]
9. Bašták Ďurán, I.; Geleyn, J.F.; Váňa, F.; Schmidli, J.; Brožková, R. A Turbulence Scheme with Two Prognostic Turbulence Energies. *J. Atmos. Sci.* **2018**, *75*, 3381–3402. [CrossRef]
10. Bougeault, P.; Lacarrere, P. Parameterization of Orography-Induced Turbulence in a Mesobeta–Scale Model. *Mon. Weather Rev.* **1989**, *117*, 1872–1890. [CrossRef]
11. Honnert, R.; Masson, V.; Lac, C.; Nagel, T. A Theoretical Analysis of Mixing Length for Atmospheric Models From Micro to Large Scales. *Front. Earth Sci.* **2021**, *8*, 537. [CrossRef]
12. Nakanishi, M. Improvement of the Mellor–Yamada Turbulence Closure Model Based on Large-Eddy Simulation Data. *Bound.-Layer Meteorol.* **2001**, *99*, 349–378. [CrossRef]
13. Nakanishi, M.; Niino, H. Development of an Improved Turbulence Closure Model for the Atmospheric Boundary Layer. *J. Meteorol. Soc. Jpn. Ser. II* **2009**, *87*, 895–912. [CrossRef]
14. Cheng, Y.; Canuto, V.M.; Howard, A.M. An Improved Model for the Turbulent PBL. *J. Atmos. Sci.* **2002**, *59*, 1550–1565. [CrossRef]
15. Lilly, D. *The Representation of Small-Scale Turbulence in Numerical Simulation Experiments*; Technical Report; National Center for Atmospheric Research: Boulder, CO, USA, 1966. [CrossRef]
16. Stull, R.B. *An Introduction to Boundary Layer Meteorology*; Kluwer Academic Publishers: Amsterdam, The Netherlands, 1988.
17. Redelsperger, J.L.; Mahé, F.; Carlotti, P. A Simple Furthermore, General Subgrid Model Suitable Both For Surface Layer Furthermore, Free-Stream Turbulence. *Bound.-Layer Meteorol.* **2001**, *101*, 375–408. [CrossRef]
18. Cedilnik, J. *Parallel Suites Documentation*; Regional Cooperation for Limited Area modeling in Central Europe: Prague, Czech Republic, 2005.
19. Mellor, G.L.; Yamada, T. A Hierarchy of Turbulence Closure Models for Planetary Boundary Layers. *J. Atmos. Sci.* **1974**, *31*, 1791–1806. [CrossRef]
20. Bougeault, P.; André, J.C. On the Stability of the THIRD-Order Turbulence Closure for the Modeling of the Stratocumulus-Topped Boundary Layer. *J. Atmos. Sci.* **1986**, *43*, 1574–1581. [CrossRef]
21. Rodier, Q.; Masson, V.; Couvreux, F.; Paci, A. Evaluation of a Buoyancy and Shear Based Mixing Length for a Turbulence Scheme. *Front. Earth Sci.* **2017**, *5*, 65. [CrossRef]
22. Brown, A.R.; Cederwall, R.T.; Chlond, A.; Duynkerke, P.; Golaz, J.C.; Khairoutdinov, M.; Lewellen, D.C.; Lock, A.P.; MacVean, M.K.; Moeng, C.H.; et al. Large-eddy simulation of the diurnal cycle of shallow cumulus convection over land. *Q. J. R. Meteorol. Soc.* **2002**, *128*, 1075–1093. [CrossRef]
23. Stevens, B.; Moeng, C.H.; Ackerman, A.S.; Bretherton, C.S.; Chlond, A.; de Roode, S.; Edwards, J.; Golaz, J.C.; Jiang, H.; Khairoutdinov, M.; et al. Evaluation of Large-Eddy Simulations via Observations of Nocturnal Marine Stratocumulus. *Mon. Weather Rev.* **2005**, *133*, 1443–1462. [CrossRef]
24. Holtslag, B. Preface: GEWEX Atmospheric Boundary-layer Study (GABLS) on Stable Boundary Layers. *Bound.-Layer Meteorol.* **2006**, *118*, 243–246. [CrossRef]

25. Beare, R.J.; Macvean, M.K.; Holtslag, A.A.M.; Cuxart, J.; Esau, I.; Golaz, J.C.; Jimenez, M.A.; Khairoutdinov, M.; Kosovic, B.; Lewellen, D.; et al. An Intercomparison of Large-Eddy Simulations of the Stable Boundary Layer. *Bound.-Layer Meteorol.* **2006**, *118*, 247–272. [CrossRef]
26. Siebesma, A.P.; Bretherton, C.S.; Brown, A.; Chlond, A.; Cuxart, J.; Duynkerke, P.G.; Jiang, H.; Khairoutdinov, M.; Lewellen, D.; Moeng, C.H.; et al. A Large Eddy Simulation Intercomparison Study of Shallow Cumulus Convection. *J. Atmos. Sci.* **2003**, *60*, 1201–1219. [CrossRef]
27. Rauber, R.M.; Stevens, B.; Ochs, H.T.; Knight, C.; Albrecht, B.A.; Blyth, A.M.; Fairall, C.W.; Jensen, J.B.; Lasher-Trapp, S.G.; Mayol-Bracero, O.L.; et al. Rain in Shallow Cumulus Over the Ocean: The RICO Campaign. *Bull. Am. Meteorol. Soc.* **2007**, *88*, 1912–1928. [CrossRef]
28. van Heerwaarden, C.C.; van Stratum, B.J.H.; Heus, T.; Gibbs, J.A.; Fedorovich, E.; Mellado, J.P. MicroHH 1.0: A computational fluid dynamics code for direct numerical simulation and large-eddy simulation of atmospheric boundary layer flows. *Geosci. Model Dev.* **2017**, *10*, 3145–3165. [CrossRef]
29. Baas, P.; de Roode, S.R.; Lenderink, G. The Scaling Behaviour of a Turbulent Kinetic Energy Closure Model for Stably Stratified Conditions. *Bound.-Layer Meteorol.* **2007**, *127*, 17–36. [CrossRef]
30. Wernli, H.; Paulat, M.; Hagen, M.; Frei, C. SAL—A Novel Quality Measure for the Verification of Quantitative Precipitation Forecasts. *Mon. Weather Rev.* **2008**, *136*, 4470–4487. [CrossRef]
31. Zilitinkevich, S.S.; Elperin, T.; Kleeorin, N.; Rogachevskii, I.; Esau, I. A Hierarchy of Energy- and Flux-Budget (EFB) Turbulence Closure Models for Stably-Stratified Geophysical Flows. *Bound.-Layer Meteorol.* **2012**, *146*, 341–373. [CrossRef]
32. van Heerwaarden, C.C.; van Stratum, B.J.H.; thijsheus. microhh/microhh: 1.0.0. 2017. Available online: https://zenodo.org/record/822842 (accessed on 26 February 2022).
33. Bašták Ďurán, I.; Reilly, S.; Anurose, T.J.; Schmidli, J. Data Accompanying the Paper Titled: An Evaluation of Algebraic Turbulence Length Scale Formulations. 2022. Available online: https://zenodo.org/record/6372434 (accessed on 26 February 2022).

atmosphere

MDPI

Article

Numerical Investigation on Impact Erosion of Aeolian Sand Saltation in Gobi

Yong Wang [1,2], Jie Zhang [1,2], Hongchao Dun [1,2,*] and Ning Huang [1,2,*]

1 Key Laboratory of Mechanics on Disaster and Environment in Western China, Lanzhou University, Lanzhou 730000, China
2 College of Civil Engineering and Mechanics, Lanzhou University, Lanzhou 730000, China
* Correspondence: dunhc@lzu.edu.cn (H.D.); huangn@lzu.edu.cn (N.H.)

Abstract: Sand drift erosion is common on aeolian landforms, particularly in the Gobi desert where sand drift is often quite strong. Sand drift erosion can lead to many types of hazards, including severe crop loss, structural damage to buildings or infrastructure, and abrasion of soil or clay components that contribute to the production of fine particulate matter. This article combines the Gobi sand flow model with the solid particles erosion model to simulate the sand drift erosion process in a variety of Gobi environments. The results show that the impact erosion of saltation particles is highly dependent on both the friction velocity and the gravel coverage. Saltation erosion amount increases with the increment of friction velocity and the gravel coverage. The vertical profile of saltating erosion rate displays a clear stratification pattern composed of a linear increasing layer, a damage layer, and a monotonic decreasing layer. The maximum value of the saltation erosion rate increases as the friction velocity increases and their curve shows a power-law relationship. The damage height caused by saltation erosion is primarily concentrated in the height range of 0.03 m to 0.15 m, and it increases approximately linearly with friction velocity.

Keywords: aeolian saltation; Gobi sand flow; wind erosion; impact erosion; erosion rate

Citation: Wang, Y.; Zhang, J.; Dun, H.; Huang, N. Numerical Investigation on Impact Erosion of Aeolian Sand Saltation in Gobi. *Atmosphere* **2023**, *14*, 349. https://doi.org/10.3390/atmos14020349

Academic Editors: Yubin Li and Jie Tang

Received: 17 January 2023
Revised: 2 February 2023
Accepted: 3 February 2023
Published: 9 February 2023

1. Introduction

Aeolian landforms are extensively distributed on Earth, Mars, and other Earth-like planets [1–3]. Wind-blown sand movement is a near-surface transport phenomenon occurring frequently in aeolian landforms such as a sand desert and the Gobi [4–6]. It is the primary cause of numerous geophysical phenomena, such as dust emissions, soil wind erosion, and desertization [7–10]. As a mode of wind erosion, impact erosion refers to the process of erosion damage to the surfaces of building structures caused by the collision and impact of airborne sand particles. Impact erosion by sand drift has long been a significant contributor to serious wear and tear on roads, houses, bridges and other mixed-concrete structures in sandy lands, which lowers the strength and safety of the mixed-concrete buildings, thus creating a potential safety threat [11–13]. Especially in the Gobi region, with strong winds and long wind periods, the concrete and steel construction around them suffer more serious impacts and damage [14,15].

Wind-blown sand movement is a complex physical process in which sand particles, airflow fields and granular beds are coupled with each other to form a two-phase flow of gas and solid [16,17]. For tens of years, considerable efforts have been devoted to the research of aeolian sediment transport [18–22]. Researchers have gradually developed and improved the mathematical model of wind-blown sand movement [23–27], and multiple wind tunnel tests and field observations have confirmed its accuracy [28–30]. Recent experimental research, in particular, parameterized the sand-bed collision mechanisms on the Gobi and granular bed utilizing natural sand bed samples and sand grains [31,32]. Those studies have laid a foundation for the establishment of the Gobi sand flow model.

Airborne particles will wear down and harm the building walls when they come in contact with them [33]. Many academics devoted themselves to studying the erosion of various target materials by sand-carrying jet impingement experiment [34–37]. Erosion rate is defined as the ratio of the lost mass of target material to the total mass of impact sand particles, and numerous empirical models are present to describe the erosion rate as a function of impact velocity and impact angle [37–40]. Particle erosion models and computational fluid dynamics (CFD) technique are frequently combined to estimate solid particle erosion in several significant areas of the oil and gas industry [41,42]. In contrast, this related technique has not been applied very often in studies on saltating sand erosion, and research on it is mainly focused on field observations and wind tunnel experiments [43–46].

Field observation on a boulder alluvial plain subjected to sand drift shows a maximum erosion at $10 \sim 12\,\text{cm}$ above the ground [47]. Liu et al. [48] gave the vertical profile of the saltation erosion rate by wind tunnel measurements of adobe abrasion. Shi and Shi [49] first combined the steady-state saltation model with the particle erosion rate model to simulate the sand drift erosion and confirmed the early conclusion that the kinetic energy is crucial to erosion. These studies, however, have been limited to qualitative analyses or simple quantitative descriptions of sand drift erosion, without providing a quantitative parametric characterization of the change in saltating erosion rate under various environmental conditions, so they are unable to provide helpful recommendations for the erosion protection of building structures in aeolian environments.

The numerical model for this investigation was built by combining the solid particle erosion model with the Gobi sand flow model. We simulate and investigate the spatial distribution of impact erosion by aeolian saltation in different Gobi environments, including varying gravel covering and friction velocity, and provide a parametric model for the variation of saltation erosion rate with friction velocity and gravel coverage.

2. Methodology

2.1. Particle Motion Governing Equations

Sand particles are assumed to be small spheres, and every particle in the air is tracked. Airborne particle motion is computed by explicitly solving the particle motion equation using an explicit temporal integration approach [50,51]. The governing equations of particle motion can be written as follows [52]:

$$m\frac{\mathrm{d}\vec{v}}{\mathrm{d}t} = \vec{f}_{\mathrm{d}} + \vec{f}_{\mathrm{l}} - m\vec{g} \tag{1}$$

$$\vec{v} = \frac{\mathrm{d}\vec{x}_{\mathrm{p}}}{\mathrm{d}t} \tag{2}$$

where $\vec{x}_{\mathrm{p}}$ and $\vec{v}$ stand for the position coordinates and particle velocity, respectively; m is the particle mass; $\vec{g}$ is the acceleration of gravity; $\vec{f}_{\mathrm{d}}$ and $\vec{f}_{\mathrm{l}}$ represent, respectively, aerodynamical drag and aerodynamical lift, and can be expressed as [22,53]:

$$\vec{f}_{\mathrm{d}} = -\frac{\pi}{8}\rho d^2 C_{\mathrm{d}} u_{\mathrm{r}}(\vec{v} - \vec{u}) \tag{3}$$

and

$$\vec{f}_{\mathrm{l}} = \frac{\pi}{8}\rho d^3 C_{\mathrm{l}}(\bigtriangledown|\vec{u}|^2) \tag{4}$$

where $u_{\mathrm{r}} = |\vec{v} - \vec{u}|$ stands for the velocity of particles relative to fluid; $\vec{u}$ is the fluid velocity; $\rho = 1.23\,\text{kg/m}^3$ is the fluid density; d is the particle diameter; $C_{\mathrm{d}} = 24/Re_{\mathrm{p}}(1 + 0.15Re_{\mathrm{p}}^{0.687})$ and $C_{\mathrm{l}} = 0.85C_{\mathrm{d}}$ represent, respectively, the drag coefficient and lift coefficient, in which $Re_{\mathrm{p}} = u_{\mathrm{r}}d/\nu$ is the particle Reynolds number and $\nu = 1.5 \times 10^{-5}\,\text{m}^2/\text{s}$ is the kinematic viscosity coefficient of the fluid.

Airborne particles are accelerated by the airflow and their motion is governed by Equations (1) and (2). There are numerous schemes available for the solution of the motion equations, including Euler schemes, Verlet schemes, and others. We use here the Runge–Kutta scheme due to its high accuracy while ensuring the efficiency of the calculation [54].

In one iteration step, after solving the particle's motion, we will scan all particles in the air and find out the particle pairs that contact each other. Because there are so many particles in midair, collisions frequently happen as they move. In this study, a collision happens when a pair of particles' centroids are closer together than the sum of their radii. Then, the post-collision velocity of two colliding particles determine directly with the collision theory, and the post-collision velocity $\vec{v}_1'$ on one of the particles after colliding is [55]:

$$\vec{v}_1' = \vec{v}_1 + \epsilon(\vec{n} \cdot \vec{v}_{12})\vec{n} + v[\vec{v}_{12} - (\vec{n} \cdot \vec{v}_{12})\vec{n}] \tag{5}$$

with

$$\epsilon = \frac{1+\varepsilon}{1+\eta} \tag{6}$$

$$v = \frac{(2/7)(1-\mu)}{1+\eta} \tag{7}$$

where the subscript 1 or 2 of the variable refers to the label of each particle in the collision pair and $\vec{v}_{12} = \vec{v}_2 - \vec{v}_1$ is the relative velocity before the collision; $\vec{n} = (\vec{x}_{p,2} - \vec{x}_{p,1})/|\vec{x}_{p,2} - \vec{x}_{p,1}|$ is a unit vector from one particle centre pointing to the centre of the other one; $\eta = m_1/m_2$ is the mess ratio of two colliding particles; ε and μ are restitution coefficients for the normal and tangential components, respectively. For quartz particles, generally set $\varepsilon = 0.9$, and μ can be calculated with the following formula [56]:

$$\mu = max\left(0, 1 - \frac{C_f(1+\varepsilon)}{2/7}\frac{u_{pn}}{u_{pt}}\right) \tag{8}$$

where $C_f = 0.4$ is the friction coefficient [57]; u_{pn} and u_{pt} are the normal and tangential relative velocities between the two particles in the colliding pair, respectively. $\vec{v}_2'$ can be calculated using the same method, thus there is no need to repeat that process here.

2.2. Aerodynamic Governing Equations

The movement of airflow can be described using the equation below:

$$\rho\frac{\partial \vec{u}}{\partial t} + \rho\vec{u} \cdot \nabla\vec{u} = \nabla T + \frac{\vec{F}_p}{1-\phi_p} \tag{9}$$

where T stands for the stress tensor of the fluid; ϕ_p is the particle volume fraction; $\vec{F}_p = -\sum_{n=1}^{N_p} \vec{f}_{d,n}/V_c$ is the reaction force of N_p particles in unit volume V_c to the flow field.

For the steady and homogeneous flow field studied in this article, the inertia and horizontal stress gradients of the fluid are neglected. If only the horizontal flow components is considered, Equation (9) can be simplified to [16,53]:

$$\frac{\partial \tau_f}{\partial z} + \frac{F_p}{1-\phi_p} = 0 \tag{10}$$

where u and F_p are, respectively, the horizontal component of $\vec{u}$ and $\vec{F}_p$; τ_f represents the stress, which contains both viscous stress and Reynolds stress. A Prandtl's mixing length model with the kinematic turbulent viscosity $\nu_t = l_m^2|\partial u/\partial z|$ is used to close Equation (10). The expression of τ_f can be given as [58]:

$$\tau_f = \rho(\nu + \nu_t)\frac{\partial u}{\partial z} = \rho\left(\nu + l_m^2\left|\frac{\partial u}{\partial z}\right|\right)\frac{\partial u}{\partial z} \tag{11}$$

with the mixing length scale $l_{\mathrm{m}}(z) = \kappa z$ and $\kappa = 0.4$ being von Karman constant. According to the boundary conditions $\tau_{\mathrm{f}}|_{z\to\infty} = \rho u_*^2$ and ignoring the effect of viscous stress, the governing equation of airflow can be obtained as:

$$\frac{\mathrm{d}u}{\mathrm{d}z} = \frac{1}{\kappa z}\left(u_*^2 - \frac{1}{\rho}\int_z^{z_{\max}} \frac{F_{\mathrm{p}}}{1-\phi_{\mathrm{p}}}\mathrm{d}z\right)^{\frac{1}{2}} \quad (12)$$

where u_* is the friction velocity.

2.3. Surface Process

When saltation sand particles hit the bed surface, complex interactions between particles and the bed happen [59,60]. In general, it is extremely challenging to accurately describe this process in sand drift [27]. The currently common method for numerical simulation is to replace it with the splash function [61–63].

The impactor, with impact velocity v_{i}, impact angle θ_{i}, and azimuthal angle φ_{i}, will return at a specified speed and angle after impacting the gravel or sand bed. We indicate the impactor's rebound speed and angle by v_{r} and θ_{r}. The velocity of rebound is given by the equation $v_{\mathrm{r}} = e_{\mathrm{r}} v_{\mathrm{i}}$, where e_{r} is the restitution coefficient. Experiments show that e_{r} and θ_{r} accord with normal distribution. When a particle impacts the sand bed, it will not only bounce back but also eject new particles from the bed. The number of sand particles ejected n_{ej}, as well as their initial velocity V_{ej} and initial angle θ_{ej} all obey lognormal distribution. As for the azimuth angle φ_{ej}, Xing and He [64] pointed out that, due to symmetry, the distribution of azimuth angle φ_{ej} satisfies normal distribution with a mean value of $\mu = 0$, The parameter of splash model are all shown in Table 1.

Table 1. Parameters of the Splash Model.

	Distribution	Mean	Std.	References
Rebound on Gravel				
e_r	Normal	$0.62 + 0.0084V_i - 0.63\sin\theta_i$	$0.19 - 0.0035V_i$	Chen et al. [31]
θ_r	Lognormal	$2.92 - 0.034V_i + 0.02\theta_i$	$0.9 - 0.049V_i$	Chen et al. [31]
Rebound on Sand Bed				
e_r	Lognormal	$0.47 + 0.015V_i - 0.02\theta_i$	$0.16 + 0.04V_i - 0.001\theta_i$	Zhang et al. [32]
θ_r	Lognormal	$0.26 + 0.04V_i - 0.032\theta_i$	$0.16 - 0.005V_i - 0.029\theta_i$	Zhang et al. [32]
Splash				
n_{ej}	Lognormal	$-0.36 + 1.35\ln V_i - 0.01\theta_i$	0.55	Chen et al. [31]
V_{ej}	Lognormal	$1.67 + 0.082V_i - 0.003\theta_i$	0.616	Chen et al. [31]
θ_{ej}	Lognormal	3.94	0.64	Chen et al. [31]
φ_{ej}	Normal	0	15	Xing and He [64]

Static particles on the granular bed are not only motivated by impactors but also are directly entrained by the airflow, which is called aerodynamic entrainment. When the friction velocity u_* is greater than the threshold friction velocity u_{*t}, the particles are motivated from the bed. The threshold friction velocity is used to determin by $u_{*t} = A\sqrt{(\rho_{\mathrm{p}}/\rho - 1)gd}$ with $A = 0.1$ [4].

The sand particles entrained by aerodynamic force will have a certain initial velocity, which is known as the initial take-off velocity. Kang et al. [65] proposes that the horizontal velocity of sand particles obeys normal distribution, while the vertical component obeys exponential distribution:

$$S(v_{x0}) = \frac{1}{\sqrt{2\pi}A}\exp\left(-\frac{(v_{x0} - B)^2}{2A^2}\right) \quad (13)$$

$$S(v_{z0}) = \frac{1}{C}\exp(-\frac{v_{z0}}{C}) \tag{14}$$

where, the values of the parameters are $A = 0.5030$, $B = 0.7135$ and $C = 0.3952$, respectively.

Aerodynamic entrainment plays an important role in the initiation stage in a sand drift. With the gradual strengthening of the saltation, the airborne shear stress on the bed surface decreases and eventually becomes smaller than the aerodynamic entrainment threshold. In this work, we study the subjects in the case of steady-state sand flow, which means that aerodynamic entrainment can be neglected during the simulation. Only Equations (13) and (14) are utilized to give the initial velocity of the particle at the first step of the sand drift.

2.4. Realization Method of Gobi Bed

In this study, the numerical simulation of wind-blown sand flow involves two types of beds, one is a sand bed and the other is a gobi bed. The sand bed is composed of sand particles, while the gobi bed is composed of sand particles covered with a certain amount of gravel.

In the numerical model, we randomly generated a certain number of circular areas on the sand bed as simulated gravel cover (Figure 1a). Referring to the previous researchers' wind tunnel simulation experiments, the diameter of each small circular area is set to 2.0 cm [66–68]. For numerical simulation, we set up the Gobi with five levels of gravel cover: 0%, 16.2%, 24.7%, 43.6% and 59.4%, as shown in Figure 1b. The four subplots in Figure 1b represent both the four gravel-covered Gobi beds used in the simulation, where the blue circles represent gravels and the white areas represent granular beds.

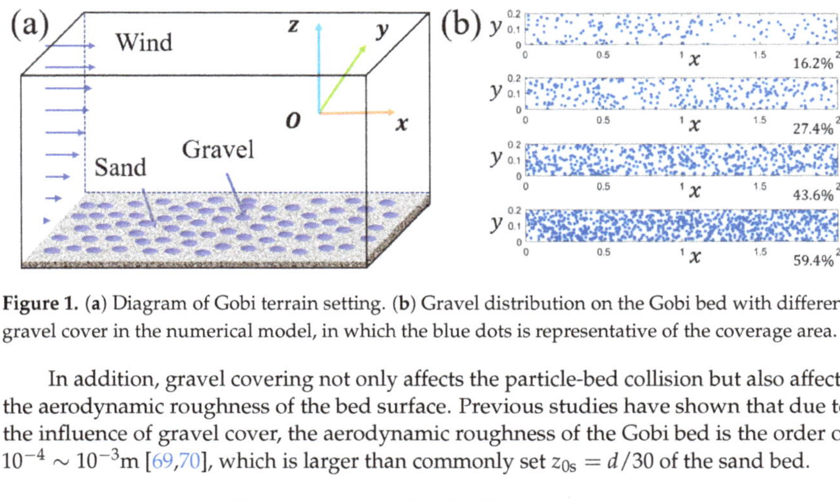

Figure 1. (**a**) Diagram of Gobi terrain setting. (**b**) Gravel distribution on the Gobi bed with different gravel cover in the numerical model, in which the blue dots is representative of the coverage area.

In addition, gravel covering not only affects the particle-bed collision but also affects the aerodynamic roughness of the bed surface. Previous studies have shown that due to the influence of gravel cover, the aerodynamic roughness of the Gobi bed is the order of $10^{-4} \sim 10^{-3}$m [69,70], which is larger than commonly set $z_{0s} = d/30$ of the sand bed.

2.5. Calculation Method of Impact Erosion by Sand Drift

Saltation sand erosion is a physical process in which the sand carrying fluid continuously strikes the solid surface, causing the solid surface materials to gradually wear away. For solid particles striking various targets, it is believed that the impact particle velocity, angle, particle size, particle type, and target strength all have an impact on the erosion rate. The parameters that have been investigated the most extensively are impact velocity and angle. Some academic researchers have also considered the impact of particle size; however, they have primarily focused on metallic targets [37,71]. For sand-impacted concrete materials, there are relatively few studies on particle size dependence. The erosion rate E_r in the model used in this study depends only on the impact velocity V_I and impact angle θ_I, and parameters such as particle size and density are considered to be constant and consistent with those in the experiment.

To study the spatial distribution characteristics of sand drift erosion, we set up a virtual wall that is 0.2 m broad and 1.0 m height in the wind-blown sand flow, as shown in Figure 2. The wall is divided into 10×100 cells, each cell size is 0.02 m $\times$ 0.01 m. The spatial distribution of E_a is obtained by calculating the accumulation of erosion in each cell during sand drift by the position coordinate y_p, z_p, impact speed V_I and impact angle θ_I of the saltating sand particles when it impacts the wall:

$$E_a(y,z)|_{y=y_p,z=z_p} = \frac{m_p E_r(V_I,\theta_I)}{\Delta y \Delta z} \tag{15}$$

where m_p is the mass of the saltating sand particles. Oka et al. [71] suggests that the variables V_I and θ_I of E_r could be separated, and it can be expressed as:

$$E_r(V_I,\theta_I) = g(\theta_I)E_{r,45}(V_I) \tag{16}$$

in which $E_{r,45}$ stand for the erosion rate at $\theta_I = 45°$.

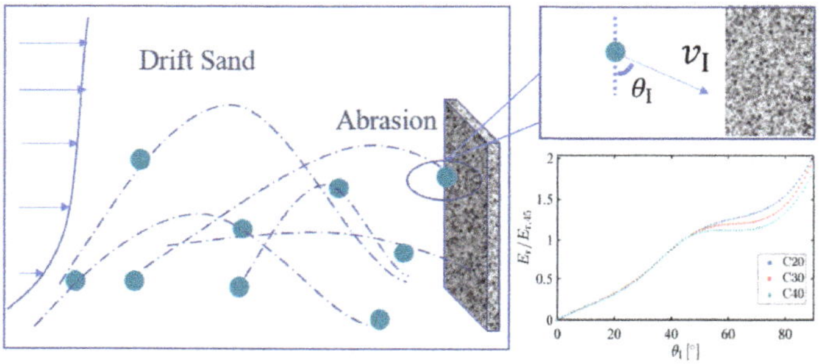

Figure 2. Impact erosion process of saltation particles to the vertical wall in a sand drift.

Table 2. Parameters of erosion model.

Material	C_{20}	C_{30}	C_{40}
K_e	0.0057	0.0037	0.0028
n_e	1.4046	1.4952	1.5555

Hao et al. [38] investigated the erosion rate of three different types of concrete materials, namely C20, C30, and C40, and propose a power empirical formula to parameterize the variation of erosion rate with impact velocity:

$$E_{r,45} = K_e V_I^{n_e} \tag{17}$$

where K_e and n_e are empirical parameters relating to concrete materials, and their values are listed in Table 2.

As was already mentioned, the velocity and angle of impactors both affect the erosion rate. Although Hao et al. [38] experimentally investigated the quantitative relationship of erosion rate with impact angle θ_I, it did not give an empirical model about $g(\theta_I)$ similar to Equation (17). Using the method of spline interpolation, we calculate the values of $g(\theta_I)$ corresponding to various angles θ_I based on the experimental results, as shown in Figure 2. As a result, the erosion rate at any impact angle can be determined.

2.6. Calculation Procedure

For a total of 10 s, two stages of numerical simulations of the sand drift were carried out. The first stage was calculated for 5 s, which is used to allow the sand flow to develop to a steady state [29,53]. The second stage begins with the saltation erosion computation, which takes also 5 s to complete. The calculation procedure is as follows:

1. Given the friction velocity u_*, the initial wind field can be calculated using Equation (10). Next, introduce 100 particles from the bed, their x, y coordinates are uniform random numbers, and initial velocity are determined by Equations (13) and (14).
2. Calculate $\vec{f}_\mathrm{d}$ and $\vec{f}_\mathrm{l}$ of each particles, and solve the motion Equations (1) and (2) and update each particle's position.
3. Search all particles and find out every collision pair. Renew the velocity of collision particles using Equation (5).
4. Find out particles below the surface and the coordinates of their impact location. The properties of bound particles and ejected particles can be inferred from the splash function in Table 1. If the rebound velocity of a particle after hitting the bed is not sufficient to move it to a height above one particle diameter, the particle is considered to have transformed into a static particle on the bed and is removed from the saltation system.
5. Identify the particles that come into contact with the wall and the corresponding collision locations. Calculate the impact erosion based on Equation (15).
6. Calculate $\vec{F}_\mathrm{p}$ and update the wind field with Equation (10).
7. Return to step (2) and start the calculation of the next step.

2.7. Model Verification

In this section, we compare the numerical simulation results of sediment transport with the experimental results to verify the reliability of the numerical model. There, the diameter and density of the sand particles were set to $d = 0.25\,\mathrm{mm}$ and $\rho_\mathrm{p} = 2650\,\mathrm{kg/m^3}$, respectively.

Figure 3 shows the change of sediment transport rate Q_s with the friction velocity u_* from simulation and experiment on the sand bed (the gravel coverage is 0%), in which the experimental results are presented by Creyssels et al. [72] and Tong and Huang [29]. It can be seen that the numerical simulation results of Q_s versus u_* are consistent with the experimental results.

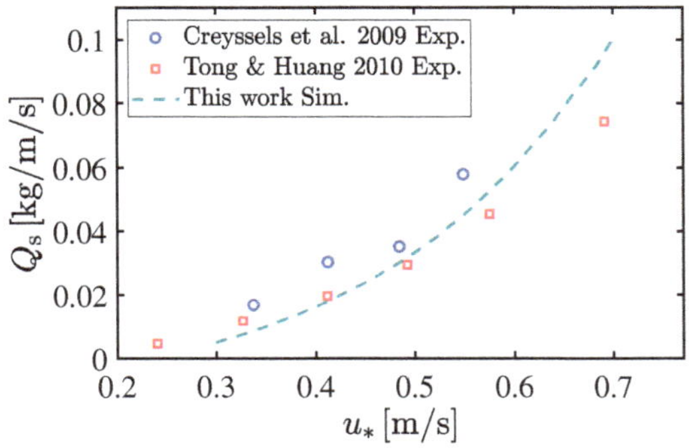

Figure 3. Comparison of the relationships between sand transport rate and the friction velocity calculated in this paper with that from experiments by Creyssels et al. [72] and Tong and Huang [29].

In addition, we contrasted the numerical results of mass flux profiles on the sand bed and on the Gobi (the gravel coverage is 60%) with the results of the experiments performed by Zhang et al. [73] and Tong and Huang [29] (Figure 4). The figure shows that, whether on the sand bed or the Gobi bed, the simulated data are found to be in good agreement with the experimental results.

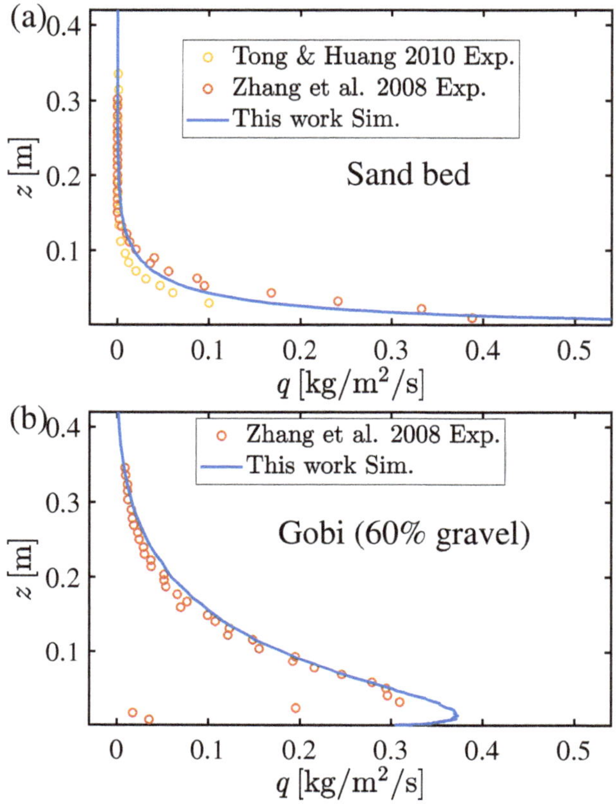

Figure 4. Comparison of the numerical results on mass flux profile with the measured data in (**a**) sand bed by Tong and Huang [29] and (**b**) Gobi with 60% gravel coverage by Zhang et al. [73].

The stratification pattern of mass flux profile, proposed by Zheng et al. [28], refers to the phenomenon that the mass flux can be divided into three layers, containing a growth layer, a saturation layer and a decay layer. which is very common in sand drift in Gobi. As shown in Figure 4b, the stratification pattern of the mass flux profile in sand drift is accurately reproduced by the simulation results on the Gobi. These numerical examples demonstrate that the numerical model used in the study can accurately simulate wind-blown sand movement on sand surfaces and the Gobi.

3. Results and Discussion

3.1. Spatial Distribution Image of Sand Drift Erosion

The erosion effect of the wind-blown sand flow causes destructive impacts on buildings next to the aeolian landform [49]. In order to reveal how gravel coverage and friction velocity affect the impact erosion of sand drift in the Gobi, and analyze the spatial distribution characteristics of saltation erosion. We calculated the sand drift erosion amount E_a[mg/m^3] with C_{20} concrete as target material under different gravel coverage and friction velocity, as shown in Figures 5 and 6.

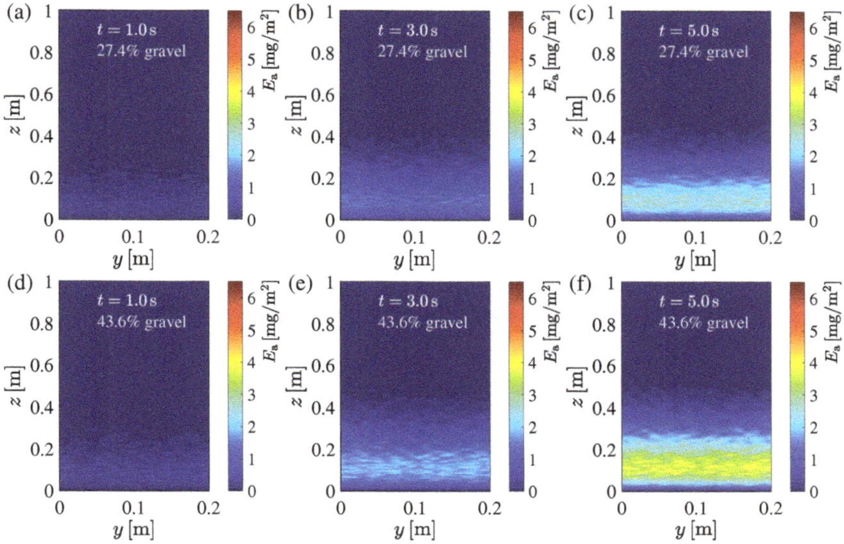

Figure 5. Spatial distribution images of impact erosion amount $E_a[\text{mg/m}^2]$ in sand drift at $u_* = 0.60\,\text{m/s}$, for (**a–c**) on Gobi with 27.4 gravel coverage and (**d–f**) on Gobi with 43.6% gravel coverage.

Sand drift creates impact erosion, which accumulates over time, causing the erosion amount E_a to rise with time, as seen in Figure 5. We can also deduce at least two other intriguing findings from Figure 5. First, the distribution of sand drift erosion along the vertical direction is not monotonic, with maximum values of impact erosion occurring at heights of about $0.05 \sim 0.15\,\text{m}$. Second, sand drift erosion is influenced by the gravel cover; On the Gobi with a 60% gravel covering, the erosion amount on the concrete wall is more than twice as great as it is on the Gobi with a 27.4% gravel coverage. This is due to the fact that an increment in gravel coverage from 16.2% to 43.6% causes a sediment flux to increase [74,75], more sand particles to impact the wall, and ultimately causes an increase of erosion amount.

In general, when the friction velocity increases, the saltating sand particle will gain more kinetic energy from the airflow. The erosion amount will be influenced by the increased wind speed since it will strengthen the wind-blown sand flow. Friction velocity u_* is used to characterize the wind speed in the investigation of aeolian sediment transport. Figure 6 shows that, with the same gravel coverage, impact erosion amount increases dramatically as friction velocity increases, with E_a being more than five times higher at $u_* = 0.65\,\text{m/s}$ compared to $u_* = 0.45\,\text{m/s}$. When compared to the same friction velocity, the Gobi covered with 60% gravel experience greater erosion amount than the Gobi covered with 16.2% gravel.

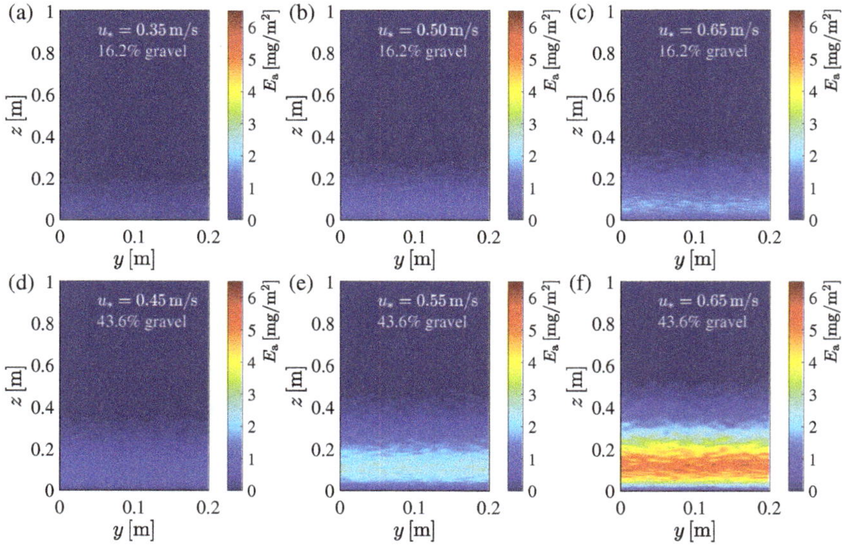

Figure 6. Spatial distribution images of impact erosion amount $E_a\,[\mathrm{mg/m^2}]$ in sand drift at $t = 5.0\,\mathrm{s}$, for (**a–c**) on Gobi with 16.2% gravel coverage and (**d–f**) on Gobi with 43.6% gravel coverage.

3.2. Vertical Structure of Saltation Erosion Rate

It is essential to study the vertical structure of saltation erosion for the design of preventative measures against infrastructure abrasion in the Gobi region. In this paper, we define the saltation erosion rate $E_s\,[\mathrm{mg/m^2/s}]$ to describe the erosion intensity of the sand drift, which characterizes the amount of sand drift erosion per unit time per unit area of the solid wall:

$$E_s = \frac{dE_a}{dt} \quad (18)$$

and E_s can be calculated by averaging E_a for 5 s.

Aeolian sediment movement differs from the pure sand-carrying jet impingement. The vertical structure of sediment movement is often complicated. Particularly in the Gobi, the presence of surface gravel further complicates the sand flow structure [67,76]. The vertical distribution profile of saltation erosion rate with C_{20} concrete as target material is shown in Figure 7. There is an obvious stratification pattern of saltation erosion rate profiles, and the profiles can be stratified into three layers along the vertical direction. The mass flux increases with height in the first layer, which is close to the surface. This is called a increasing layer. Then, as the height increases, the saltation erosion rate then reaches its maximum value. The high erosion rate will cause this interval to be the most severely eroded area of the vertical wall and the first to suffer wear damage. It makes sense to call this layer the damage layer as a result, and the height corresponding to the maximum saltation erosion rate is called as damage height z_d. The saltation erosion rate will rapidly drop as the height rises further, and this portion of the profile is referred to as the decreasing layer.

The saltation erosion rate is significantly influenced by the friction velocity. In the Gobi sand flow with same gravel coverage, the increment of friction velocity not only causes a saltation erosion rate increase, but also leads to the rise of damage height. Figure 7 clearly shows that, as friction velocity increases from 0.35 m/s to 0.60 m/s, the saltation erosion rate increases by more than 10 times, and the damage height increases by about 0.05 m. It can also be seen from Figure 7 that the value of the saltation erosion rate increases as the gravel coverage increases from 16.2% to 59.4%.

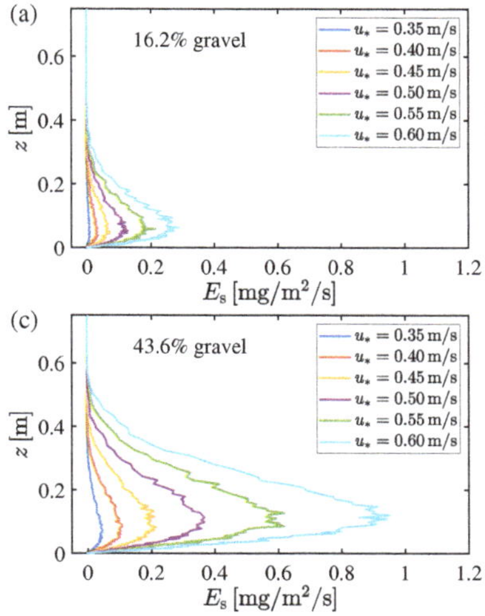

Figure 7. Vertical profiles of saltation erosion rate E_s, corresponding to different friction velocity, for (**a**) on Gobi with 16.2% gravel coverage, (**b**) on Gobi with 27.4% gravel coverage, (**c**) on Gobi with 43.6% gravel coverage, and (**d**) on Gobi with 59.4% gravel coverage.

The vertical wall's material strength has a significant impact on the profile of saltation erosion rate as well. Figure 8 depicts the results of saltation erosion that we performed on three types of concrete, C_{20}, C_{30} and C_{40}, in the 43.6% gravel cover Gobi conditions. As can be seen from the figure, at the same friction velocity, when concrete strength increases, the saltation erosion rate will decrease. It reveals that concrete materials with higher strengths also have greater impact on erosion resistance.

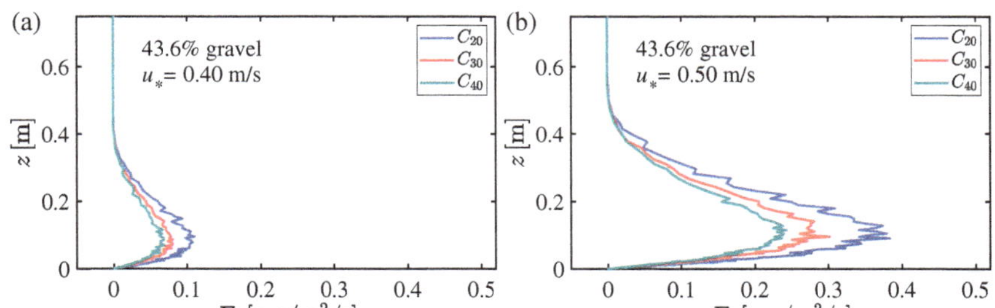

Figure 8. Vertical profiles of saltation erosion rate E_s, corresponding to three types of concrete walls C_{20}, C_{30} and C_{40} in Gobi with 43.6% gravel coverage, for (**a**) at $u_* = 0.4\,\mathrm{m/s}$ and (**b**) at $u_* = 0.5\,\mathrm{m/s}$.

As shown in Figure 8b, the extreme erosion rate for concrete C20 is around $3.8\,\mathrm{mg/m^2/s}$, while the higher strength concrete C40 has an extreme value of only nearly $2.3\,\mathrm{mg/m^2/s}$. The extreme value of the saltation erosion rate decreases obviously with the increase of the material strength. The impact of material strength on damage height, however, is not very noticeable. At a friction velocity of $u_* = 0.5\,\mathrm{m/s}$, the damage heights of the saltation erosion rate profiles of the three types of concrete all range from 0.5 m to 0.7 m.

3.3. Scaling Rate of Saltation Erosion

Effective prediction of protection height and protection intensity based on the environmental conditions facilitates the design of protective measures in the erosion protection of building structures. This necessitates a parameterization scheme that describes the extreme value of saltation erosion rate and damage height variations with environmental variables, such as wind speed and gravel coverage. Since C_{20} concrete has the weakest saltation erosion resistance, the parametric study was carried out under this most unfavourable condition.

The maximum value of saltation erosion rate E_m is correlated with both the friction velocity and gravel coverage. Figure 9 reveals the power relationship between E_m and u_* with different gravel coverage:

$$E_m = \lambda(u_* - u_{*t})^\gamma \tag{19}$$

and the scalar rate γ and scalar coefficient λ are provided in Table 1.

As seen in Figure 9b, the damage height z_d caused by saltation erosion is primarily concentrated between $0.03 \sim 0.15$ m, and it increases with increasing friction velocity. There is an approximate linear scalar relationship between damage height and friction velocity:

$$z_d = \alpha u_* + \beta \tag{20}$$

although the linear scale factor is small, as illustrated in Table 3. The value of parameter β also vary for various gravel coverage.

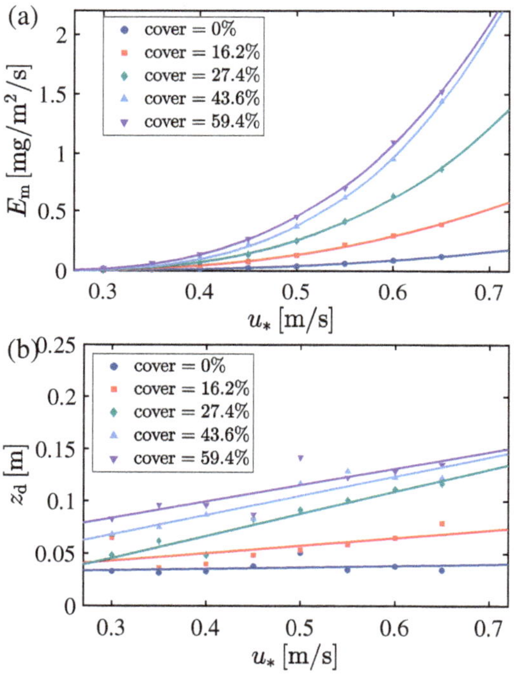

Figure 9. (**a**) The relationship between parameter E_m and friction velocity u_* on Gobi with different gravel coverage. (**b**) The relationship between parameter z_d and friction velocity u_* on Gobi with different gravel coverage. The solid line represents the fitting result.

Table 3. Parameters of fit formula.

Gravel Coverage	λ	γ	α	β
0%	0.9841	2.6207	0.0150	0.02945
16.2%	3.3752	2.6917	0.0725	0.02132
27.4%	10.214	3.0666	0.2120	0.00790
43.6%	19.984	3.3035	0.1848	0.01278
59.4%	16.653	2.9912	0.1580	0.03632

Sharp [47] recorded the maximum erosion height via field observation on a boulder alluvial plain in Kotchera Valley, California, USA. It ranged from $0.10 \sim 0.15$ m, which qualitatively reflected the vertical curve shape of impact erosion. Our research directly gives the stratification pattern of the saltation erosion rate profiles in Gobi, and the damage height obtained by numerical simulation is consistent with the peak wear height reported by Sharp [47]. The eroded flute of the Gobi region's ventifact also reflects this phenomenon. The maximum erosion on ventifacts in Gobi areas occurs at some finite height above the ground. Tan et al. [15] found a ventifact with an eroded flute in a field observation station in Gobi, and its maximum erosion occurred at about 0.10 m above the ground.

Equations (19) and (20) give the relationship between the extreme erosion rate and damage height with friction velocity for different gravel cover. It presents a theoretical tool for the erosion protection of building structures in aeolian landforms. Especially in the Gobi region, where sand drift is frequent and strong, the building structures nearby need to be protected from erosion at a height of 0.05 m to 0.15 m with some effective measures, such as adding anti-erosion coatings or additional high-strength materials. Meanwhile, in order to improve the protection efficiency, specific protection measures should be designed according to the local surface characteristics and dominant wind speed combined with the parameterized model provided in this study.

4. Conclusions

In this paper, we coupled the sediment transport model with the solid particle erosion model and conducted numerous numerical simulation investigations on the impact erosion process of aeolian saltation on the vertical wall of the Gobi surface. The results indicate that friction velocity and gravel coverage both affect saltation erosion. The erosion amount increases with friction velocity, and it also increases with increasing gravel coverage. A distinct stratification pattern can be seen in the vertical profile of saltation erosion rate, which includes increasing, damage and decreasing layers.

The damage height caused by saltation erosion is primarily concentrated in the height range of 0.03 m to 0.15 m, and it increases approximately linearly as friction velocity increases. The Gobi surface's gravel layer also significantly affects the damage height. Therefore, in order to obtain greater protection for these construction facilities against impact erosion, proper design should be performed in accordance with the local desert or Gobi gravel coverage characteristics and the primary distribution range of local wind speed.

Author Contributions: Conceptualization, Y.W. and N.H.; methodology, Y.W. and J.Z.; software, Y.W.; validation, Y.W. and H.D.; formal analysis, Y.W. and H.D.; investigation, Y.W.; resources, H.D.; data curation, Y.W.; writing—original draft preparation, Y.W.; writing—review and editing, J.Z., H.D. and N.H.; visualization, Y.W.; supervision, N.H.; project administration, J.Z.; funding acquisition, N.H. All authors have read and agreed to the published version of the manuscript.

Funding: This research was funded by the Young Scientists Fund of National Natural Science Foundation of China (42106218), Key Programme of National Natural Science Foundation of China (41931179), the Major Science and Technology Project of Gansu Province (21ZD4FA010), Innovation Star Project of Excellent Postgraduates of Gansu Province (2022CXZX-123).

Institutional Review Board Statement: Not applicable.

Informed Consent Statement: Not applicable.

Data Availability Statement: Not applicable.

Conflicts of Interest: The authors declare no conflict of interest.

References

1. Carr, M.H. The Geology of the Terrestrial Planets. *Rev. Geophys.* **1983**, *21*, 160–172. [CrossRef]
2. Titus, T.N.; Lancaster, N.; Hayward, R.; Fenton, L.; Bourke, M. Priorities for Future Research on Planetary Dunes Planetary Dunes: Workshop: A Record of Climate Change; Alamogordo, New Mexico, 28 April to 2 May 2008. *Eos Trans. Am. Geophys. Union* **2008**, *89*, 447–448. [CrossRef]
3. Lancaster, N.; Nickling, W.G.; Gillies, J.A. Sand Transport by Wind on Complex Surfaces: Field Studies in the McMurdo Dry Valleys, Antarctica. *J. Geophys. Res. Earth Surf.* **2010**, *115*, F03027. [CrossRef]
4. Bagnold, R.A. *The Physics of Blown Sand and Desert Dunes*; Springer: Dordrecht, The Netherlands, 1941.
5. Owen, P.R. Saltation of Uniform Grains in Air. *J. Fluid Mech.* **1964**, *20*, 225–242. [CrossRef]
6. Williams, G. Some aspects of the eolian saltation load. *Sedimentology* **1964**, *3*, 257–287. [CrossRef]
7. Barndorff-Nielsen, O.E.; Willetts, B.B. (Eds). *Aeolian Grain Transport 1; Volume 1, Acta Mechanica Supplementum*; Springer: Vienna, Austria, 1991. [CrossRef]
8. Pye, K.; Tsoar, H. *Aeolian Sand and Sand Dunes*; Springer: Berlin/Heidelberg, Germany, 2009. [CrossRef]
9. Kok, J.F.; Parteli, E.J.R.; Michaels, T.I.; Karam, D.B. The Physics of Wind-Blown Sand and Dust. *Rep. Prog. Phys.* **2012**, *75*, 106901. [CrossRef]
10. Valance, A.; Rasmussen, K.R.; Ould El Moctar, A.; Dupont, P. The Physics of Aeolian Sand Transport. *Comptes Rendus Phys.* **2015**, *16*, 105–117. [CrossRef]
11. Woodruff, N.P. Wind-Blown Soil Abrasive Injuries to Winter Wheat Plants. *Agron. J.* **1956**, *48*, 499–504. [CrossRef]
12. Armbrust, D.V. Recovery and Nutrient Content of Sandblasted Soybean Seedlings1. *Agron. J.* **1972**, *64*, 707–708. [CrossRef]
13. Heshmati, G.A.; Squires, V.R. Introduction to Deserts and Desertified Regions in China. In *Combating Desertification in Asia, Africa and the Middle East: Proven Practices*; Heshmati, G.A., Squires, V.R., Eds.; Springer: Dordrecht, The Netherlands, 2013; pp. 3–20. [CrossRef]
14. Qu, J.; Huang, N.; Dong, G.; Zhang, W. The Role and Significance of the Gobi Desert Pavement in Controlling Sand Movement on the Cliff Top near the Dunhuang Magao Grottoes. *J. Arid. Environ.* **2001**, *48*, 357–371. [CrossRef]
15. Tan, L.; Qu, J.; Wang, T.; Zhang, K.; An, Z. Field Observation Evidence for Kink Points in the Vertical Kinetic Energy Flux Profiles of Wind-Blown Sand Over Gobi and Its Significance. *Geophys. Res. Lett.* **2021**, *48*, e2020GL091224. [CrossRef]
16. Werner, B.T. A Steady-State Model of Wind-Blown Sand Transport. *J. Geol.* **1990**, *98*, 1–17. [CrossRef]
17. Wu, J.J.; Luo, S.H.; He, L.H. The Characteristic of Streamwise Mass Flux of Windblown Sand Movement. *Geomorphology* **2012**, *139–140*, 188–194. [CrossRef]
18. Bagnold, R.A.; Taylor, G.I. The Movement of Desert Sand. *Proc. R. Soc. London. Ser. A-Math. Phys. Sci.* **1936**, *157*, 594–620. [CrossRef]
19. White, B.R.; Schulz, J.C. Magnus Effect in Saltation. *J. Fluid Mech.* **1977**, *81*, 497–512. [CrossRef]
20. Rumpel, D.A. Successive Aeolian Saltation: Studies of Idealized Collisions. *Sedimentology* **1985**, *32*, 267–280. [CrossRef]
21. Shao, Y.; Li, A. Numerical Modelling of Saltation in the Atmospheric Surface Layer. *Bound.-Layer Meteorol.* **1999**, *91*, 199–225. .:1001816013475. [CrossRef]
22. Kok, J.F.; Renno, N.O. A Comprehensive Numerical Model of Steady State Saltation (COMSALT). *J. Geophys. Res.* **2009**, *114*, D17204. [CrossRef]
23. Ungar, J.E.; Haff, P.K. Steady State Saltation in Air. *Sedimentology* **1987**, *34*, 289–299. [CrossRef]
24. Anderson, R.S.; Sørensen, M.; Willetts, B.B. A Review of Recent Progress in Our Understanding of Aeolian Sediment Transport. In *Aeolian Grain Transport 1*; Barndorff-Nielsen, O.E., Willetts, B.B., Eds.; Springer: Vienna, Austria, 1991; Volume 1, pp. 1–19. [CrossRef]
25. Huang, N.; Ren, S.; Zheng, X. Effects of the Mid-Air Collision on Sand Saltation. *Sci. China Ser. Phys. Mech. Astron.* **2008**, *51*, 1416–1426. [CrossRef]
26. Huang, N.; Wang, C.; Pan, X. Simulation of Aeolian Sand Saltation with Rotational Motion. *J. Geophys. Res.* **2010**, *115*, D22211. [CrossRef]
27. Durán, O.; Andreotti, B.; Claudin, P. Numerical Simulation of Turbulent Sediment Transport, from Bed Load to Saltation. *Phys. Fluids* **2012**, *24*, 103306. [CrossRef]
28. Zheng, X.; He, L.; Wu, J. Vertical Profiles of Mass Flux for Windblown Sand Movement at Steady State: VERTICAL PROFILES OF MASS FLUX. *J. Geophys. Res. Solid Earth* **2004**, *109*. [CrossRef]
29. Tong, D.; Huang, N. Numerical Simulation of Saltating Particles in Atmospheric Boundary Layer over Flat Bed and Sand Ripples: SIMULATION OF PARTICLES OVER TWO BEDS. *J. Geophys. Res. Atmos.* **2012**, *117*, n/a. [CrossRef]
30. Dupont, S.; Bergametti, G.; Marticorena, B.; Simoëns, S. Modeling Saltation Intermittency. *J. Geophys. Res. Atmos.* **2013**, *118*, 7109–7128. [CrossRef]
31. Chen, Y.; Zhang, J.; Huang, N.; Xu, B. An Experimental Study on Splash Functions of Natural Sand-Bed Collision. *J. Geophys. Res. Atmos.* **2019**, *124*, 7226–7235. [CrossRef]

32. Zhang, C.; Huang, N.; Dun, H.C. Experimental Study on Sand/Bed Collision Over the Gobi Surface. *J. Geophys. Res. Atmos.* **2022**, *127*, e2021JD035766. [CrossRef]
33. Wang, Z.; Liu, L.; Li, X.; Zhao, L. An Experimental Method for Analyzing Environmental Effects of Blowing Sands on Glass Abrasion. *Procedia Environ. Sci.* **2010**, *2*, 207–217. [CrossRef]
34. Molinari, J.F.; Ortiz, M. A Study of Solid-Particle Erosion of Metallic Targets. *Int. J. Impact Eng.* **2002**, *27*, 347–358. [CrossRef]
35. López, D.; Congote, J.P.; Cano, J.R.; Toro, A.; Tschiptschin, A.P. Effect of Particle Velocity and Impact Angle on the Corrosion–Erosion of AISI 304 and AISI 420 Stainless Steels. *Wear* **2005**, *259*, 118–124. [CrossRef]
36. Bousser, E.; Martinu, L.; Klemberg-Sapieha, J.E. Solid Particle Erosion Mechanisms of Protective Coatings for Aerospace Applications. *Surf. Coatings Technol.* **2014**, *257*, 165–181. [CrossRef]
37. Tarodiya, R.; Levy, A. Surface Erosion Due to Particle-Surface Interactions—A Review. *Powder Technol.* **2021**, *387*, 527–559. [CrossRef]
38. Hao, Y.; Feng, Y.; Fan, J. Experimental Study into Erosion Damage Mechanism of Concrete Materials in a Wind-Blown Sand Environment. *Constr. Build. Mater.* **2016**, *111*, 662–670. [CrossRef]
39. Goretta, K.C.; Burdt, M.L.; Cuber, M.M.; Perry, L.A.; Singh, D.; Wagh, A.S.; Routbort, J.L.; Weber, W.J. Solid-Particle Erosion of Portland Cement and Concrete. *Wear* **1999**, *224*, 106–112. [CrossRef]
40. Vieira, R.E.; Mansouri, A.; McLaury, B.S.; Shirazi, S.A. Experimental and Computational Study of Erosion in Elbows Due to Sand Particles in Air Flow. *Powder Technol.* **2016**, *288*, 339–353. [CrossRef]
41. Mbabazi, J.G.; Sheer, T.J. Computational Prediction of Erosion of Air Heater Elements by Fly Ash Particles. *Wear* **2006**, *261*, 1322–1336. [CrossRef]
42. Zhang, Y.; Reuterfors, E.P.; McLaury, B.S.; Shirazi, S.A.; Rybicki, E.F. Comparison of Computed and Measured Particle Velocities and Erosion in Water and Air Flows. *Wear* **2007**, *263*, 330–338. [CrossRef]
43. Greeley, R.; Leach, R.N.; Williams, S.H.; White, B.R.; Pollack, J.B.; Krinsley, D.H.; Marshall, J.R. Rate of Wind Abrasion on Mars. *J. Geophys. Res. Solid Earth* **1982**, *87*, 10009–10024. [CrossRef]
44. Lancaster, N. Characteristics and Occurrence of Wind Erosion Features in the Namib Desert. *Earth Surf. Process. Landforms* **1984**, *9*, 469–478. [CrossRef]
45. Ta, W. Study of the Energy Abrasion Rates of Five Soil Types Subject to Oblique Impacts. *Geoderma* **2007**, *140*, 97–105. [CrossRef]
46. Raffaele, L.; Bruno, L. Windblown Sand Action on Civil Structures: Definition and Probabilistic Modelling. *Eng. Struct.* **2019**, *178*, 88–101. [CrossRef]
47. Sharp, R.P. Wind-Driven Sand in Coachella Valley, California. *GSA Bull.* **1964**, *75*, 785–804. .[785:WSICVC]2.0.CO;2. [CrossRef]
48. Liu, L.Y.; Gao, S.Y.; Shi, P.J.; Li, X.Y.; Dong, Z.B. Wind Tunnel Measurements of Adobe Abrasion by Blown Sand: Profile Characteristics in Relation to Wind Velocity and Sand Flux. *J. Arid. Environ.* **2003**, *53*, 351–363. [CrossRef]
49. Shi, X.; Shi, X. Numerical Prediction on Erosion Damage Caused by Wind-Blown Sand Movement. *Eur. J. Environ. Civ. Eng.* **2014**, *18*, 550–566. [CrossRef]
50. Huo, X.; Dun, H.; Huang, N.; Zhang, J. 3D Direct Numerical Simulation on the Emergence and Development of Aeolian Sand Ripples. *Front. Phys.* **2021**, *9*, 662389. [CrossRef]
51. Huo, X.; Huang, N.; Zhang, J. A Numerical Research on the Relationship between Aeolian Sand Ripples and the Sand Flux. *Processes* **2022**, *10*, 354. [CrossRef]
52. Shao, Y. *Physics and Modelling of Wind Erosion*; Springer: Dordrecht, The Netherlands, 2008.
53. Anderson, R.S.; Haff, P.K. Wind Modification and Bed Response during Saltation of Sand in Air. In *Aeolian Grain Transport 1*; Barndorff-Nielsen, O.E., Willetts, B.B., Eds.; Springer: Vienna, Austria, 1991; Volume 1, pp. 21–51. [CrossRef]
54. Zheng, L.; Zhang, X. Chapter 8—Numerical Methods. In *Modeling and Analysis of Modern Fluid Problems*; Zheng, L., Zhang, X., Eds.; Mathematics in Science and Engineering; Academic Press: Cambridge, MA, USA, 2017; pp. 361–455. [CrossRef]
55. Brilliantov, N.V.; Poschel, T. *Kinetic Theory of Granular Gases*, illustrated edition; Oxford University Press: Oxford, UK, 2010.
56. Schwager, T.; Becker, V.; Pöschel, T. Coefficient of Tangential Restitution for Viscoelastic Spheres. *Eur. Phys. J. E* **2008**, *27*, 107–114. [CrossRef]
57. Fohanno, S.; Oesterlé, B. Analysis of the Effect of Collisions on the Gravitational Motion of Large Particles in a Vertical Duct. *Int. J. Multiph. Flow* **2000**, *26*, 267–292. [CrossRef]
58. Versteeg, H.K.; Malalasekera, W. *An Introduction to Computational Fluid Dynamics: The Finite Volume Method*, 2nd ed.; Pearson Education Ltd: Harlow, UK; New York, NY, USA, 2007.
59. Mitha, S.; Tran, M.Q.; Werner, B.T.; Haff, P.K. The Grain-Bed Impact Process in Aeolian Saltation. *Acta Mech.* **1986**, *63*, 267–278. [CrossRef]
60. McEWAN, I.K.; Willetts, B.B.; Rice, M.A. The Grain/Bed Collision in Sand Transport by Wind. *Sedimentology* **1992**, *39*, 971–981. [CrossRef]
61. Oger, L.; Ammi, M.; Valance, A.; Beladjine, D. Study of the Collision of One Rapid Sphere on 3D Packings: Experimental and Numerical Results. *Comput. Math. Appl.* **2008**, *55*, 132–148. [CrossRef]
62. Ammi, M.; Oger, L.; Beladjine, D.; Valance, A. Three-Dimensional Analysis of the Collision Process of a Bead on a Granular Packing. *Phys. Rev. E* **2009**, *79*, 021305. [CrossRef]
63. Lämmel, M.; Dzikowski, K.; Kroy, K.; Oger, L.; Valance, A. Grain-Scale Modeling and Splash Parametrization for Aeolian Sand Transport. *Phys. Rev. E* **2017**, *95*, 022902. [CrossRef]

64. Xing, M.; He, C. 3D Ejection Behavior of Different Sized Particles in the Grain-Bed Collision Process. *Geomorphology* **2013**, *187*, 94–100. [CrossRef]
65. Kang, L.; Guo, L.; Liu, D. Reconstructing the Vertical Distribution of the Aeolian Saltation Mass Flux Based on the Probability Distribution of Lift-off Velocity. *Geomorphology* **2008**, *96*, 1–15. [CrossRef]
66. Dong, Z.; Wang, H.; Liu, X.; Wang, X. A Wind Tunnel Investigation of the Influences of Fetch Length on the Flux Profile of a Sand Cloud Blowing over a Gravel Surface. *Earth Surf. Process. Landforms* **2004**, *29*, 1613–1626. [CrossRef]
67. Zhang, W.; Tan, L.; Zhang, G.; Qiu, F.; Zhan, H. Aeolian Processes over Gravel Beds: Field Wind Tunnel Simulation and Its Application atop the Mogao Grottoes, China. *Aeolian Res.* **2014**, *15*, 335–344. [CrossRef]
68. Li, H.; Zou, X.; Zhang, C.; Kang, L.; Cheng, H.; Liu, B.; Liu, W.; Fang, Y.; Yang, D.; Wu, X. Effects of Gravel Cover on the Near-Surface Airflow Field and Soil Wind Erosion. *Soil Tillage Res.* **2021**, *214*, 105133. [CrossRef]
69. Dong, Z.; Liu, X.; Wang, X. Aerodynamic Roughness of Gravel Surfaces. *Geomorphology* **2002**, *43*, 17–31. [CrossRef]
70. Zhang, K.; Zhang, W.; Tan, L.; An, Z.; Zhang, H. Effects of Gravel Mulch on Aeolian Transport: A Field Wind Tunnel Simulation. *J. Arid. Land* **2015**, *7*, 296–303. [CrossRef]
71. Oka, Y.I.; Okamura, K.; Yoshida, T. Practical Estimation of Erosion Damage Caused by Solid Particle Impact: Part 1: Effects of Impact Parameters on a Predictive Equation. *Wear* **2005**, *259*, 95–101. [CrossRef]
72. Creyssels, M.; Dupont, P.; El Moctar, A.O.; Valance, A.; Cantat, I.; Jenkins, J.T.; Pasini, J.M.; Rasmussen, K.R. Saltating Particles in a Turbulent Boundary Layer: Experiment and Theory. *J. Fluid Mech.* **2009**, *625*, 47–74. [CrossRef]
73. Zhang, K.; Qu, J.; Zu, R.; Ta, W. Characteristics of Wind-Blown Sand on Gobi/Mobile Sand Surface. *Environ. Geol.* **2008**, *54*, 411–416. [CrossRef]
74. Zhang, W.; Wang, T.; Wang, W.; Liu, B. Wind Tunnel Experiments on Vertical Distribution of Wind-Blown Sand Flux and Change of the Quantity of Sand Erosion and Deposition above Gravel Beds under Different Sand Supplies. *Environ. Earth Sci.* **2011**, *64*, 1031–1038. [CrossRef]
75. Tan, L.; Zhang, W.; Qu, J.; Zhang, K.; An, Z.; Wang, X. Aeolian Sand Transport over Gobi with Different Gravel Coverages under Limited Sand Supply: A Mobile Wind Tunnel Investigation. *Aeolian Res.* **2013**, *11*, 67–74. [CrossRef]
76. Liu, J.; Kimura, R.; Wu, J. Vertical Profiles of Wind-Blown Sand Flux over Fine Gravel Surfaces and Their Implications for Field Observation in Arid Regions. *Atmosphere* **2020**, *11*, 1029. [CrossRef]

atmosphere

MDPI

Article

The Impacts of Wind Shear on Spatial Variation of the Meteorological Element Field in the Atmospheric Convective Boundary Layer Based on Large Eddy Simulation

Hailiang Zhang [1,2], Jinfang Yin [3], Qing He [2,*] and Minzhong Wang [2]

[1] College of Geography and Remote Sensing Sciences, Xinjiang University, Urumqi 830046, China
[2] Institute of Desert Meteorology, China Meteorological Administration, Urumqi 830002, China
[3] State Key Laboratory of Severe Weather, Chinese Academy of Meteorological Sciences, Beijing 100081, China
* Correspondence: qinghe@idm.cn

Abstract: As wind shear increases, the quasi-two-dimensional structure of flows becomes more significant in the convective boundary layer (CBL), indicating that wind shear plays an essential role in the variation of the field of atmospheric flow. Therefore, sensitive numerical experiments based on Large Eddy Simulation (LES) techniques were conducted to comprehensively investigate the effects of wind shear on the spatial variations in the velocity and potential temperature (θ) horizontal fields. Under the constant surface heat flux condition, the main findings are summarized. Firstly, in the CBL, the variances of the streamwise velocity (u), cross-stream velocity (v), and θ enhance as wind shear increases, whereas the variance of vertical velocity (w) is insensitive to wind shear. Secondly, in the CBL, with increasing wind shear, low-wavenumber Power Spectrum Densities (PSDs) of u, v, w, and θ increase significantly, suggesting that the increasing wind shear always enhances the large-scale motions of the atmosphere (i.e., low-wavenumber PSD). Therefore, it is more likely that some mesoscale weather processes will be triggered. Thirdly, generally, in the high-wavenumber range, with increasing wind shear, the PSDs of u, v, and θ increase slightly, whereas the PSD of w decreases slightly. This study provides a new perspective for understanding the role of wind shear in the spatial variations of the horizontal fields of meteorological elements under the same conditions of surface heat flux.

Keywords: Large Eddy Simulation; convective boundary layer; Power Spectrum Density; wind shear; spatial variation; horizontal fields of velocity; velocity variance

Citation: Zhang, H.; Yin, J.; He, Q.; Wang, M. The Impacts of Wind Shear on Spatial Variation of the Meteorological Element Field in the Atmospheric Convective Boundary Layer Based on Large Eddy Simulation. *Atmosphere* **2022**, *13*, 1567. https://doi.org/10.3390/atmos13101567

Academic Editors: Yubin Li and Jie Tang

Received: 31 August 2022
Accepted: 22 September 2022
Published: 25 September 2022

1. Introduction

Under moderate surface heat fluxes and weak to negligible wind shear, convection can form into a quasi-two-dimensional structure known as an open cell. It is similar to the coherent turbulence structure of Rayleigh–Bénard convection in the laboratory, which consists of narrow spokes of strong updrafts and broad zones of compensating downdrafts [1,2]. Nevertheless, when mean wind shear becomes increasingly strong, convective updrafts tend to organize into horizontal rolls aligned within 10–20° of the geostrophic wind direction in the northern hemisphere [3,4]. All the above suggests that wind shear plays an essential role in the spatial variation of the convective boundary layer (CBL) flow field.

Understanding wind shear's effects on the spatial variation of the flow field of the CBL has important implications for microscale and mesoscale meteorology. This is because wind shear produces perturbations in velocity, temperature, and the water vapor mixing ratio, which can influence the initiation of deep and moist convection [5–7]. In addition, it is known that vertical wind shear contributes to storms [8] and tornado formation [9]. Using a 10-year mesoscale convective system (MCS) dataset, Baidu et al. [10] showed that a strong vertical shear is associated with long-lived, moderate speed, moderate size, and cold (deep) storms with high rain rates over West and Central Africa. Furthermore, based on four years of daily Special Sensor Microwave Imager (SSMI) and Tropical Rainfall Measuring Mission (TRMM) Microwave Imager (TMI) satellite passive microwave retrievals, Back et al. [11] demonstrated that at a high column relative humidity, faster winds are associated with substantially more precipitation throughout the Pacific Intertropical Convergence Zone (ITCZ). All the above may be due to the enhancement of the quasi-two-dimensional structure of the flow field induced by the strong wind shear.

Generally, turbulence within the CBL is generated and maintained by two forces, i.e., buoyancy (equivalent to heat flux) and wind shear [12]. Therefore, many studies have primarily focused on the relationships between the coherent structure and atmospheric instability $-z_i/L$, a synthetic index of buoyancy and wind shear [13–15]. Here, z_i is the CBL's depth, and L is the Monin–Obukhov length. However, as illustrated by Moeng et al. [12], due to the different mechanisms of buoyancy and wind shear forces, the flow structure patterns and turbulence statistics of the flow field are quite different. In fluid mechanics research, there are numerous outstanding works investigating the effects of wind shear on the structure and statistics of turbulent flows. For example, through direct numerical simulations of homogeneous turbulent flows, Lee et al. [16] show that high shear rates alone are sufficient to produce streak-like structures and that the presence of solid boundaries is unnecessary. However, in the CBL, the situation is more complex, with shear turbulence always accompanied by buoyant turbulence. To the authors' knowledge, in meteorological research, relatively few studies have been carried out on the effects of wind shear on atmospheric motion at different scales in the CBL. Therefore, exploring the wind shear effect under identical heat flux conditions is valuable with respect to more comprehensively understanding the spatial variation of the horizontal field of meteorological elements in the CBL and discussing its impact on mesoscale weather processes.

Therefore, this study focuses on the effect of wind shear on the spatial variation of meteorological elements' (i.e., velocity and potential temperature) fields in the CBL by using the Large Eddy Simulation (LES) model, Kernel Smoothing Function Estimation (KSFE) method, and two-dimensional (2-d) Power Spectral Density (PSD) techniques [15]. This study introduces a new perspective to more comprehensively understand the role of wind shear in the spatial variations of the atmospheric flow field.

2. Data and Methods

In recent years, the Weather Research and Forecasting (WRF) model framework embedded with an LES model (WRF-LES) has been used to analyze the behavior of the PBL under different stability conditions and to implement and develop new turbulence parametrizations [17,18]. Given LES's prominent and important role in studying boundary layer dynamics, this study uses the WRF-LES version 4.0 to reproduce the turbulent flow data.

Three benchmark simulations are run on a domain of 12.8 × 12.8 × 3 km, with a 512 × 512 × 120 grid (horizontal grid spacing $\Delta x = \Delta y = 25$ m; vertical grid spacing $\Delta z = 25$ m), and a timestep of $\Delta t = 0.15$ s. At a speed of 20 m s^{-1}, we can calculate that the Courant–Friedrichs–Lewy (CFL) number is equal to 0.12, which satisfies the requirement for convergence of the numerical calculation. The initial Convective Boundary Layer (CBL) depth z_i is set at about 1000 m. To evaluate the impact of wind shear on the CBL flow's self-organization, the LES was forced by different zonal pressure gradients expressed in terms of the zonal geostrophic wind speed U_g and a constant surface heat flux (Q_0). The increase in U_g represents the increase in wind shear. The Coriolis parameter was set to $f = 1.0 \times 10^{-4}$ s^{-1}, corresponding to latitude $\phi = 43.3°$N, and the roughness length was set to $z_0 = 0.1$ m. The 1.5-order turbulent kinetic energy (TKE) model was chosen as the WRF-LES sub-grid closure scheme [19,20]. Periodic boundary conditions are used in the x- and y- directions. The rigid-lid upper boundary condition is used, and Rayleigh damping is applied for the top 1000 m of the simulation domain.

According to Moeng et al. [12], the dynamic flow field typically takes about 6 large-eddy turnover times (represented by $t_0 = 6 \times \tau_*$) to reach a statistically quasi-steady state. Here, τ_* is the large-eddy turnover time estimated by z_i/w_*, approximately 500 s in the WRF-LES simulations, which agrees with Shin et al. [21]. Simulations are performed for 3 h of physical time, much greater than the minimum duration of 0.83 h, to obtain the fully developed turbulence data. Since the horizontal grid number (512 × 512) is sufficient for statistics, the 3-d simulation data at 3 h of physical time are used in the study for analysis. Following the derivation employed by Troen et al. [22], the boundary layer height is expressed as Equation (1)

$$z_i = Rib_{cr}\frac{\theta_{va} \times |U(z_i)|^2}{g \times [\theta_v(z_i) - \theta_s]} \tag{1}$$

where Rib_{cr} is the critical bulk Richardson number, $U(z_i)$ is the horizontal wind speed at z_i, θ_{va} is the virtual potential temperature at the lowest model level, $\theta_v(z_i)$ is the virtual potential temperature at z_i, and θ_s is the surface potential temperature. It is generally accepted that when the Richardson number is less than 0.25, turbulence is usually considered fully developed; therefore, Rib_{cr} = 0.25 is used to calculate z_i [23].

A summary of the characteristics of the simulations, including the forcings (U_g and Q_0), characteristic length (L and z_i), and velocity scales (friction velocity—u_* and convective velocity—w_*) can be found in Table 1.

Table 1. Properties of numerical simulations, including geostrophic velocity (U_g), surface heat flux (Q_0), CBL depth (z_i), Monin–Obukhov length (L), friction velocity (u_*), and convective velocity scale (w_*).

Runs	Q_0	U_g	z_i	L	w_*	z_i/L	u_*	u_*/w_*
-	Km s^{-1}	m s^{-1}	m	m	m s^{-1}	-	m s^{-1}	-
Shear-free	0.24	00	1130	−3.6	2.04	−1047	0.21	0.102
Shear-10	0.24	10	1138	−47.2	2.05	−33	0.51	0.251
Shear-20	0.24	20	1213	−181.6	2.09	−8	0.82	0.390

The atmospheric convective boundary layer height is typically about 1 to 1.5 km, so we use the same initial potential temperature profile as Shin et al. [20], as shown in Equation (2).

$$\theta = \begin{cases} 300m \ : 0 < z \le 925m \\ 300K + (z - 925m) \times 0.0536K{\cdot}m^{-1} \ : 925 < z \le 1075m \\ 308.5K + (z - 1075m) \times 0.003K{\cdot}m^{-1} \ : z > 1075m \end{cases} \tag{2}$$

Since the horizontal gradient can effectively represent the spatial variation of physical quantities, to perceive the spatial variation of the horizontal fields of the streamwise wind component (u), cross-stream wind component (v), vertical velocity (w), and potential temperature (θ) with the mean wind shear, the horizontal spatial gradients of u, v, w, and θ, i.e., ∇u, ∇v, ∇w, and $\nabla\theta$, are used to directly and approximately account for their spatial variation.

The KSFE (based on a normal kernel function) is adopted to calculate the Probability Density Function (PDF) of the meteorological elements with the default bandwidth of the smoothing window (which is optimal for normal densities). The kernel distribution is suitable when a parametric distribution cannot properly describe the data or when the user wants to avoid making assumptions about the data distribution and create a smooth, continuous probability density function for the data set.

This study emphasizes the spatial PSD characteristics of u, v, w, and θ horizontal fields calculated by the 2-dimension Fourier Transform. In the CBL, velocity and scalar fields can approach isotropy in the horizontal plane, and the power spectral density depends only on horizontal wavenumber's magnitude [24]. Thus, Peltier et al. [25] defined and introduced a 1-dimension PSD $F_{cc}(k_h)$ by integrating 2-dimension PSD $\phi_{cc}(k_1, k_2)$ over circular rings of wavenumber radius k_h (Equation (3))

$$F_{cc}(k_h) = \int_0^{2\pi} \phi_{cc}(k_1, k_2) \times k_h d\theta \tag{3}$$

where $k_h = \sqrt{k_1^2 + k_2^2}$; k_1 and k_2 are x-direction and y-direction wavenumbers, respectively. We adopt the more precise Peltier et al. [25] method to calculate the PSD of the instantaneous velocity and horizontal temperature fields.

3. Results

3.1. Effects of Wind Shear on the Horizontal Fields of Meteorological Elements

To better understand the evolution of the coherent structure of velocity and scalar fields with wind shear, we examine the horizontal structures of the flow velocity and potential temperature. Figure 1 presents the instantaneous snapshots of the flow velocity and potential temperature fields on the horizontal plane at z/z_i = 0.5 and 3 h of physical time.

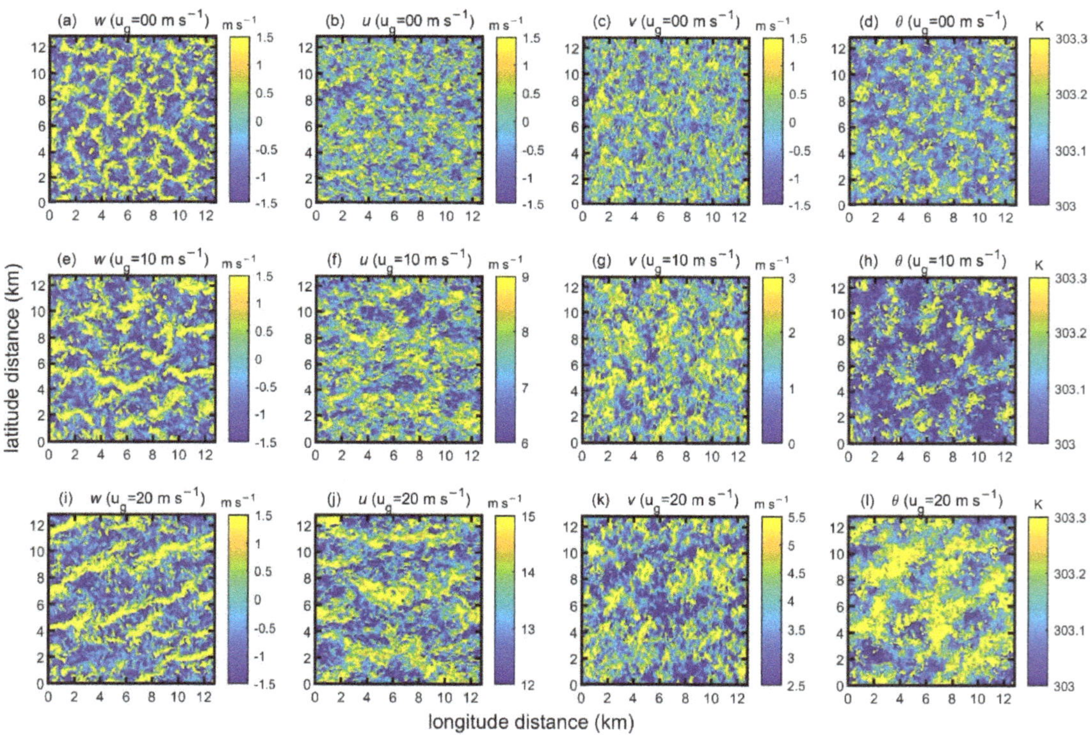

Figure 1. Instantaneous snapshots of velocity component w (**first column**), velocity component u (**second column**), velocity component v (**third column**), and potential temperature θ (**last column**) from LES simulations at z/z_i = 0.5 and 3 h of physical time across the range of geostrophic wind speed. First row U_g = 0 ms^{-1}; second row U_g = 10 ms^{-1}; third row U_g = 20 ms^{-1}. Note that the color scale on each plot is different for contrast and clarity.

As shown in the first column of the plots in Figure 1, w's horizontal field transforms from cellular structures to roll-type organizations as U_g increases from 0 to 20 ms^{-1}, which is consistent with Young et al.'s findings [4]. The horizontal structures of w consist of rectangular cells with narrow updrafts along the edges and broad downdrafts in the center when U_g = 0 ms^{-1} (Figure 1a); as U_g = 20 ms^{-1} (Figure 1e), they are characterized by long and linear updrafts bent slightly from the x-direction; as U_g = 10 ms^{-1} (Figure 1i), they are in an intermediate transition state. The transition of w's horizontal structures corresponds to Salesky et al.'s [26] and Zhou et al.'s [27] findings.

Furthermore, from the remaining columns of the plots in Figure 1, we also find that as U_g increases, the organized large-scale structures of u, v, w, and θ grow significantly, implying that wind shear might play a key role in the organization of the flow velocity and other scalar horizontal fields, and enhanced wind shear would, in turn, produce more organized and larger structures in the CBL. Our next step is to prove and explain this effect based on statistics.

3.2. *Effects of Wind Shear on the Vertical Mean Profiles of the Meteorological Elements*

The vertical profiles of the horizontal averaged u, v, w, and θ are initially investigated at 3 h of physical time. As displayed in Figure 2a–c, u, v, and θ increase monotonically with an increasing U_g, whereas the differences in θ's vertical profiles are not significant with an increasing U_g throughout the CBL's depth, which is consistent with the studies of Park et al. [14]. The magnitude of u and v increase near the ground and reach a quasi-constant value in the middle of the CBL; then, u begins to rise rapidly to the U_g, and v declines rapidly to 0 near the inversion (around z/z_i = 1.0), which is consistent with the studies of Salesky et al. [26].

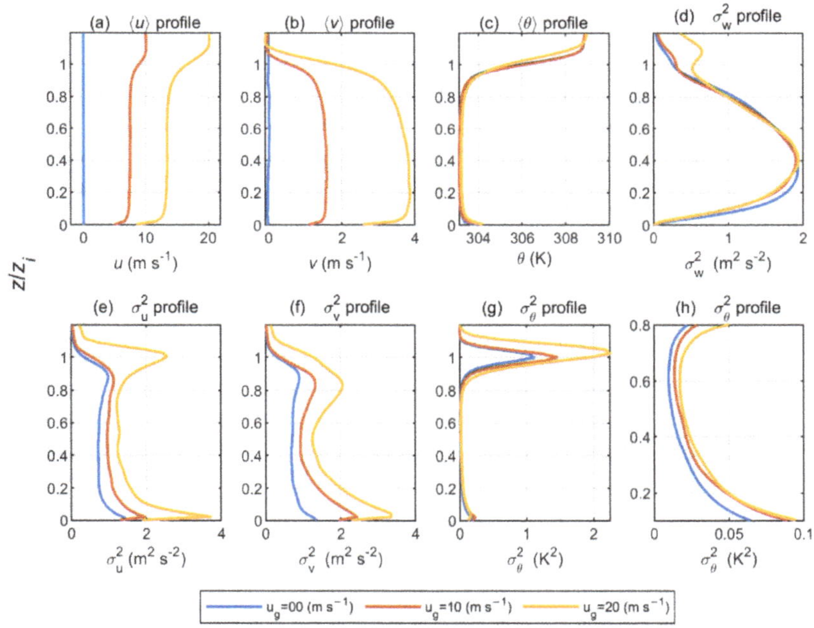

Figure 2. The vertical profiles of horizontal averaged u (**a**), v (**b**), and θ (**c**), and variances of w (**d**), u (**e**), v (**f**), and θ (**g**,**h**) from LES simulations at z/z_i = 0.5 and 3 h of physical time. The blue, red, and yellow lines represent the geostrophic wind forcing of U_g = 0, 10, and 20 ms^{-1}, respectively.

The w variance (σ_w^2, Figure 2d), u variance (σ_u^2, Figure 2e), v variance (σ_v^2, Figure 2f), and θ variance (σ_θ^2, Figure 2g) are examined throughout the height of the CBL. For a fixed value of z/z_i and a constant Q_0, σ_w^2 differs insignificantly, whereas σ_u^2 and σ_v^2 increase monotonically with an increasing U_g, suggesting that the TKE vertical component, σ_w^2, remains nearly identical, and the TKE horizontal components, σ_u^2 and σ_v^2, become larger with an increasing U_g, which is consistent with the results of Salesky et al. [26].

As a forcing factor, Q_0 remains constant, and so σ_w^2 remains stable regardless of wind shear; otherwise, an increase in σ_w^2, i.e., an increase in w fluctuations, will lead to an increase in Q_0, which contradicts the condition that Q_0 remains constant. In addition, the profiles of σ_w^2 present a mono-peak structure with a peak near the middle of the CBL (z/z_i = 0.4). Meanwhile, the profiles of σ_u^2 present a dual-peak structure with a peak near the surface and the other near the top of the CBL, similar to the theoretical variance vertical profiles of Rayleigh–Bénard convection [27,28]. However, the θ variance, σ_θ^2, increases in the entrainment layer (around z/z_i = 1.0) significantly (Figure 2g), whereas it increases in the middle CBL (around z/z_i = 0.5) only slightly with an increasing wind shear (Figure 2h). Nonetheless, the σ_θ^2 vertical profiles in Figure 2g are generally consistent with those reported by Sorbjan et al. [29].

3.3. *Effects of Wind Shear on the Spatial Gradients of Horizontal Fields of the Meteorological Elements*

In Figure 3, the distributions of u, v, w, and θ's horizontal gradient magnitudes are similar to the structures of u, v, w, and θ's horizontal fields at z/z_i = 0.5 (Figure 1). Moreover, u, v, and θ's horizontal gradient magnitude distributions (Figure 3) demonstrate that the horizontal gradient magnitudes increase with an increasing wind shear at z/z_i = 0.5, which means that the spatial variation of u, v, and θ strengthens as the wind shear increases in the middle of the CBL. On the other hand, the variational tendency of the horizontal gradient magnitude of w with wind shear is not obvious at first glance and requires further investigation.

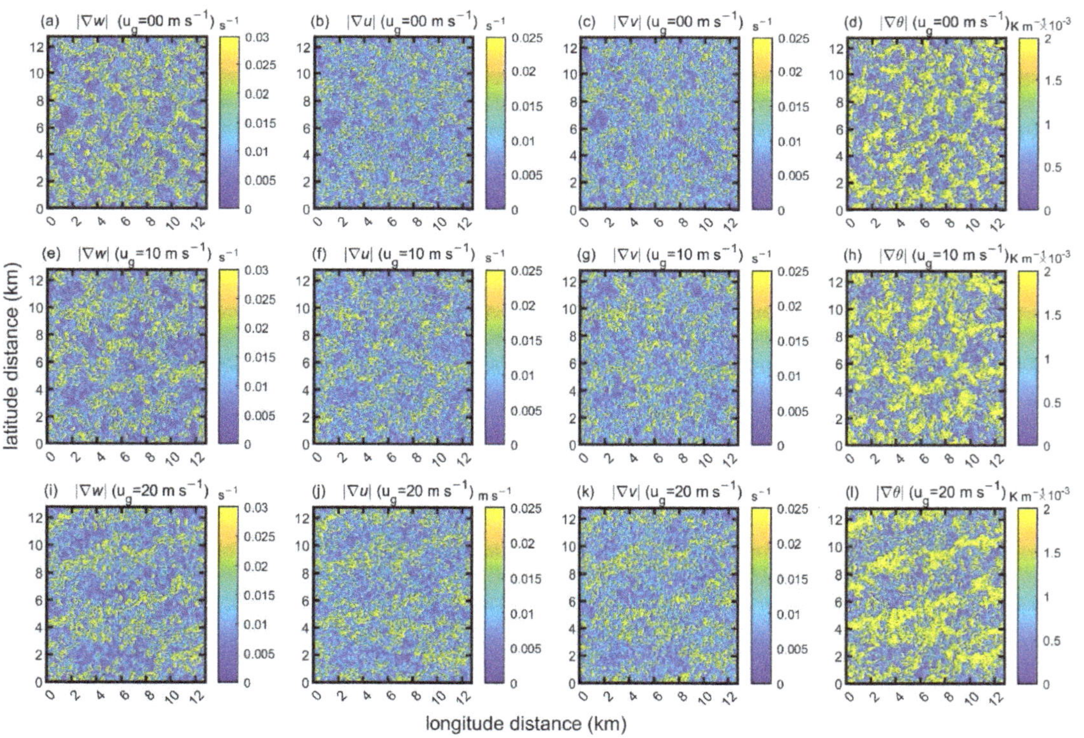

Figure 3. Instantaneous snapshots of w's gradient magnitude ($|\nabla w|$, **the first column**), u's gradient magnitude ($|\nabla u|$, **second column**), v's gradient magnitude ($|\nabla v|$, **the third column**), and θ's gradient magnitude ($|\nabla\theta|$, the last column) from LES simulations at z/z_i = 0.5 at 3 h of physical time across the range of U_g. First row U_g = 0 ms^{-1}; second row U_g = 10 ms^{-1}; third row U_g = 20 ms^{-1}. Note that the color scales of $|\nabla w|$, $|\nabla u|$, $|\nabla v|$, and $|\nabla\theta|$ are adjusted separately for clarity and contrast.

To represent the spatial variation characteristics of the meteorological elements with wind shear more precisely and quantitatively, we applied the KSFE to estimate the PDFs of the horizontal gradient vector magnitudes of u, v, w, and θ. However, it is reasonable and feasible to use the union of the real and imaginary parts of gradient vectors to estimate the PDFs (Figure 4) for u, v, and θ's gradients. At $z/z_i = 0.5$, the PDFs of the gradient vector magnitudes decrease with an increasing wind shear near the point of gradient value = 0 in the x-coordinate and increase relatively far away from the point as wind shear increases (Figure 4b–d). Since the sum of the samples is identical for all cases, the small gradient sample number decreases, and the large gradient sample number increases with increasing wind shear. In other words, the spatial variations of u, v, and θ become more pronounced as wind shear increases in the middle of the CBL, which is consistent with the result that the variances of u, v, and θ increase with an increasing wind shear at $z/z_i = 0.5$ (Figure 2e–h).

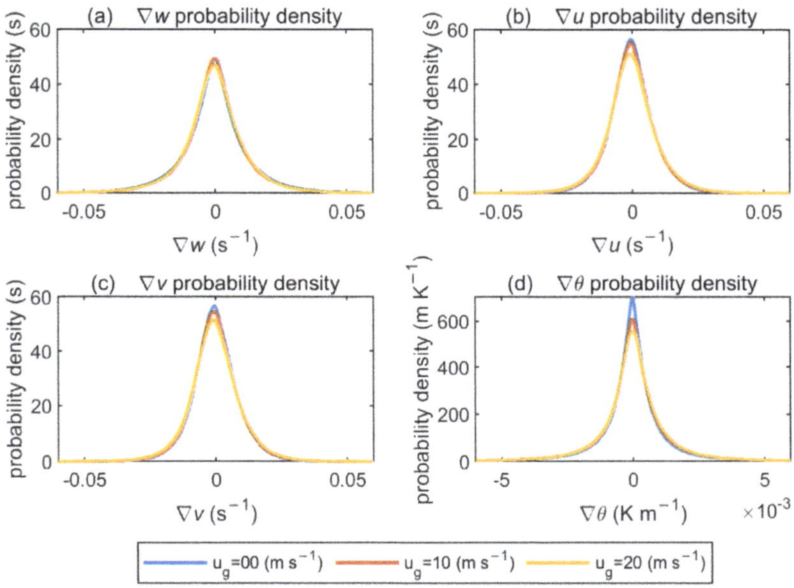

Figure 4. Comparison of w (**a**), u (**b**), v (**c**), and θ (**d**) spatial horizontal gradient PDFs from LES simulations for $z/z_i = 0.5$ at 3 h of physical time across the range of U_g. Blue line $U_g = 0$ ms^{-1}; Red line $U_g = 10$ ms^{-1}; Yellow line $U_g = 20$ ms^{-1}. The x-coordinate in the plots represent w (**a**), u (**b**), v (**c**), and θ (**d**) gradient vector values. The y-coordinate represents the probability density.

For w, at $z/z_i = 0.5$. However, with an increasing wind shear, there is no obvious variational tendency in the PDFs of the horizontal gradient magnitudes. Furthermore, the PDFs with $U_g = 0$ ms^{-1} and $U_g = 10$ ms^{-1} are almost identical, whereas the PDF with $U_g = 20$ ms^{-1} is minimum near the point of gradient value = 0 in the x-coordinate (Figure 4a), which is consistent with the result that the variances of w almost remain identical with $U_g = 10$ ms^{-1} and $U_g = 20$ ms^{-1} at $z/z_i = 0.5$ (Figure 2d).

The distributions of the horizontal gradients of u, v, w, and θ near the surface ($z/z_i = 0.1$) are also investigated. As shown in Figures 5 and 6, with an increasing wind shear, the evolutions of w, u, v, and θ's horizontal gradient magnitudes at $z/z_i = 0.1$ are similar to those at $z/z_i = 0.5$, i.e., with increasing wind shear, at $z/z_i = 0.1$, the horizontal gradient magnitudes of u, v, and θ increase, whereas the horizontal gradient magnitudes of w still present an insignificant variation trend (Figure 5). This is due to the results that, with increasing wind shear, w's variances differ a little and almost remain identical throughout the CBL's depth (Figure 2d). Hence, at $z/z_i = 0.1$, the PDFs of horizontal gradient magnitudes of w also present an insignificant variation trend with wind shear, such as reaching a maximum with U_g = 10 ms^{-1} and a minimum with U_g = 0 ms^{-1} near the point of gradient value = 0 in the x-coordinate (Figure 6a).

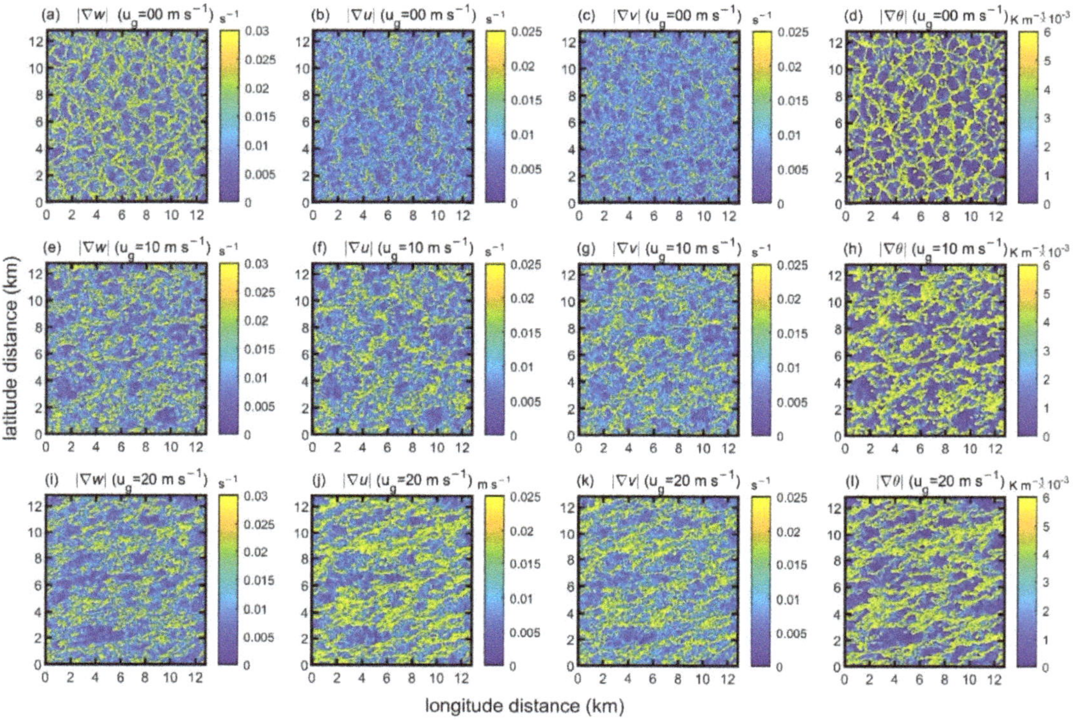

Figure 5. As in Figure 3, but at z/z_i = 0.1.

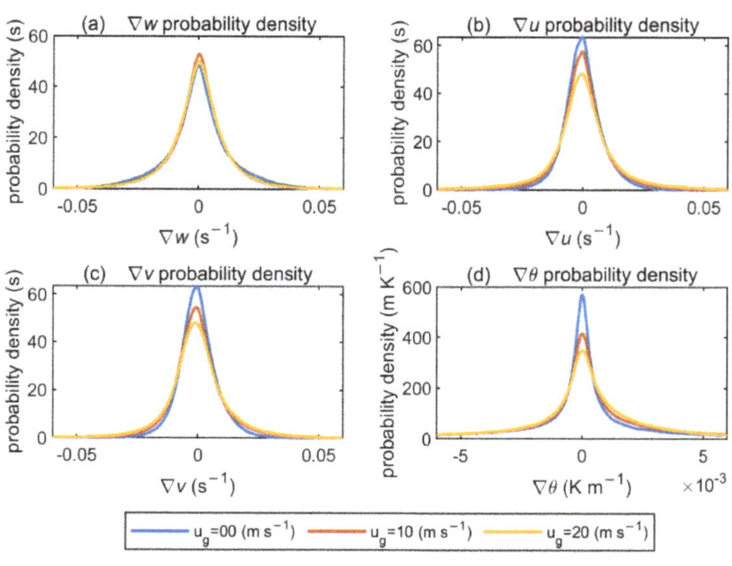

Figure 6. As in Figure 4, but at z/z_i = 0.1.

However, based on the gradient and variance analysis, it can be inferred that, under the condition of identical surface heat flux, the spatial variations of u, v, and θ's horizontal fields are enhanced as the wind shear increases. Nevertheless, the spatial variations of w's horizontal fields are insensitive to wind shear throughout the CBL.

3.4. Effects of Wind Shear on 2-d PSDs of Meteorological Element Fields (2 Dimensions)

The study of the horizontal gradient of meteorological elements indicates that the spatial variations of u, v, and θ's horizontal fields are significantly influenced by wind shear, whereas the spatial variations of w's horizontal fields are insensitive to wind shear. Nonetheless, the horizontal gradients cannot precisely and comprehensively describe the details of the impact. Better mathematical tools are required to achieve it. As the 2-d PSD can provide more reliable scale information than the 1-d PSD, Peltier et al. [25] applied the 2-d PSD to study the unstable layer. Gibbs et al. [30] also applied the technique to compare the velocity data produced by two models. Therefore, the 2-d PSD is adopted to reveal the spatial variation characteristics of u, v, w, and θ fields with wind shear. However, u is taken as an example to demonstrate the calculating process of 2-d PSD (Equations (4)–(5). The u_m^n represents u at (m, n); m and n are the x-direction and y-direction grid indexes, respectively; M and N are the x-direction and y-direction grid numbers, respectively; U_p^q is the 2-d Fourier Coefficient at (p, q); $\phi_{uu}(k_1, k_2)$ is the PSD at (k_1, k_2); k_1 and k_2 are the x-direction and y-direction wavenumbers, respectively; and K_1 and K_1 are the x-direction and y-direction sampling wavenumbers, respectively. In this study, $K_1 = K_2 = 1/25$. Before calculating the PSD, the mean values of u, v, w, and θ fields at each vertical level are removed, respectively.

$$U_p^q = \sum_{m=0}^{M} \sum_{n=0}^{N} u_m^n e^{-2\pi i\left(\frac{pm}{M} + \frac{qn}{N}\right)} p \in [0, M-1], q \in [0, N-1] \quad (4)$$

$$\phi_{uu}(k_1, k_2) = \phi_{uu}\left(\frac{p \times K_1}{M}, \frac{q \times K_2}{N}\right) = \frac{\left|U_p^q\right|^2}{M \times N \times K_1 \times K_2} \quad (5)$$

As the 2-d PSD represents the harmonic power, Figures 7 and 8 show the power distributions of the harmonic constituents with different wavenumbers. The harmonic powers peak near the center of the 2-d wavenumber plane (around k_h = 0) and decrease rapidly toward the edge of the 2-d wavenumber plane (toward high $|k_1|$ and $|k_2|$), which indicates that a low wavenumber harmonic carries more power than high wavenumber harmonics, and harmonic power decreases with an increasing wavenumber with respect to only one harmonic being inspected.

Notably, u 2-d PSDs are elongated in the k_2-direction (i.e., y-direction; Figures 7b and 8b), and the v 2-d PSDs are elongated in the k_1-direction (i.e., x-direction; Figures 7c and 8c). This is consistent with the turbulence theory, which suggests that the ratio of longitudinal to transversal spectra in the inertial subrange should be larger than one. For isotropic turbulence, this ratio is equal to 4/3 [31]. According to the theory, as the x and y directions are transverse directions for w, the ratios of the isotropic inertial-subrange 2-d PSDs of w along the wavenumber coordinates (k_1, k_2) should approximately equal unity. As illustrated in Figures 7a and 8a, the 2-d PSD of the w field presents a geometric distribution of center symmetry consistent with the simulation results of Gibbs et al. [30]. Since θ has a strong correlation with w, the 2-d PSDs of θ present identical geometric distributions (Figures 7d and 8d).

As displayed in Figures 7 and 8, the 2-d PSDs of u, v, w, and θ increase monotonically with increasing wind shear in the low wavenumber region (the center of the 2-d wavenumber plane) but almost remain stable in the high wavenumber region (far away from the center of the 2-d wavenumber plane), regardless of z/z_i = 0.1 or z/z_i = 0.5. To better analyze the variation trend of the PSDs with wind shear, the 2-d PSD is converted to the 1-d PSD (Figures 9 and 10) by following Peltier et al. [25] and Sullivan et al. [32] (introduced in Section 2).

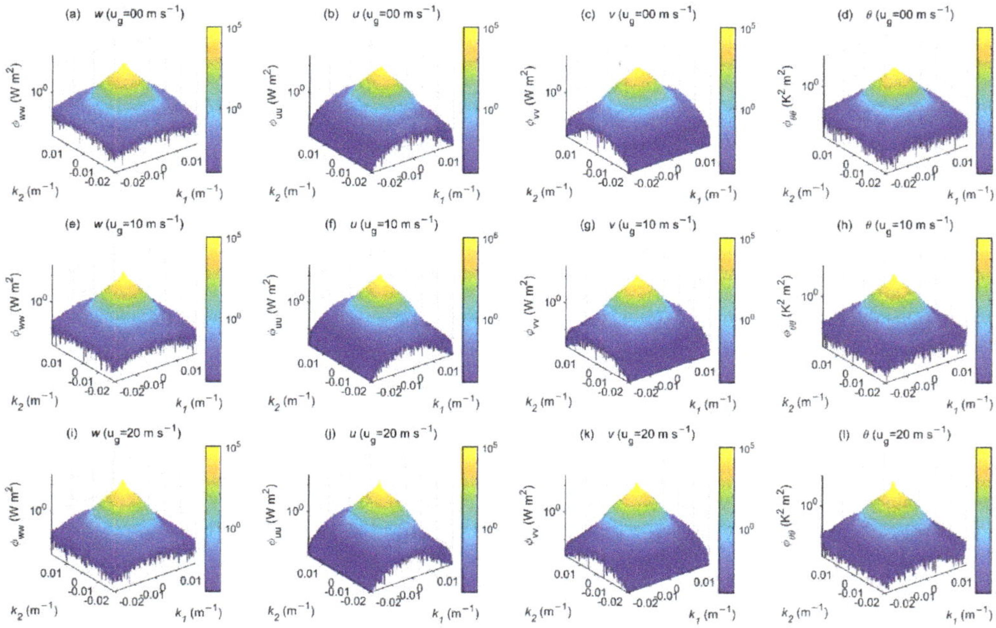

Figure 7. Instantaneous snapshots of w's PSD (ϕ_{ww}, **the first column**), u's PSD (ϕ_{uu}, **the second column**), v's PSD (ϕ_{vv}, **the third column**), and potential temperature θ PSD ($\phi_{\theta\theta}$, the **last column**) from LES simulations at z/z_i= 0.5 and 3 h of physical time across the range of U_g. First row $U_g = 0\ \mathrm{ms}^{-1}$; second row $U_g = 10\ \mathrm{ms}^{-1}$; third row $U_g = 20\ \mathrm{ms}^{-1}$. The x-coordinate represents the k_1 wavenumber, y-coordinate the k_2 wavenumber, and the z-coordinate the PSD. Note that the PSDs of ϕ_{ww}, ϕ_{uu} ϕ_{vv}, and $\phi_{\theta\theta}$ are scaled by the logarithm for clarity and contrast (i.e., $\log_{10}\phi_{ww}$, $\log_{10}\phi_{uu}$, $\log_{10}\phi_{vv}$, $\log_{10}\phi_{\theta\theta}$).

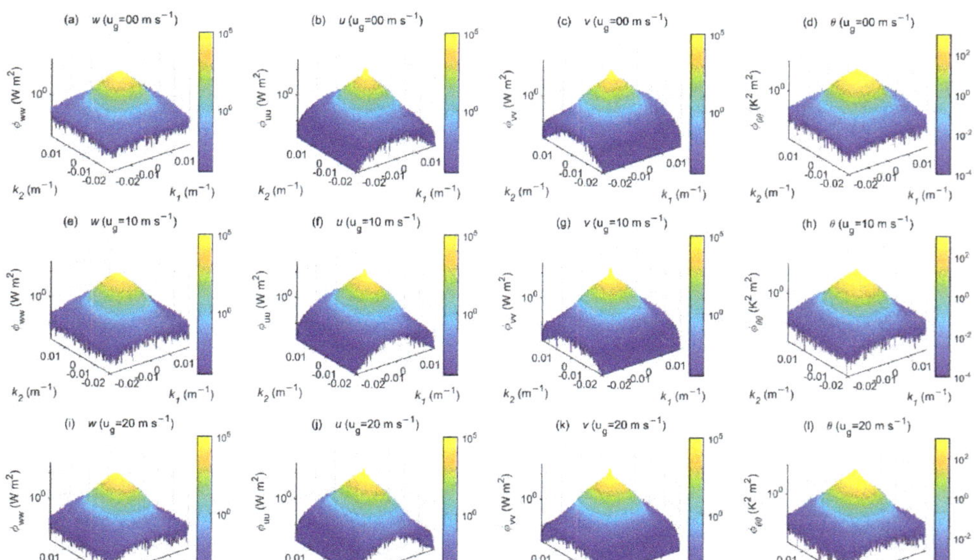

Figure 8. As in Figure 7 at $z/z_i = 0.1$.

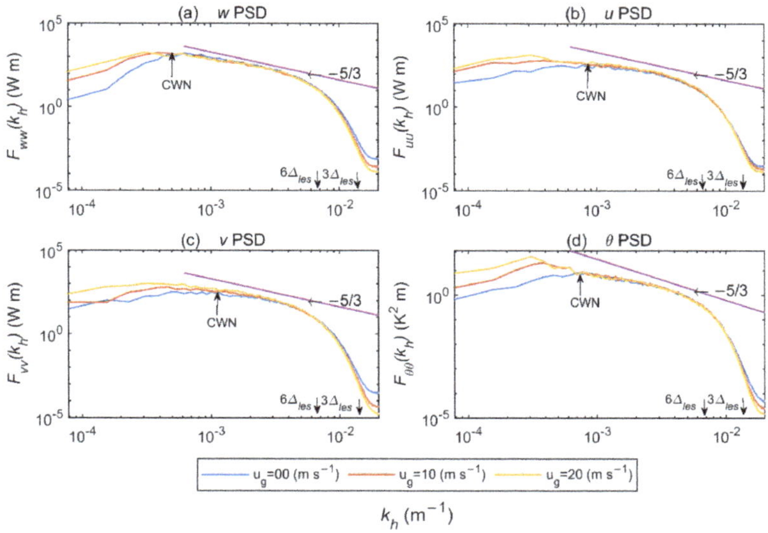

Figure 9. The instantaneous snapshots of w 1-d PSD $F_{ww}(k_h)$ (**a**), u 1-d PSD $F_{uu}(k_h)$ (**b**), v 1-d PSD $F_{vv}(k_h)$ (**c**), and θ 1-d PSD $F_{\theta\theta}(k_h)$ (**d**) from LES simulations at z/z_i = 0.5 and 3 h of physical time with U_g = 0 (blue line), 10(red line), and 20 ms^{-1} (yellow line). The x-coordinate represents the k_h wavenumber ($k_h = \sqrt{k_1^2 + k_2^2}$). y-coordinate represents the PSD value. The inclined straight lines have a slope of $k_h^{-5/3}$.

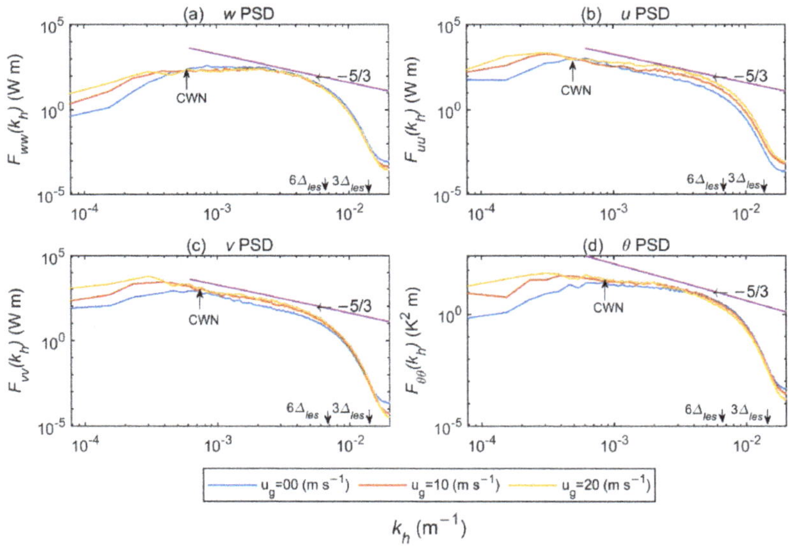

Figure 10. As in Figure 9 at z/z_i = 0.1.

3.5. *Effects of Wind Shear on 2-d PSDs of Meteorological Element Fields (1 Dimension)*

The 1-d PSDs of u, v, w, and θ horizontal fields at z/z_i = 0.5 and z/z_i = 0.1 are investigated. As shown in Figures 9 and 10, the PSDs demonstrate the transfer process of energy from large scales (in the energy-containing range) of motion to the small scales (inertial subrange), capture the peak reasonably, and exhibit the $k^{-5/3}$ slope in the inertial subrange [33]. The peak in the w PSDs broadens and shifts to higher wavenumbers at z/z_i = 0.1, which is in agreement with the results of Sullivan et al. [32] and occurs due to inviscid blocking by the presence of walls.

However, it should be noted that the PSD damps more rapidly than $k_h^{-5/3}$ in the very high wavenumber space ($\lambda < 6\Delta_{les}$, approximately), as described by Shin et al. [21], where Δ_{les} represents the grid spacing of the LES and λ is the wavelength. Zhou et al. [34] also reported the rapid decay of the Advanced Regional Prediction System (ARPS) LES simulations. Skamarock et al. [35] suggest that the rapid decay of the PSD of small-scale eddies ($\lambda < 6\Delta_{les}$) is caused by the formulation and application of explicit and implicit filters in numerical schemes, which diffuse and attenuate physical variables. Moreover, in the extremely high wavenumber space ($\lambda < 3\Delta_{les}$), the PSD decreases with the increasing wind shear (Figures 9 and 10). This is because the odd-ordered schemes are dissipative and possess a dissipation term with a coefficient proportional to the Courant number [36]. Therefore, as the larger horizontal wind speed entails a larger numerical dissipation, the PSDs decrease with the increasing wind shear in the extremely high wavenumber space ($\lambda < 3\Delta_{les}$).

However, the numerical scheme's dissipation causes the PSDs to decay more rapidly than the $k_h^{-5/3}$ slope in the very high wavenumber space ($\lambda < 6\Delta_{les}$). Therefore, in this study, the very high wavenumber spaces ($\lambda < 6\Delta_{les}$) were excluded to obtain the correct conclusion.

For convenience, we define a critical point on the wavenumber x-coordinate, named the Critical Wavenumber (CWN), to describe the location where the PSD curves almost collapse into each other for the first time. We define the space consisting of wavenumbers less than the CWN as the low-wavenumber range (k_h < CWN) and the space consisting of wavenumbers greater than the CWN as the high-wavenumber range (k_h > CWN).

At $z/z_i = 0.5$, the slope of the PSD curve of w is steeper than those of u, v, and θ, which is more consistent with the $k_h^{-5/3}$ law (Figure 9). However, with the increasing wind shear, the PSD increases significantly in the low-wavenumber range, while in the high-wavenumber range, for u and v, the PSD increases slightly with increasing wind shear, while for w, the PSD decreases slightly with increasing wind shear (Figure 9a–c).

According to Parseval's Theorem, the integration of a PSD over the entire wavenumber range is equal to the variance of the signals. Therefore, under identical surface heat flux conditions, with increasing wind shear, the PSDs of u and v nearly increase in the entire wavenumber range, resulting in the variance of u and v increasing significantly, whereas the PSDs of w in the high-wavenumber range decrease slightly, resulting in the variance of w remaining stable at $z/z_i = 0.5$.

At $z/z_i = 0.1$, some new changes can occur because of the shear forces strengthening near the surface and the presence of walls.

For u and v, as $U_g = 0$ ms^{-1}, in the inertial subrange, the slopes of u and v PSD are more consistent with the $k_h^{-5/3}$ law, while as $U_g = 10$ ms^{-1} and $U_g = 20$ ms^{-1}, the slopes of u and v PSD deviate from the $k_h^{-5/3}$ law (Figure 10b–c). This may be because the turbulent eddies in the inertial subrange lose their isotropy due to the presence of walls. The PSDs of u and v increase with increasing wind shear in the entire wavenumber range, resulting in the variance of u and v increasing significantly with increasing wind shear. However, the PSD curves of u and v still collapse into each other at the CWN point.

For w, at $z/z_i = 0.1$, the PSD curves tend to be flatter in the entire wavenumber space range than those at $z/z_i = 0.5$ (Figure 9a vs. Figure 10a), due to the w fields at $z/z_i = 0.1$ being blocked by the presence of walls [32]. It should be noted that in the high-wavenumber range, the PSD of w slightly decreases with increasing wind shear. Thus, at $z/z_i = 0.1$, because (1) the low-wavenumber PSD values are much smaller than those of u and v, and (2) the PSD decreases slightly with increasing wind shear in the high-wavenumber range, resulting in the variance of w being the largest as $U_g = 0$ ms^{-1} (Figure 2d).

For θ, the PSD curves are flatter over the entire wavenumber space at $z/z_i = 0.1$ than at $z/z_i = 0.5$ (Figure 9d vs. Figure 10d). With increasing wind shear, the PSD of θ increases significantly in the low-wavenumber range and slightly in the high-wavenumber range.

However, the magnitude of the PSD of θ is significantly smaller than that of u and v (Figures 9 and 10), which leads to a slight increase in the variance of θ with increasing wind shear at z/z_i = 0.1 (Figure 2h).

3.6. The Effects of Filtering the Energy of Large-Scale Atmospheric Motions Out

The Butterworth Filter is a signal-processing filter designed to have a frequency response that is as flat as possible in the passband [37]. To further investigate the evolution of the PSDs of meteorological element fields with wind shear, the 2-d Butterworth High Pass Filter (2-d BHPF) is introduced here to remove low wavenumber components from u, v, w, and θ horizontal fields and preserve high wavenumber components. According to Equation (6), the transfer function is defined as $H(p,q)$, where D_0 represents the cut-off wavenumber. The 2-d BHPF passes all the wavenumbers greater than the D_0 value without attenuation and discards all the wavenumbers less than it. $D(p,q)$ is the Euclidean distance from any point (p, q) to the origin of the wavenumber plane, i.e., $D(p,q) = \sqrt{p^2 + q^2}$. The p and q represent the wavenumbers in the x-direction and y-direction, respectively. The n is the order of the filter and is set as six to reduce the transition process from passband to stopband sharply. The approach to obtaining the filtered data follows these steps: first, obtain the 2-d Fourier Transform of u, v, w, and θ horizontal fields; second, acquire the product of the 2-d Fourier spectrum and the Transfer Function in the wavenumber space; and finally, take the Inverse 2-dimension Fourier Transform of the product data of u, v, w, and θ to obtain the filtered data.

$$H(u,v) = \frac{1}{1 + (D_0/D(u,v))^{2n}} \tag{6}$$

The D_0 is set as 9.375×10^{-4} m^{-1}, which means the harmonics with wavenumbers less than 9.375×10^{-4} (i.e., $\lambda > 1067$ m) are filtered out. As wind shear increases, the snapshots of the w, u, v, and θ horizontal fields at z/z_i = 0.5, which are filtered out of harmonics with $\lambda > 1067$ m, present no obvious large-scale organized structures, preferring the distribution of random white noise, as shown in Figure 11. In contrast, as displayed in Figure 1, the organized large-scale structures of the u, v, w, and θ horizontal fields are more pronounced, indicating that large-scale atmospheric motions (i.e., a low-wavenumber PSD) play an essential role in the organization of the large-scale structures of the CBL. Therefore, it is reasonable to expect that as the mean wind shear increases, the large-scale atmospheric motions will also increase. Therefore, the organized large-scale structures in the CBL become apparent with increasing wind shear.

To investigate the effects of low wavenumber harmonics on the variances of the u, v, w, and θ horizontal fields, the D_0 is set as 3.125×10^{-4} m^{-1} (λ = 3200 m), 6.25×10^{-4} m^{-1} (λ = 1600 m), 9.375×10^{-4} m^{-1} (λ = 1067 m), and 1.25×10^{-4} m^{-1} (λ = 800 m) to remove the harmonics with λ greater than 3200, 1600, 1067, and 800 m, respectively. The vertical variance profiles of the filtered u, v, w, and θ horizontal fields are plotted in Figure 12.

As D_0 increases, increasingly more low wavenumber harmonics are removed from the u, v, w, and θ horizontal fields, and the vertical variances of the u, v, w, and θ decrease significantly at all elevations of the CBL. This indicates that the large-scale motions (i.e., low-wavenumber PSD) contribute significantly to the magnitudes of the horizontal field variances of the u, v, w, and θ.

In terms of the u, in the range of 0.3 to 0.9 z/z_i, as D_0 increases, the variances with U_g = 20 ms^{-1} decrease sharply and rapidly approach the variances with U_g = 0 and U_g = 10 ms^{-1} (Figure 12, second column). This is because (1) the low-wavenumber PSD is large enough to affect the magnitude of the variance significantly, and (2) the low-wavenumber PSD increases with increasing wind shear, resulting in the variance with a large wind shear rapidly approaching that of a small wind shear as increasingly more low-wavenumber harmonics are filtered out (Figure 12r). However, in the near-surface layer, about 0 to 0.3 z/z_i above ground level, even if D_0 is set to the largest value (i.e., $k_h = 1.25 \times 10^{-4}$ m^{-1}), the differences between the vertical profiles of variance are still noticeable (Figure 12r). This is because the PSDs increase with increasing wind shear in the high-wavenumber range at z/z_i = 0.1 (Figure 10b), resulting in the vertical profiles of variance constantly diverging from each other in the near-surface layer (0.0–0.3 z/z_i), regardless of the wind shear.

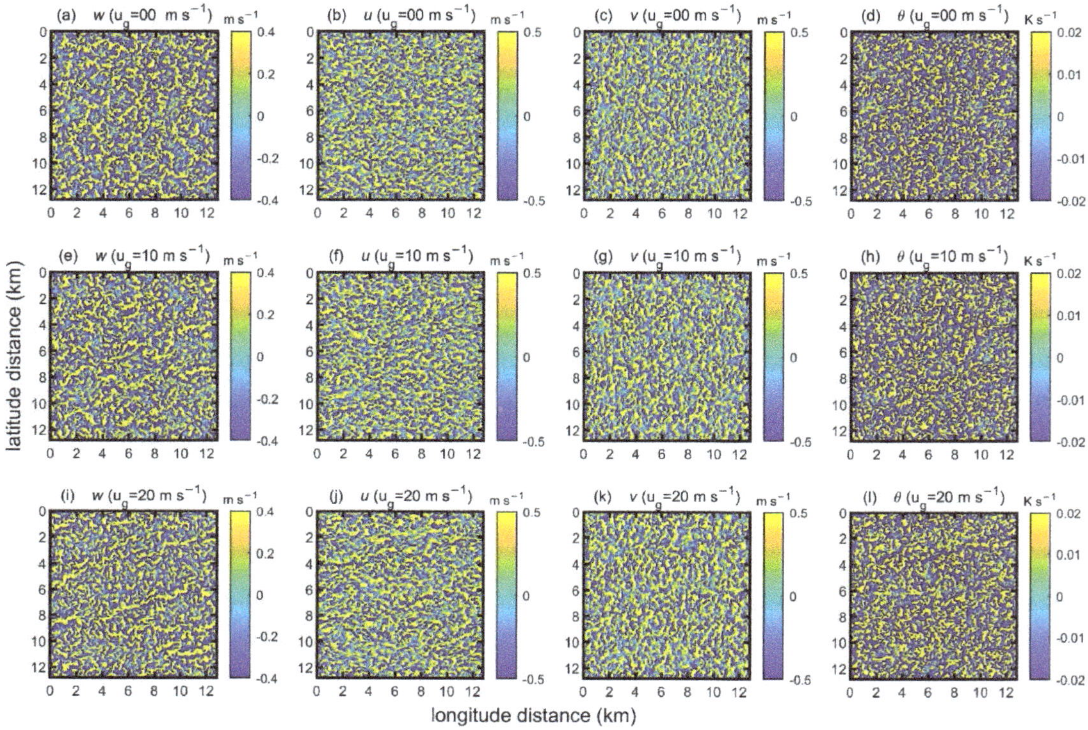

Figure 11. Filtered snapshots of w (**first column**), u (**second column**), v (**third column**), and θ (**last column**) from LES simulations at z/z_i = 0.5 and 3 h of physical time with U_g = 0 ms^{-1} (**First row**), U_g = 10 ms^{-1} (**second row**), and U_g = 20 ms^{-1} (**third row**). D_0 is set as 9.375 × 10^{-4} m^{-1}.

In terms of v, the situation is similar to that of the u (Figure 12, third column).

In terms of w, in the range of 0.0–1.0 z/z_i, as D_0 increases, the variances with U_g = 10 ms^{-1} and U_g = 20 ms^{-1} decrease more rapidly than those with U_g = 0 ms^{-1} (Figure 12, first column). This implies that the increases in the PSD in the low-wavenumber range compensate for the slight decreases in the high-wavenumber range with the increasing wind shear (Figures 9a and 10a). Therefore, as the low wavenumber harmonics are filtered out, the vertical variance profiles of w with U_g = 10 ms^{-1} and U_g = 20 ms^{-1} decrease more rapidly than those with U_g = 0 ms^{-1}.

In terms of θ, the variances increase with increasing wind shear in the range of 0.1–0.8 z/z_i (Figure 12d). However, as D_0 increases, the variances with U_g = 10 ms^{-1} and U_g = 20 ms^{-1} sharply decrease and approach those with U_g = 0 ms^{-1} (Figure 12h,l,p,t). This indicates that as wind shear increases, the increases in PSD in the low-wavenumber range contribute significantly to the increases in the variance of the θ horizontal fields. Furthermore, because the PSD magnitude of θ is significantly smaller than that of u and v (Figures 9 and 10), for example, the PSD magnitudes of θ are less than 10^{-2} while the PSD magnitudes of u and v are much larger than 10^{-2} (Figures 9 and 10), the variance of the θ horizontal field is much smaller than that of u and v (Figure 12).

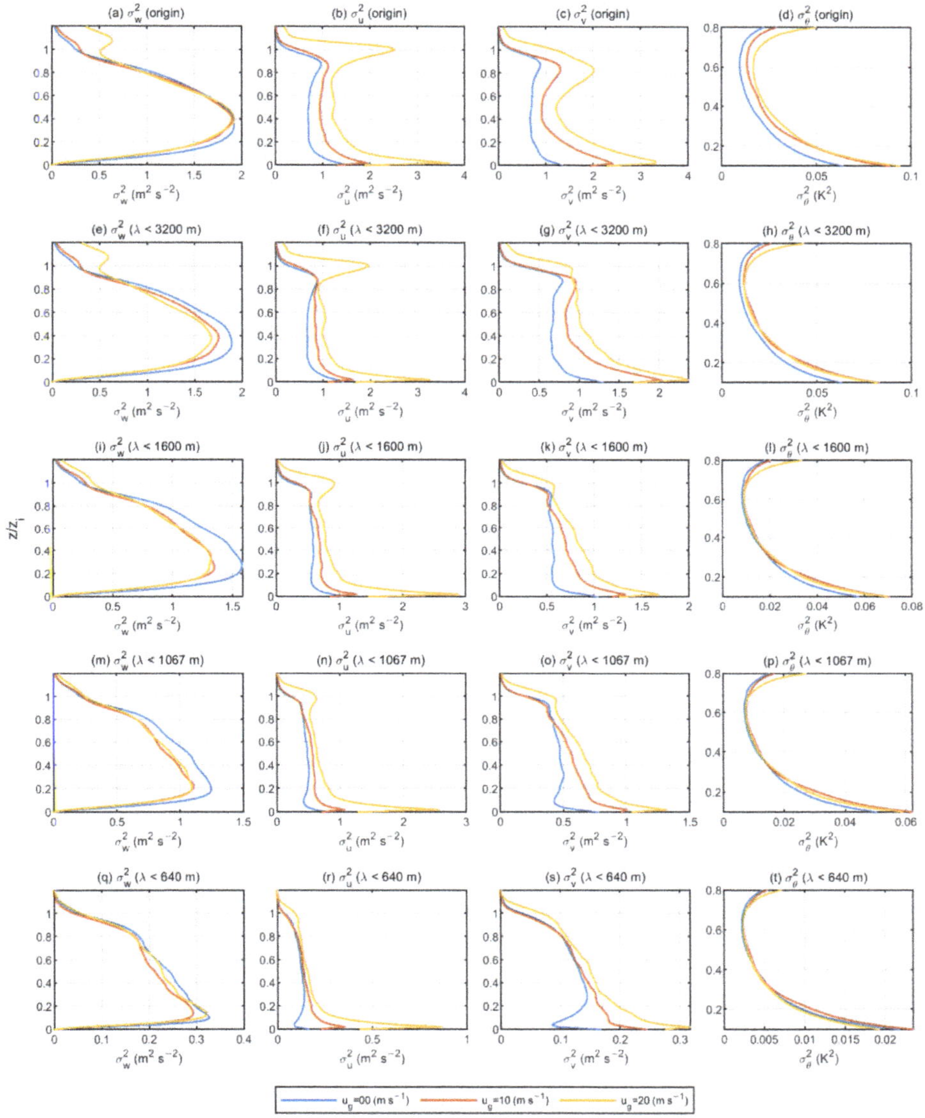

Figure 12. The vertical profiles of horizontal variances of w (**first column**), u (**second column**), v (**third column**), and θ (**last column**) data filtered by 2-d BHPF with cut-off wavenumber (D_0) set as 3.125×10^{-4} m^{-1} (i.e., λ = 3200 m; **second row**), 6.25×10^{-4} m^{-1} (i.e., λ = 1600 m; **third row**), 9.375×10^{-4} m^{-1} (i.e., λ = 1067 m; **fourth row**), and 1.25×10^{-4} m^{-1} (i.e., λ = 800 m; **last row**) from LES simulations at 3 h of physical time. The blue, red, and yellow lines represent U_g = 0, 10, and 20 ms^{-1}, respectively. Note that the first row represents original vertical profiles of horizontal variances of w, u, v, and θ without filtering.

4. Discussion and Conclusions

For clarity, the main findings can be summarized as follows. It should be noted that all conclusions are drawn under the same conditions of surface heat flux.

(1) In the CBL, as wind shear increases, the variance of w differs insignificantly, u and v increase monotonically, and the variance of θ increases slightly, which implies that the spatial variations of u, v, and θ are enhanced as the wind shear increases. The spatial variation of w is insensitive to wind shear throughout the CBL depth. This is consistent with the results of the horizontal gradients of u, v, w, and θ.
(2) In the middle CBL (about 0.2–0.8z/z_i), with increasing wind shear, the low-wavenumber PSDs of u, v, w, and θ increase significantly. In addition, in the high-wavenumber range, with increasing wind shear, the PSDs of u and v increase slightly, and the PSD of w decreases slightly, while the PSD of θ almost remains stable.
(3) In the surface layer CBL (about 0.0–0.2 z/z_i), low-wavenumber PSDs of u, v, w, and θ increase significantly with the increasing wind shear. Moreover, in the high-wavenumber range, with increasing wind shear, the PSDs of u and v increase, and the PSD of w decreases slightly, while the PSD of θ increases slightly.
(4) However, the low-wavenumber PSDs of u, v, w, and θ increase significantly with increasing wind shear in the CBL, indicating that the large-scale motions of the atmosphere are constantly enhanced with increasing wind shear, which means the large-scale coherent 2-d structures of the atmospheric flows in the CBL become more significant and ordered with increasing wind shear.
(5) The PSDs can more precisely elucidate the spatial variation of u, v, w, and θ with wind shear. Generally, with increasing wind shear, the PSDs of u and v increase in nearly the entire wavenumber range, resulting in u and v having increased variances. On the contrary, with increasing wind shear, the PSD of w increases in the low-wavenumber range and decreases slightly in the high-wavenumber range, which results in the variance of w differing insignificantly. Since the magnitude of the PSD of θ is much smaller than those of u and v, the variance of θ increases with increasing wind shear, but it is much smaller than the variances of u and v.

The finding that a low-wavenumber PSD increases significantly with increasing wind shear can explain the phenomenon that self-organized convection strengthens with the increase in wind shear. This is because the large-scale motions of the atmosphere (i.e., low-wavenumber PSD) always increase as wind shear increases and are, therefore, more likely to trigger some mesoscale weather processes. This interpretation can cover some findings of mesoscale meteorological phenomena. For example, horizontal convective rolls influence the initiation of deep and moist convection [5–7] and the formation of storms [8] and tornados [9]. In addition, faster winds are associated with substantially more precipitation [11]. This may help forecasters to speculate on the likelihood of strong convective weather processes based on the mean wind shear intensity of the CBL.

Moreover, z_i/L is used to define the turbulent state of the atmosphere and determine the degree of stability or instability [38]. For example, in Table 1, with increasing wind shear, z_i/L increases from -1047 to -8. This means that as wind shear increases, the atmosphere experiences a transition from extremely unstable to nearly neutral. Therefore, it is reasonable to expect that with increasing wind shear, the increase in large-scale motions (low-wavenumber PSD) can increase the atmospheric stability of the CBL. With increasing wind shear, the enhanced large-scale motions strengthen the large-scale coherent flow structures in the CBL. In addition, with increasing wind shear, the atmospheric flow structures become more orderly; in this sense, the atmosphere is more stable.

The results of this study are based on the same surface heat flux preconditions. However, the surface heat flux cannot remain constant as the wind shear increases. Therefore, we will further investigate the interaction between wind shear and surface heat flux under variable surface heat flux conditions. This study provides a new perspective for understanding the role of wind shear in the spatial variations of horizontal fields of meteorological elements and provides a reference for other researchers to conduct further studies.

Author Contributions: Conceptualization, H.Z., Q.H. and M.W.; methodology, H.Z. and J.Y.; software, H.Z.; validation, H.Z. and J.Y.; formal analysis, J.Y.; investigation, M.W.; resources, Q.H.; data curation, J.Y.; writing—original draft preparation, H.Z.; writing—review and editing, M.W.; visualization, H.Z.; supervision, Q.H.; project administration, Q.H.; funding acquisition, Q.H. All authors have read and agreed to the published version of the manuscript.

Funding: This research was funded by the Open Grants of the State Key Laboratory of Severe Weather (grant number 2022LASW-B04), the Xinjiang Science Fund for Distinguished Young Scholars (grant number 2022D01E07), the Flexible Talents Introducing Project of Xinjiang (grant number 2021-49) the National Natural Science Foundation of China (grant number 42030612), the National Natural Science Foundation of China (grant number 41830968), the Second Tibetan Plateau Scientific Expedition and Research (STEP) program (grant number 2019QZKK010206), the Third Xinjiang Scientific Expedition and Research program (grant number 2021xjkk030501), and the Special Project for Innovation and Development of China Meteorological Administration (grant number CXFZ2022P074).

Institutional Review Board Statement: Not applicable.

Informed Consent Statement: Not applicable.

Data Availability Statement: The original code of WRF version 4.0 (including LES) can be downloaded at https://www2.mmm.ucar.edu/wrf/users/download/get_source.html#current. (accessed on 8 June 2018). The data can be generated by the WRF-LES numerical experiments in this paper or obtained from the corresponding authors.

Acknowledgments: We gratefully acknowledge the suggestions for the study from Bowen Zhou of Nanjing University. We also thank Xu Zhang for our insightful conversations.

Conflicts of Interest: The authors declare no conflict of interest.

References

1. Hunt, J.C.R.; Kaimal, J.C.; Gaynor, J.E. Eddy Structure in the Convective Boundary Layer—New Measurements and New Concepts. *Q. J. R. Meteorol. Soc.* **1988**, *114*, 827–858.
2. Mesoscale Shallow Convection in the Atmosphere-Atkinson-1996-Reviews of Geophysics-Wiley Online Library. Available online: https://agupubs.onlinelibrary.wiley.com/doi/abs/10.1029/96RG02623 (accessed on 29 August 2022).
3. Brown, R.A. Longitudinal Instabilities and Secondary Flows in the Planetary Boundary Layer: A Review. *Rev. Geophys.* **1980**, *18*, 683–697. [CrossRef]
4. Young, G.S.; Kristovich, D.A.; Hjelmfelt, M.R.; Foster, R.C. Rolls, Streets, Waves, and More: A Review of Quasi-Two-Dimensional Structures in the Atmospheric Boundary Layer. *Bull. Am. Meteorol. Soc.* **2002**, *83*, 997–1002.
5. Weckwerth, T.M.; Wilson, J.W.; Wakimoto, R.M. Thermodynamic Variability within the Convective Boundary Layer Due to Horizontal Convective Rolls. *Mon. Weather Rev.* **1996**, *124*, 769–784. [CrossRef]
6. Weckwerth, T.M. The Effect of Small-Scale Moisture Variability on Thunderstorm Initiation. *Mon. Weather Rev.* **2000**, *128*, 4017–4030. [CrossRef]
7. Xue, M.; Martin, W.J. A High-Resolution Modeling Study of the 24 May 2002 Dryline Case during IHOP. Part I: Numerical Simulation and General Evolution of the Dryline and Convection. *Mon. Weather Rev.* **2006**, *134*, 149–171. [CrossRef]
8. Trier, S.B.; Parsons, D.B.; Clark, J.H. Environment and Evolution of a Cold-Frontal Mesoscale Convective System. *Mon. Weather Rev.* **1991**, *119*, 2429–2455. [CrossRef]
9. Wakimoto, R.M.; Wilson, J.W. Non-Supercell Tornadoes. *Mon. Weather Rev.* **1989**, *117*, 1113–1140. [CrossRef]
10. Baidu, M.; Schwendike, J.; Marsham, J.H.; Bain, C. Effects of Vertical Wind Shear on Intensities of Mesoscale Convective Systems over West and Central Africa. *Atmos. Sci. Lett.* **2022**, *23*, e1094. [CrossRef]
11. Back, L.E.; Bretherton, C.S. The Relationship between Wind Speed and Precipitation in the Pacific ITCZ. *J. Clim.* **2005**, *18*, 4317–4328. [CrossRef]
12. Moeng, C.-H.; Sullivan, P.P. A Comparison of Shear-and Buoyancy-Driven Planetary Boundary Layer Flows. *J. Atmos. Sci.* **1994**, *51*, 999–1022. [CrossRef]
13. Schmidt, H.; Schumann, U. Coherent Structure of the Convective Boundary Layer Derived from Large-Eddy Simulations. *J. Fluid Mech.* **1989**, *200*, 511–562. [CrossRef]
14. Park, J.; Basu, S.; Manuel, L. Large-Eddy Simulation of Stable Boundary Layer Turbulence and Estimation of Associated Wind Turbine Loads: LES of SBL Turbulence and Wind Turbine Loads. *Wind Energy* **2014**, *17*, 359–384. [CrossRef]
15. Park, S.-B.; Baik, J.-J.; Han, B.-S. Role of Wind Shear in the Decay of Convective Boundary Layers. *Atmosphere* **2020**, *11*, 622. [CrossRef]
16. Lee, M.J.; Kim, J.; Moin, P. Structure of Turbulence at High Shear Rate. *J. Fluid Mech.* **1990**, *216*, 561–583. [CrossRef]

17. Skamarock, W.C.; Klemp, J.B.; Dudhia, J.; Gill, D.O.; Barker, D.M.; Wang, W.; Powers, J.G. A Description of the Advanced Research WRF Version 3. NCAR Technical Note-475+ STR. 2008. Available online: https://citeseerx.ist.psu.edu/viewdoc/summary?doi=10.1.1.484.3656 (accessed on 30 August 2008).
18. Noh, Y.; Cheon, W.G.; Hong, S.Y.; Raasch, S. Improvement of the K-Profile Model for the Planetary Boundary Layer Based on Large Eddy Simulation Data. *Bound.-Layer Meteorol.* **2003**, *107*, 401–427. [CrossRef]
19. Deardorff, J.W. Stratocumulus-Capped Mixed Layers Derived from a Three-Dimensional Model. *Bound.-Layer Meteorol.* **1980**, *18*, 495–527. [CrossRef]
20. Moeng, C.-H. A Large-Eddy-Simulation Model for the Study of Planetary Boundary-Layer Turbulence. *J. Atmos. Sci.* **1984**, *41*, 2052–2062. [CrossRef]
21. Shin, H.H.; Hong, S.-Y. Analysis of Resolved and Parameterized Vertical Transports in Convective Boundary Layers at Gray-Zone Resolutions. *J. Atmos. Sci.* **2013**, *70*, 3248–3261. [CrossRef]
22. Troen, I.B.; Mahrt, L. A Simple Model of the Atmospheric Boundary Layer; Sensitivity to Surface Evaporation. *Bound.-Layer Meteorol.* **1986**, *37*, 129–148. [CrossRef]
23. Shikhovtsev, A.Y.; Kovadlo, P.G.; Khaikin, V.B.; Nosov, V.V.; Lukin, V.P.; Nosov, E.V.; Torgaev, A.V.; Kiselev, A.V.; Shikhovtsev, M.Y. Atmospheric Conditions within Big Telescope Alt-Azimuthal Region and Possibilities of Astronomical Observations. *Remote Sens.* **2022**, *14*, 1833. [CrossRef]
24. Batchelor, G.K. *The Theory of Homogeneous Turbulence*; Cambridge University Press: Cambridge, UK, 1953.
25. Peltier, L.J.; Wyngaard, J.C.; Khanna, S.; Brasseur, J.O. Spectra in the Unstable Surface Layer. *J. Atmos. Sci.* **1996**, *53*, 49–61. [CrossRef]
26. Salesky, S.T.; Chamecki, M.; Bou-Zeid, E. On the Nature of the Transition between Roll and Cellular Organization in the Convective Boundary Layer. *Bound.-Layer Meteorol.* **2017**, *163*, 41–68. [CrossRef]
27. Zhou, B.; Simon, J.S.; Chow, F.K. The Convective Boundary Layer in the Terra Incognita. *J. Atmos. Sci.* **2014**, *71*, 2545–2563. [CrossRef]
28. Drazin, P.G.; Reid, W.H. *Hydrodynamic Stability*; Cambridge University Press: Cambridge, UK, 2004.
29. Sorbjan, Z. Effects Caused by Varying the Strength of the Capping Inversion Based on a Large Eddy Simulation Model of the Shear-Free Convective Boundary Layer. *J. Atmos. Sci.* **1996**, *53*, 2015–2024. [CrossRef]
30. Gibbs, J.A.; Fedorovich, E. Comparison of Convective Boundary Layer Velocity Spectra Retrieved from Large-Eddy-Simulation and Weather Research and Forecasting Model Data. *J. Appl. Meteorol. Climatol.* **2014**, *53*, 377–394. [CrossRef]
31. Tennekes, H.; Lumley, J.L.; Lumley, J.L. *A First Course in Turbulence*; MIT Press: Cambridge, MA, USA, 1972.
32. Sullivan, P.P.; Patton, E.G. The Effect of Mesh Resolution on Convective Boundary Layer Statistics and Structures Generated by Large-Eddy Simulation. *J. Atmos. Sci.* **2011**, *68*, 2395–2415. [CrossRef]
33. Jiménez, M.A.; Cuxart, J. Large-Eddy Simulations of the Stable Boundary Layer Using the Standard Kolmogorov Theory: Range of Applicability. *Bound.-Layer Meteorol.* **2005**, *115*, 241–261. [CrossRef]
34. Zhou, B.; Sun, S.; Sun, J.; Zhu, K. The Universality of the Normalized Vertical Velocity Variance in Contrast to the Horizontal Velocity Variance in the Convective Boundary Layer. *J. Atmos. Sci.* **2019**, *76*, 1437–1456. [CrossRef]
35. Skamarock, W.C. Evaluating Mesoscale NWP Models Using Kinetic Energy Spectra. *Mon. Weather Rev.* **2004**, *132*, 3019–3032. [CrossRef]
36. Wicker, L.J.; Skamarock, W.C. Time-Splitting Methods for Elastic Models Using Forward Time Schemes. *Mon. Weather Rev.* **2002**, *130*, 2088–2097. [CrossRef]
37. Selesnick, I.W.; Burrus, C.S. Generalized Digital Butterworth Filter Design. *IEEE Trans. Signal Process.* **1998**, *46*, 1688–1694. [CrossRef]
38. Golder, D. Relations among Stability Parameters in the Surface Layer. *Bound.-Layer Meteorol.* **1972**, *3*, 47–58. [CrossRef]

atmosphere

MDPI

Article

Assessing Spatial Variation of PBL Height and Aerosol Layer Aloft in São Paulo Megacity Using Simultaneously Two Lidar during Winter 2019

Gregori de Arruda Moreira [1,2,3,*], Amauri Pereira de Oliveira [2], Georgia Codato [2], Maciel Piñero Sánchez [2], Janet Valdés Tito [2], Leonardo Alberto Hussni e Silva [2], Lucas Cardoso da Silveira [2], Jonatan João da Silva [3,4], Fábio Juliano da Silva Lopes [3] and Eduardo Landulfo [3]

1 Federal Institute of Education, Science and Technology of São Paulo (IFSP), Campus Registro, São Paulo 11900-000, Brazil
2 Instituto de Astronomia, Geofísica e Ciências Atmosféricas, Universidade de São Paulo, São Paulo 05508-090, Brazil; apdolive@usp.br (A.P.d.O.); gecodato@usp.br (G.C.); maciel.sanchez@iag.usp.br (M.P.S.); janet.valdes@iag.usp.br (J.V.T.); leohussni@hotmail.com (L.A.H.e.S.); lucas.silveira@usp.br (L.C.d.S.)
3 Center of Lasers and Applications (CELAP), Instituto de Pesquisas Energéticas e Nucleares, São Paulo 05508-000, Brazil; silvajonatan@ufob.edu.br (J.J.d.S.); fabiolopes@alumni.usp.br (F.J.d.S.L.); elandulf@ipen.br (E.L.)
4 Center for Exact Sciences and Technologies (CCET), Campus Barreiras, Federal University of Western Bahia (UFOB), Barreiras 47810-047, Brazil
* Correspondence: gregori.moreira@ifsp.edu.br

Citation: Moreira, G.d.A.; Oliveira, A.P.d.; Codato, G.; Sánchez, M.P.; Tito, J.V.; Silva, L.A.H.e.; Silveira, L.C.d.; Silva, J.J.d.; Lopes, F.J.d.S.; Landulfo, E. Assessing Spatial Variation of PBL Height and Aerosol Layer Aloft in São Paulo Megacity Using Simultaneously Two Lidar during Winter 2019. *Atmosphere* **2022**, *13*, 611. https://doi.org/10.3390/atmos13040611

Academic Editors: Yubin Li and Jie Tang

Received: 11 February 2022
Accepted: 29 March 2022
Published: 11 April 2022

Abstract: This work presents the use of two elastic lidar systems to assess the horizontal variation of the PBL height (PBLH) and aerosol layer aloft in the São Paulo Megacity. These two lidars performed simultaneous measurements 10.7 km apart in a highly urbanized and relatively flat area of São Paulo for two winter months of 2019. The results showed that the PBLH differences display diurnal variation that depends on the PBL during daytime growth phases. Cloud and sea breeze effects control most of PBLH variation. In the absence of cloud and sea breeze, the maximum difference (~300 m) occurs in the rapid development stage and is due to topographic effects. When the PBL approaches its maximum daily value, it tends to level off with respect to the topography. In addition, it was presented a method that combines elastic lidar (to detect an aerosol layer) and satellite data (to classify such a layer from Aerosol Optical Depth (AOD) and Aerosol Index (AI) information) for the detection of biomass burning events. This methodology demonstrated that the variations caused by Biomass Burning in AOD and AI enable both the detection of aerosol plumes originating from biomass burning and the identification of their origin.

Keywords: atmospheric boundary layer height; horizontal homogeneity; elastic lidar; biomass burning

1. Introduction

In the last decade, lidar systems have been widely applied to analyze a large variety of tropospheric properties [1–4]. Among them, the following stand out: aerosol optical properties [5,6], vertical displacements of aerosol layers [7], cloud microphysics [8], Planetary Boundary Layer Height (PBLH) estimation [1–4,9], and detection of extreme aerosol intrusion events [10–12].

In the case of elastic lidars [13], the PBLH detection methods are based on the abrupt reduction in the backscattered signal intensity that occurs in the transition layer between the PBL and Free Atmosphere (FA). Unfortunately, such a drop in the backscattered signal intensity occurs in ideal conditions, observed during the convective period in the absence of low clouds and decoupled aerosol layers. As ideal conditions are rarely observed, several algorithms of PBLH detection have been developed to cope with the most-frequently-observed not-ideal conditions [1,2,14,15].

The spatial variation of the PBL responds, at large, to the modulation of turbulence intensity induced by topography and land use variation [16,17]. These effects are particularly important in urban areas where observations indicate a significant PBLH variation associated with the urban-rural contrast [18]. Although lidar systems have a high cost, the most appropriate way to monitor the spatial variation of PBLH is by carrying out simultaneous measurements with multiples lidar systems over more than one measuring point or with lidar systems capable of scanning large atmospheric volumes, such as Plan Position Indicator (PPI) and Range Height Indicator (RHI) techniques. Although most algorithms applied to retrieve PBLH from one-point lidar measurements have displayed reliable results for different atmospheric scenarios [1–3,9,14], there are still uncertainties related to horizontal representativity of one-point analysis, mainly in urban regions over non-flat terrains [19] that requires to be addressed.

In recent years, the increase in the Biomass Burning [BB] frequency caused by anthropogenic activities has been followed by a significant improvement in the quality of monitoring systems, allowing a more accurate description of extreme aerosol intrusion events [11,20,21]. In the last three years, an unprecedented number of wildfires have been detected in the Brazilian central west [22] and Amazon [23] regions. There is observational evidence indicating the plumes produced by these fires have been advected to São Paulo City [10,24,25] so that they are considered the main external anthropogenic sources of BB for this region from May to October [26,27].

Despite the predominance of urban land use, 44% of São Paulo city territory is endowed with unequally distributed green areas [28,29]. Such land-use heterogeneity, combined with topographic diversity [30] and a high frequency of sea breeze [31,32], are likely to induce a horizontal variation of the PBL during daytime [19]. Furthermore, the presence of BB plumes from the Brazilian central west and Amazon regions during May and October may also be affecting the PBLH in the city of São Paulo, aggravating the air quality during winter.

Considering both scenarios, this paper proposes to assess the spatial variation of the PBLH in São Paulo city using simultaneous measurements of two elastic lidars systems located 10.7 km apart and characterize the aerosol-layer aloft the PBL produced by BB activities by deploying a new method that explores the synergy between lidars and other remote sensing systems data. The results presented here are based on 12 days selected from lidar measurements performed in São Paulo city during winter, from 22 July to 29 August 2019.

This article is organized as follows: in Section 2, the study area and instruments are described. Section 3 deals with detection methods of PBLH and BB plumes. In Section 4, the spatial variation of the PBLH and the detection of the aerosol layer aloft the PBL are discussed. Finally, the conclusions can be found in Section 5.

2. Materials and Methods

2.1. *São Paulo Megacity*

São Paulo is the most populous Brazilian city, with around 12.3 million inhabitants distributed in an urban area of 1521 km^2 [33]. It is situated in the "Paulista" Plateau at 700 m above sea level (asl) and about 60 km from the Atlantic Ocean (Figure 1a). São Paulo has a high elevation subtropical humid (Cwb) climate, where summers (December–February) are warm and wet, while winters (June–August) are dry and mildly cold [34].

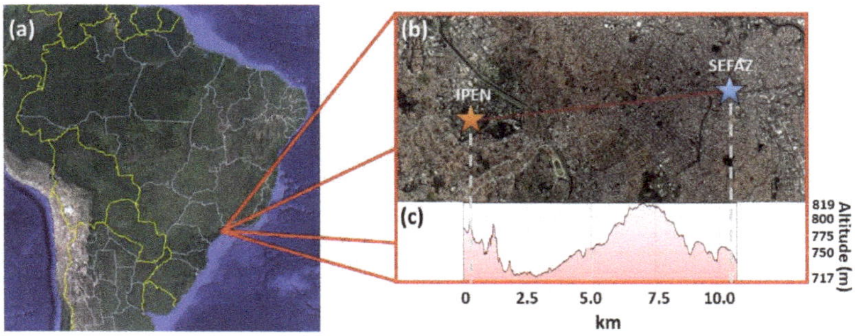

Figure 1. (**a**) Brazil map. The red dot indicates the location of São Paulo City. (**b**) Location of the two elastic lidar systems (orange star-IPEN and blue star (SEFAZ)) and (**c**) elevation profile.

The measurement campaign was held in São Paulo city from 22 July to 29 August 2019. During this period, two elastic lidar systems, approximately 10.7 km apart (Figure 1b), were operating from 09 to 18 Local Time (LT). The urban site, located downtown in São Paulo city (23°32′ S, 46°37′ W, 739 m above sea level), was named SEFAZ. The suburban site, located west of São Paulo city (23°34′ S, 46°43′ W, 782 m above sea level), was named IPEN.

2.2. Instrumentation

2.2.1. Metropolitan São Paulo Lidar 1 (MSP1) System

The Metropolitan São Paulo Lidar 1 (MSP1) system is a coaxial ground-based multi-wavelength Raman lidar located at the Nuclear and Energy Research Institute (IPEN) in the suburban region of São Paulo city (23°34′ S, 46°43′ W, 782 m asl) (Figure 1b). MSP1 operates with a pulsed Nd:YAG laser pointed towards the zenith direction, emitting radiation at 355 nm, 532 nm, and 1064 nm with a repetition frequency of 10 Hz. This system detects three Raman-shifted channels (387, 408, and 530 nm) and three elastic channels (355, 532, and 1064 nm), reaching the full overlap at 300 m above ground level [7]. During the field campaign, MSP1 was run with a temporal and spatial resolution of 1 min and 7.5 m, respectively, from 09 to 18 LT.

2.2.2. Metropolitan São Paulo Lidar (MSP) 2 System

The Metropolitan São Paulo Lidar 2 (MSP2) is a mobile biaxial ground-based multiwavelength Raman lidar system. MSP2 operates with a pulsed Nd: YAG laser, which emits radiation at 355 nm and 532 nm in the zenith direction and detects one elastic channel (532 nm) and one Raman-shifted channel (387 nm). Such a system reaches the full overlap at 180 m above ground level. During the field campaign, this system was allocated in the Department of the Treasury of the São Paulo State (SEFAZ) in the urban region of São Paulo city (23°32′ S, 46°37′ W, 739 m asl) (Figure 1b), operating continuously with a temporal and spatial resolution of 1 min and 7.5 m, respectively.

2.2.3. Suomi National Polar-Orbiting Partnership (Suomi NPP) Data

The Suomi National Polar-orbiting Partnership (Suomi NPP) is a weather satellite operated by the National Oceanic and Atmospheric Administration (NOAA). Launched in 2011, Suomi NPP is equipped with the Visible Infrared Imaging Radiometer Suite (VIIRS) and the Ozone Mapping and Profiling Suite Nadir-Mapper (OMPS-NM). The VIIRS is a whiskbroom scanner radiometer, which passively observes reflectance at visible and infrared wavelengths [35].

In this work, Aerosol Optical Depth (AOD) and Aerosol Index (AI) were obtained respectively from VIIRS data and normalized radiances using two-wavelength pairs at 340 and 378.5 nm of OMPS-NM. AI is a qualitative index that indicates the presence of aerosol layers, such as biomass-burning, desert-dust, and volcanic-ash plumes, monitoring the absorption in the above-mentioned wavelength pairs [36]. Both datasets are available at: https://earthdata.nasa.gov (accessed on 1 March 2021).

2.2.4. AERONET Sunphotometer

The AErosol RObotic NETwork (AERONET) [37] is the NASA sunphotometer global network that supplies automatic sun and sky scanning measurements. Using direct sun measurements, AERONET supplies both Aerosol Optical Depth (AOD) and the Ångström Exponent (Å), which gives the wavelength dependence of the AOD. Using multiangular and multispectral measurements of atmospheric radiance and applying a flexible inversion algorithm [38], the AERONET data can also supply several additional aerosol optical and microphysical parameters, such as size distribution, single-scattering albedo, and refractive index. The operating principle of this system is to acquire aureole and sky radiance observations using a large number of solar scattering angles through a constant aerosol profile and thus retrieve the aerosol size distribution, the phase function, and the AOD [39]. In this work, the AERONET sunphotometer data were measured in the SP-EACH station located in São Paulo city (23°48′ S, 46°49′ W, 754 m asl). These data were used to derive the aerosol size distribution and Angstrom Matrix during the BB event from 17 to 19 August 2019.

3. Methodology

3.1. PBLH Detection

The PBLH was estimated from the 532 nm backscattered signal, so the algorithm is divided into two parts: raw lidar data pre-processing [40] and Wavelet Covariance Transform (WCT) algorithm [14]. The data pre-processing begins with the subtraction of the dark current signal (DC(z)) of the raw lidar signal (P) to reduce the influence of electrical noise. Then, the background radiation signal (BG) is removed to attenuate the influence of external sources. Finally, due to the attenuation of the lidar signal with the height, the result of the previous steps is multiplied by the square of the corresponding height (z), resulting in the Range Corrected Signal (RCS_{532}), as indicated in Equation (1):

$$RCS_{532}(z) = (P(z) - DC(z) - BG) \times z^2, \tag{1}$$

After the raw signal pre-processing, the second part of the algorithm is performed, where the WCT algorithm is applied. Firstly, the covariance ($W(a,b)$) between the average $RCS_{532}(z)$ obtained for one hour ($\overline{RCS}_{532}(z)$) and Haar function ($h\left(\frac{z-b}{a}\right)$), the mother-wavelet is done:

$$W(a,b) = \frac{1}{a}\int_{z_i}^{z_f} \overline{RCS}_{532}(z) h\left(\frac{z-b}{a}\right) dz, \tag{2}$$

where z is the height above the ground, z_i and z_f are the lower and upper limit of the $\overline{RCS}_{532}(z)$, respectively, and the respective values of dilatation and transition-related to mother-wavelet are given by a and b. Considering a previous study by Moreira et al. [41], which was held in São Paulo city, the parameters a and b received the values 200 and 40 m, respectively. Then, the height z where the maximum $W(a,b)$ occurs is classified as PBLH:

$$PBLH = Max(W(a,b)), \tag{3}$$

3.2. PBLH Levelness

The Levelness number (L) was defined by Stull [42] as:

$$L = \Delta z_i / \Delta z_T, \tag{4}$$

where Δz_i and Δz_T are PBLH and topography differences, respectively. The ratio L provides information about how the PBLH varies horizontally with respect to the topographic variations. If $L < 0$, the PBLH varies in an opposite way to the topography, that is, the higher terrains have lower PBLH values and vice-versa. $L = 0$ indicates that PBLH remains level concerning the surface. When L = 1, the PBLH follows topography, in other words, higher terrains have higher PBLH values and vice-versa. Finally, L > 1 indicates that the amplitude (absolute value) of PBLH differences is larger than topographic ones, so higher (lower) terrain has much higher (lower) PBLH.

In the case of São Paulo city $\Delta z_i = PBLH_{SEFAZ} - PBLH_{IPEN}$ and $\Delta z_T = -57$ m is the difference between the altitude of SEFAZ and IPEN lidar sites. To take advantage of the high spatial resolution of lidar systems (~7.5 m) used in São Paulo, cases L = 1 and L = 0 are replaced by L~1 and L~0, respectively. L~1 represents cases where L is closer to 1 than to 0, and PBLH tends to follow topography. As the difference between SEFAZ and IPEN altitudes $\Delta z_T = -57$ m, L~1 occurs when $57 \text{ m} \geq |\Delta z_i| \geq 29$ m (Table 1). On the other hand, L~0 indicates the cases where L is closer to 0 than to 1, and the PBLH tends to level when $|\Delta z_i| < 29$ m. Table 1 summarizes how L and Δz_i are related in the case of São Paulo city.

Table 1. Behavior classes of PBLH with respect to topography based on Levelness number (L) for São Paulo city. Classes of L are estimated in terms of PBLH (Δz_i) and topography (Δz_T) differences for SEFAZ and IPEN lidar sites. The background colors for each line represent the colors that will be used to illustrate the behavior of L in the next figures.

L	$\Delta z_i = PBLH_{SEFAZ} - PBLH_{IPEN}$	Behavior of PBLH with Respect to Topography
L < 0	$\Delta z_i > 29$ m	PBLH varies opposite to topography
L~0	$|\Delta z_i| < 29$ m	PBLH tends to level
L~1	$-57 \text{ m} \geq \Delta z_i \geq -29$ m	PBLH tends to follow the topography
L > 1	$\Delta z_i < -57$ m	PBLH differences are larger than topographic ones

3.3. Detection Algorithm of BB Events

Earlier studies show that in São Paulo city, the BB events are, in general, associated with the presence of high-concentration aerosol layers in the first 5 km of the atmosphere [10,25]. They occur between May and October when the number of forest fires in the center-west Brazil and Amazon regions and the burning of organic matter produced in the sugarcane crops in the countryside of São Paulo State [10,25] increases.

The intense aerosol loading combined with high solar radiation absorption capacity let these BB events be identified by the simultaneous increase in the daily values of Aerosol Optical Depth (AOD) and Aerosol Index (AI) [43,44]. Therefore, in this work, a new detection algorithm for BB events was developed and applied to São Paulo city. As displayed schematically in Figure 2, this algorithm combines key properties of lidar and satellite data. First, lidar data are used to find the aerosol layers. In the case of a positive outcome, daily values of AOD and AI are estimated for the lidar site and compared with corresponding daily values in the previous day.

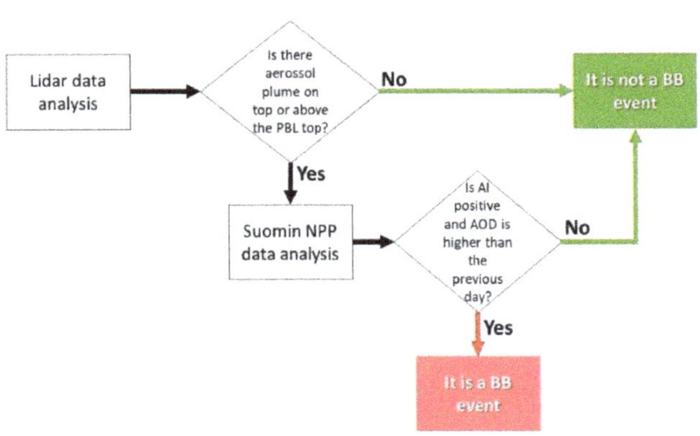

Figure 2. Scheme to BB event detection from the combination of elastic lidar and VIIRS data.

As there is no evidence that São Paulo city is affected by dust outbreaks, identifying the BB event is necessary only to verify whether daily values of AI remain positive during two consecutive days. To track the origin of the BB event is necessary to verify whether AOD daily values are increasing during these two consecutive days when AI values were positive. If both conditions are satisfied, it is plausible to infer that the aerosol layer detected in São Paulo was produced by a BB event and VIIRS data can be used to identify its origin.

4. Results

4.1. PBLH Horizontal Variation

In this section, the diurnal evolution of differences between PBLH retrieved at IPEN ($PBLH_{IPEN}$) and SEFAZ ($PBLH_{SEFAZ}$) from lidar measurements is analyzed. Firstly, two case studies illustrate how the absence (23 July 2019) and presence (8 August 2019) of low clouds affect the relationship between PBLH and topographic difference during the day. Then, a statistical analysis of the PBLH behavior is discussed, considering observations carried out for 12 days selected is also presented. It is important to emphasize that sea-breeze circulation is observed with frequency in São Paulo city [31]; however, sea breeze circulation was not observed on 23 July and 8 August 2019.

4.1.1. Case 1: Absence of Low Clouds

The diurnal evolution of PBLH is shown (by black stars) in the RCS-Intensity curtain plots of 23 July 2019 at IPEN (Figure 3a) and SEFAZ (Figure 3b). Due to technical differences in the lidar systems and possible spatial variation in the aerosol concentrations, the RCS intensities observed in both sites are not the same. However, the vertical structure of the aerosol layers is quite similar. This day is characterized by the absence of low clouds and the aerosol layer adjacent to the surface is well mixed, consequently a well-defined PBL in both sites. There is a thin aerosol layer (purple dashed boxes in Figure 3a,b) above the PBL, between 2000 and 3000 m, in both sites. As demonstrated, the Section 4.2 aerosol layers aloft the PBL may result from BB events advected into São Paulo city.

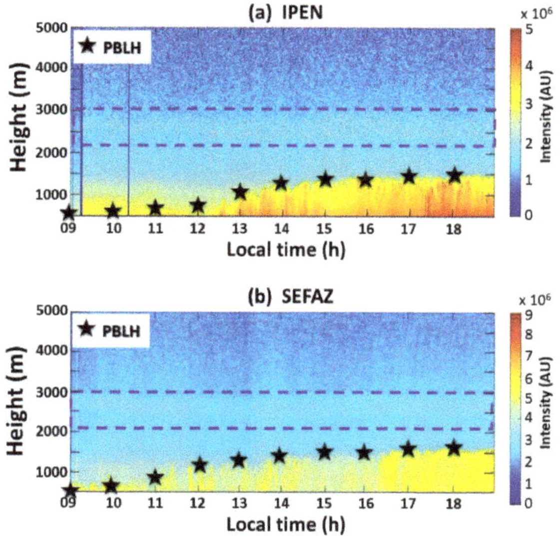

Figure 3. Diurnal evolution of (**a**) $PBLH_{IPEN}$ and (**b**) $PBLH_{SEFAZ}$ and corresponding RCS-Intensity curtain plot on 23 July 2019. The purple dashed boxes indicate an aerosol layer aloft the PBLH. Hourly values of PBLH are indicated by black stars.

At IPEN (Figure 3a), the PBLH remains practically constant from 09 to 11 LT, and it has a fast-growth stage from 12 to 14 LT reaching 1400.0 ± 7.5 m. Then, from 15 to 18 LT, PBLH remains practically constant and equal to 1452.0 ± 7.5 m.

At SEFAZ (Figure 3b), the PBLH displayed a longer fast-growth stage from 09 to 13 LT, reaching 1390.0 ± 7.5 m. At 15 LT, the PBLH reaches its maximum value of 1460.0 ± 7.5 m and maintains constant until 18 LT.

Figure 4 display the diurnal evolution of $PBLH_{SEFAZ}$-$PBLH_{IPEN}$ (Δz_i) and PBLH Levelness (L) on 23 July 2019. The higher differences occur during the growth stage because $PBLH_{SEFAZ}$ and $PBLH_{IPEN}$ displayed different growth rates (156.7 and 153.0 $m \cdot h^{-1}$, respectively), which is probably associated with the land use of each region (IPEN–suburban, SEFAZ–urban). Comparatively, on most mornings, $PBLH_{SEFAZ}$ is higher and has larger growth rate values than $PBLH_{IPEN}$. Such behavior is expected because the sensible heat flux in the urban region (SEFAZ) is expected to be higher than in the suburban region (IPEN). The maximum difference (Δz_i = 300.0 ± 10.6 m) was observed at 12 LT when $PBLH_{SEFAZ}$ reached the mature stage and $PBLH_{IPEN}$ just started the fast-growth stage. After the fast growth stage, at 14 LT, $PBLH_{IPEN}$ and $PBLH_{SEFAZ}$ displayed similar behavior, reaching a remarkably similar maximum value and remaining almost constant until the end of the day. During this period, an intense reduction in the difference between the PBLH values occurs (Δz_i = 8.0 ± 10.6 m).

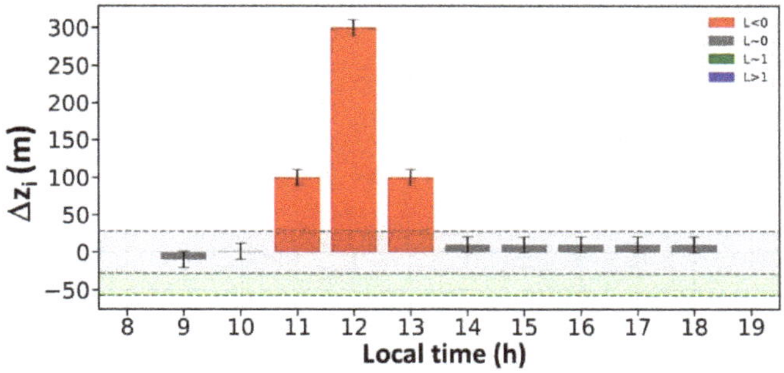

Figure 4. Diurnal evolution of Δz_i and L on 23 July 2019. Gray and green shadows represent the regions where L tends to 0 and 1, respectively.

Regarding L, during the first two hours (09 to 10 LT), while the fast PBL-growth stage has not started, the PBLH tends to level (L~0). During the fast-growth stage (11 to 13 LT), PBLH varies horizontally opposite to topography because the growth rate at SEFAZ is higher than at IPEN (L < 0). However, at 14 LT, when both PBL are almost fully developed, the PBLH tends to level again (L~0). This result agrees with Stull [42], which indicates that the PBLH tends to level when it is fully developed.

4.1.2. Case 2: Presence of Low Clouds

The diurnal evolution of PBLH is shown (by black stars) in the RCS-Intensity curtain plots on 8 August 2019 at IPEN (Figure 5a) and SEFAZ (Figure 5b). This day is characterized by the presence of scattered low clouds at both sites.

At IPEN, $PBLH_{IPEN}$ displays a fast-growth stage from 09 to 15 LT, reaching its maximum value of 2600 ± 7.5 m. The presence of scattered low clouds affected the performance of the WTC method, which tends to retrieve PBLH as the base of the cloud [15]. During the afternoon, the presence of clouds became more frequent in both sites, and PBLH was retrieved below the actual PBLH, indicating that the cloud base was below the actual PBLH in the IPEN site. Between 16 and 18 LT, PBLH was retrieved below the actual PBLH, indicating that the cloud base was below the actual PBLH.

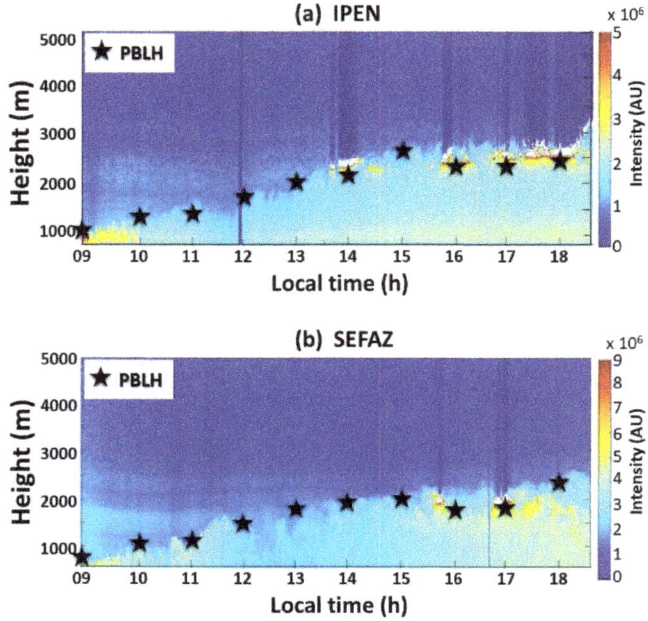

Figure 5. Diurnal evolution of (**a**) $PBLH_{IPEN}$ and (**b**) $PBLH_{SEFAZ}$ and corresponding RCS-Intensity curtain plot on 8 August 2019. Hourly values of PBLH are indicated by black stars.

At SEFAZ, $PBLH_{SEFAZ}$ displays a fast-growth stage from 09 to 14 LT. Between 15:30 and 17 LT, the presence of scattered low clouds affected the performance of the PBLH-detection algorithm so that hourly values of PBLH retrieved from the WCT method were misplaced below the actual PBLH. At 18 LT, $PBLH_{SEFAZ}$ reaches its maximum daily value of 2400 ± 7.5 m.

Like the previous case, $PBLH_{IPEN}$ and $PBLH_{SEFAZ}$ display different growth rates (266.7 and 166.7 $m{\cdot}h^{-1}$, respectively). However, in this case, the $PBLH_{IPEN}$ grows faster than $PBLH_{SEFAZ}$ from 9 to 14 LT (Figure 5), indicating that other causes rather than land use are affecting the diurnal evolution of PBLH in both sites.

Regarding L, it remains >1 during the fast-growth stage (09–13 LT) and most of the mature stage (15–17 LT), indicating that the amplitude of the absolute values of PBLH differences are systematically larger than topographic ones. The higher Δz_i value (-500 ± 10.6 m) is observed during the mature stage period. Comparatively to the fast-growth stage, the frequency of low clouds is higher during the mature stage in both sites (Figure 5). At 14 LT, no clouds were identified by both lidars sites (Figure 5), the growth rate of $PBLH_{IPEN}$ and $PBLH_{SEFAZ}$ are equally attenuated, and the PBLH difference decreases significantly ($\Delta z_i = -48 \pm 10.6$ m). As indicated by L~1 at 14 LT (Figure 6), the PBLH tends to follow the topography. At 18 LT, the presence of low clouds was observed only in the IPEN region. As a result, PBLH differences become smaller than Δz_i (56 ± 10.6 m), and, as indicated by $L < 0$, PBLH difference varies in the opposite way to the topography one.

Based on the above analysis, it seems plausible to infer that the presence of clouds has affected the diurnal evolution of PBLH in both sites by keeping the PBLH differences between SEFAZ and IPEN negative, which contradicts the expected land-use effect ng and by increasing it in absolute terms during the afternoon.

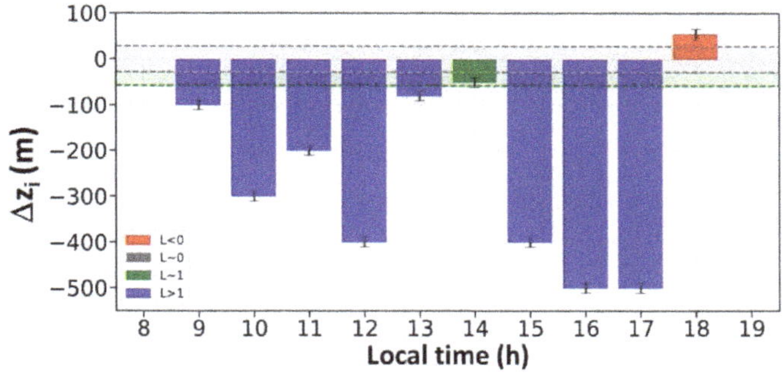

Figure 6. Diurnal evolution of Δz_i and L on 8 August 2019. Gray and green shadows represent the regions where L tends to 0 and 1, respectively.

The presence of low clouds intensifies the vertical mixing, mainly inside the cloud, reducing or even eliminating the presence of aerosol layers aloft at the PBL site. Several aerosol layers present in the first 3 km of the atmosphere at the IPEN (Figure 5a) were diluted after 13 LT and incorporated the PBL totally at 15 LT. While at SEFAZ (Figure 5b), higher aerosol layers remain above the PBL during the 17 LT. As will be shown in Section 4.2 below, these aerosol layers are composed of black carbons produced by BB events advected to São Paulo city during winter. During the daytime, they absorb incoming solar radiation, heating the atmosphere locally between 2 and 3 km. This localizing heat generates elevated inversion layers that, in turn, inhibit the vertical evolution of the PBL. This mechanism seems to be acting on 23 July 2019, when no low clouds were detected, restricting the vertical evolution of PBL to 1450 m in IPEN (Figure 3a) and SEFAZ (Figure 3b). On 8 August 2019, the presence of low clouds disrupted the process of stabilization induced by the aerosol layers from BB events, allowing a much deeper vertical evolution of the PBL, but as indicated in the Figure 5a,b, clouds were more frequent over IPEN than SEFAZ, so that the PBL differences remain negative during almost the entire daytime and becomes more intense during the afternoon when the presence of low clouds are more intense at IPEN site.

4.1.3. All Campaign

The diurnal evolution of statistical properties of $PBLH_{IPEN}$ and $PBLH_{SEFAZ}$, as well as their respective mean values ($\overline{PBLH}_{IPEN}$ and $\overline{PBLH}_{SEFAZ}$), retrieved from lidar performed in SEFAZ and IPEN during the 12 days of the field campaign, are presented in Figure 7.

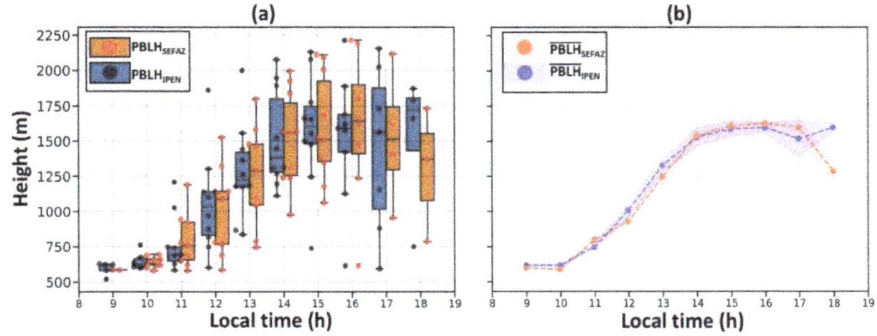

Figure 7. Diurnal evolution of (**a**) hourly values of $PBLH_{IPEN}$ (black dots) and $PBLH_{SEFAZ}$ (orange dots) and (**b**) average values of $PBLH_{IPEN}$ ($\overline{PBLH}_{IPEN}$, blue dots) and $PBLH_{SEFAZ}$ ($\overline{PBLH}_{SEFAZ}$, orange dots). PBLH values were retrieved from the WCT method during the 12-day field campaign of 2019. Symbols in (**a**) obey the boxplot standard. In (**b**), the statistical error is indicated by blue and red shadow bands.

From 09 to 10 LT, $PBLH_{IPEN}$ and $PBLH_{SEFAZ}$ have low variability and similar evolutions. At 11 LT, $PBLH_{SEFAZ}$ has greater variability and mean value. Between 12 and 13 LT, $PBLH_{SEFAZ}$ has the greatest variability, but $\overline{PBLH}_{IPEN}$ is higher than $\overline{PBLH}_{SEFAZ}$. Between 14 and 16 LT, both distributions and mean values of $PBLH_{IPEN}$ and $PBLH_{SEFAZ}$ have very similar behavior so considering the statistical error, $\overline{PBLH}_{IPEN}$ and $\overline{PBLH}_{SEFAZ}$ are equal. The maximum values of $\overline{PBLH}_{IPEN}$ (1600.0 ± 38.7 m) and $\overline{PBLH}_{SEFAZ}$ (1630.0 ± 54.2 m) are observed at 16 LT. At 17 LT, $PBLH_{IPEN}$ has high variability and an average value less than $PBLH_{SEFAZ}$. However, considering the statistical error, $\overline{PBLH}_{IPEN}$ (1520.0 ± 129.0 m) and $\overline{PBLH}_{SEFAZ}$ (1600.0 ± 77.4 m).

At 18 LT, $PBLH_{SEFAZ}$ has greater variability than $PBLH_{IPEN}$ and a predominance of smaller values. Such an effect is caused by the sea breeze, which systematically penetrates more than 50% of all days of the year in São Paulo city [31]. The sea breeze brings colder and more moist air from the coast, changing the vertical aerosol distribution and causing a high aerosol concentration in the lower part of the PBL. Then, methods based on the vertical aerosol gradient (e.g., WCT) tend to underestimate the PBLH in this situation [7]. Analyzing high-resolution data from numerical simulation with WRF Model, Ribeiro et al. [32] also show that the passage of sea breeze during the afternoon induces a significant drop in the PBLH in São Paulo city.

Figure 8 presents the diurnal evolution of the levelness parameter and the difference between $\overline{PBLH}_{SEFAZ}$ and $\overline{PBLH}_{IPEN}$ ($\Delta\overline{z_i}$). From 09 to 10 LT, the average Δz_i remains below $|28|$ m so that during this period, the PBLH tends to level (L~0). At 11 LT, $PBLH_{SEFAZ}$ display a large growth rate, Δz_i intensifies, reaching 55 ± 30.7 m and the average horizontal variation of PBLH is opposite to the topography (L < 1). Between 12 and 13 LT, the average PBLH displays the most intense growth rate in both sites and Δz_i becomes greater, so that the PBLH varies horizontally, amplifying topographic differences (L > 1). From 14 to 16 LT, the period in which $\overline{PBLH}_{SEFAZ}$ and $\overline{PBLH}_{IPEN}$ have little variation, Δz_i is again smaller than 28 m, and PBLH tends to levelling (L~0). At 17 LT, due to the reduction in the mean value of $PBLH_{IPEN}$, PBLH varies horizontally as opposed to topography (L < 0). Then, at 18 LT, the $PBLH_{SEFAZ}$ reduces sharply so that the PBLH varies, intensifying the topographical differences (L > 1). Such a phenomenon occurs due to the presence of the sea breeze, which arrives first at the SEFAZ, later spreading to the IPEN region. However, due to slow sea breeze displacement in urban areas [32,42], at 18 LT, only the SEFAZ region was reached by it. So, $PBLH_{SEFAZ}$ has a significant drop (Figure 7), becoming lower than $PBLH_{IPEN}$.

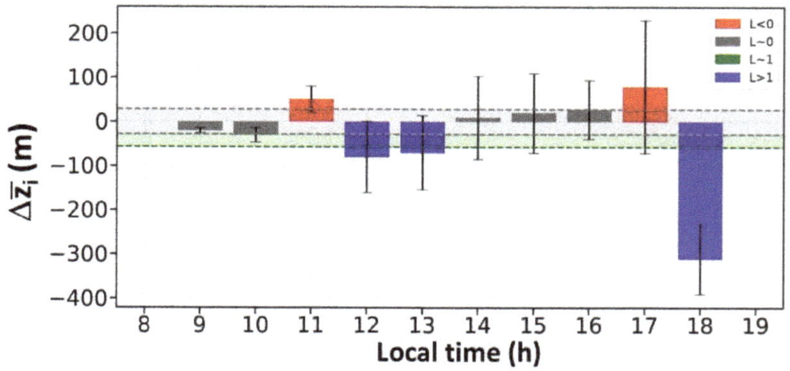

Figure 8. Diurnal evolution of $\overline{z_i}$ and L. Gray and green horizontal bands represent the regions where L tends to 0 and 1, respectively.

It is important to emphasize that the effects of thermal stabilization associated with the presence of aerosol layers associated with BB events on average seem to be canceled out by the effect of low clouds so that PBLH differences between SEFAZ and IPEN are very small.

4.2. Biomass Burning Detection

Figure 9 shows the RCS-Intensity curtain plots measured by the lidar at the SEFAZ on 19 August 2019. From 06 to 13 LT, an intense aerosol plume (dashed violet box) extends from 1000 m to the $PBLH_{SEFAZ}$. In the presence of intense aerosol plumes, such as the BB event on 19 August, the lidar methods based on the aerosol gradient tend to retrieve PBLH at the top of the aerosol layer. Therefore, the behavior of $PBLH_{SEFAZ}$ between 09 and 12 LT does not necessarily represent the real top of the PBL [2,13]. Indeed, on average, the PBLH displays a fast-growth stage during this period (Figure 7b). From 22 LT until the end of the measurements at 24 LT, a shallow layer with high aerosol concentration appeared near the surface, causing an intense backscattering and impeding laser signal to reach above this shallow layer.

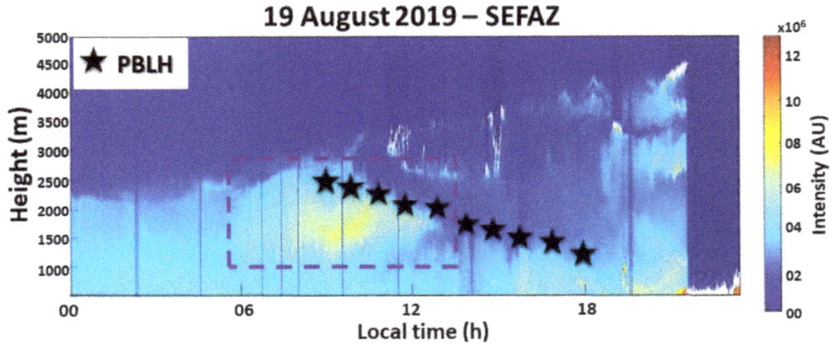

Figure 9. RCS-Intensity curtain plot obtained on 19 August 2019 at SEFAZ.

Figure 10 presents the spatial distribution of AI in Brazil, Paraguay, Bolivia, northern Argentina, and Uruguay from 17 to 19 August 2019. On 17 August, an aerosol plume, with AI values ranging between 2 and 5, was observed near the Bolivia–Brazil frontier. With the predominance of AI values higher than 3, the aerosol plume occupied a large horizontal extension on 18 August, spreading from the Brazil–Bolivia frontier to the Midwest region of Brazil. On 19 August, the aerosol plume was displaying AI values around 5. Located entirely in Brazil, this aerosol plume covered the States of São Paulo, Mato Grosso do Sul, Minas Gerais, and Rio de Janeiro.

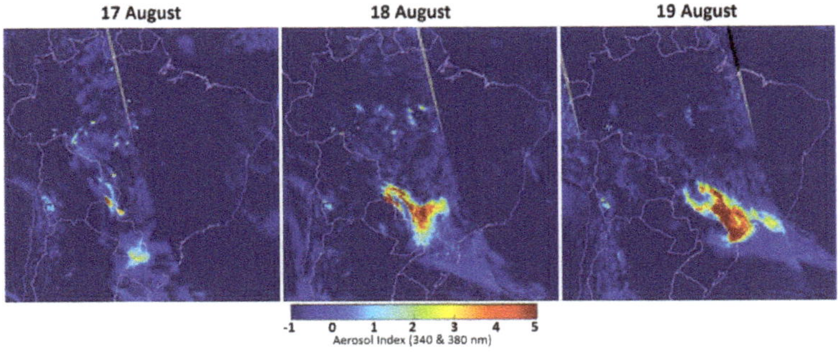

Figure 10. AI map values in Brazil, Paraguay, Bolivia, and northern Argentina and Uruguay from 17 to 19 August 2019. Such values were obtained from the OMPS-NM instrument on the Suomi-NPP satellite.

Figure 11 presents the variation of AOD in Brazil, Paraguay, Bolivia, and the northern region of Argentina and Uruguay, from 17 to 19 August 2019. On 17 August, while low values of AOD could be observed over southeastern Brazil, where São Paulo city is located, high AOD values (AOD > 0.8) were detected in the Brazil–Bolivia frontier. On 18 August, the area with high AOD values (>0.8) moved to the region between Brazil–Bolivia frontier and the Brazilian Midwest, coinciding with the aerosol plume shown in Figure 10. On the other hand, the AOD value for the visible region (without clouds) in the São Paulo city area remains below 0.2. On 19 August, despite the presence of clouds, it is possible to observe that AOD values are approximately 1 from the Brazilian Midwest to the São Paulo city area. Therefore, AOD values increased significantly in São Paulo city on 19 August in comparison to the previous two days.

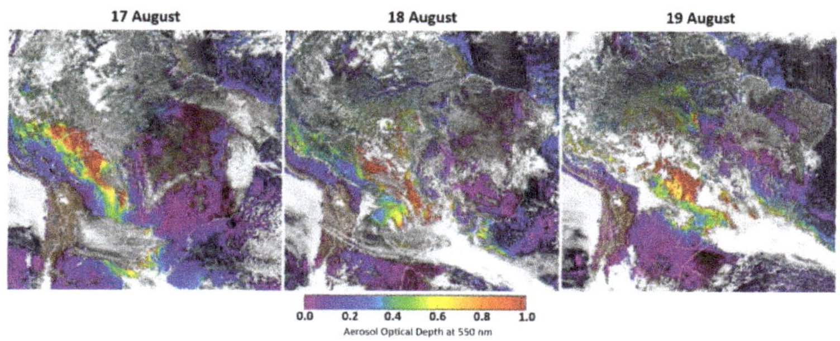

Figure 11. AOD map indicating BB plume over Brazil, Paraguay, Bolivia, and northern Argentina and Uruguay from 17 to 19 August 2019. AOD values were obtained from the VIIRS on Suomi NPP satellite data.

Inversion data from SP-EACH AERONET sunphotometer station were employed to derive the log-normal size distribution of aerosol from 17 to 19 August 2019. As shown in Figure 12, the log-normal distribution on 17 August changes during the day. However, fine mode aerosols are dominant during the whole period and there is strong evidence that most aerosols come from the BB event produced by the wildfires on the Brazil–Bolivia frontier.

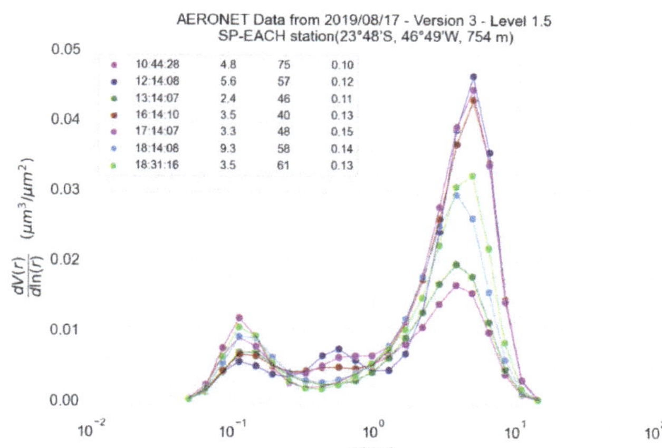

Figure 12. Aerosol size distribution was retrieved from AERONET sunphotometer measurements at the EACH station in São Paulo city (23°48′ S, 46°49′ W, 754 m asl) on 17 August 2019. Solid circle-lines numbers in the legend (top-left) indicate local time, particle radius (μm), number, and volume density ($\mu m^3\ \mu m^{-2}$). $dV(r)/dln(r)$ is the particle volume density and r is the particle radius.

According to Cazorla et al., 2012 [45], the chemical composition of aerosol can be identified from the Ångström Matrix. It consists of a dispersion diagram of Scattering Ångström Exponent (SAE) by Absorption Ångström Exponent (AAE). The sub-sections of this diagram define aerosol composition by grouping absorbing aerosol types [46,47]. During the aerosol transport event that occurred between 17 and 19 August over MRSP, the Ångström Matrix in Figure 13 shows the presence of small particles with low absorption, with some cases of black carbon aerosol composition.

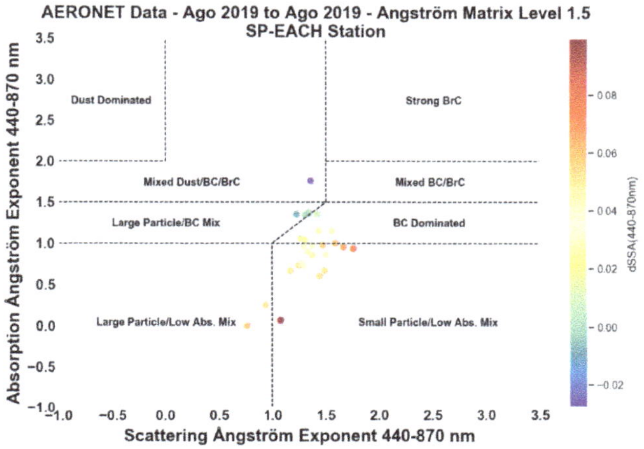

Figure 13. Dispersion diagram of Scattering Ångström Exponent by Absorption Ångström Exponent (Ångström Matrix Level 1.5). Ångström Exponent retrieved from AERONET sunphotometer measurements in São Paulo city (23°48′ S, 46°49′ W, 754 m asl) during August 2019. BrC, BC, and Abs indicate Brown Carbon, Black Carbon, and Absorption, respectively.

Therefore, based on the methodology presented in Section 3.3, it is possible to conclude that the intense aerosol layer observed on 19 August 2019 represents a BB event. Such a result is in accordance with Pereira et al. (2021) [48], who from a combination of back trajectory analyses (from Hybrid Single-Particle Lagrangian Integrated Trajectory [49]) and images from satellite (GOES-16 Advanced Baseline Instrument and Moderate Resolution Imaging Spectroradiometer) observed air masses transporting plumes of biomass burning aerosols from Bolivia and the Amazon Basin to São Paulo state. Such a BB event was widely publicized in the press due to its unprecedented effects, such as "black rain" [48], which could be detected in several regions of the São Paulo state.

In the 12 days of measurements, strong aerosol plumes above or on the PBL top were found in 3 of them (23 July, 10 and 19 August 2019), but only on 10 August variations in AOD and AI could be associated with the presence of material from biomass burning.

5. Conclusions

Elastic lidar systems have been widely applied in studies related to PBL structure. The high spatial and temporal resolutions of this type of remote sensing system allow it to be used in several environmental applications. In this work, data from two elastic LIDAR systems, which operated simultaneously at approximately 10.7 km of distance, were used to analyze the homogeneity of the PBLH in the city of São Paulo. In addition, a new procedure based on elastic lidar and satellites data were used to identify the impact of BB events in the aerosol layer detected above the PBL.

Except for the cases with the presence of sea breezes, in general, higher differences in the PBLH horizontal homogeneity (around 100 m) are observed during its growth period. Before the PBL growth stage and when the PBLH is fully developed, it was observed that PBLH tends to levelling (L~0). Sea breeze moves slowly in urban areas, so its presence makes the PBLH vary horizontally, amplifying topographical differences (L > 1). Furthermore, a stabilization of the atmosphere caused by the presence of aerosol layers associated with BB events was observed. However, such a result needs a more intensive investigation using radiosonde data to identify the presence of thermal inversion induced by these aerosol layers.

In addition, the synergy between the elastic lidar and satellites data enabled the detection of a BB event, as well as identifying its origin. Such a result demonstrates that the proposed methodology is an efficient and easy-applicable tool to detect BB events in São Paulo city. However, it is necessary to extend the application to a larger set of cases, using AE as a benchmark.

Therefore, these results reinforce the great importance and applicability of elastic lidar systems in PBL studies. So that the combination of a set of lidar systems or a synergy use of lidars and other remote sensing systems can provide a better understanding of certain phenomena.

Author Contributions: Conceptualization, G.d.A.M. and A.P.d.O.; methodology, G.d.A.M.; software, G.d.A.M.; formal analysis, G.d.A.M.; investigation, G.d.A.M., A.P.d.O., G.C., M.P.S., J.V.T., L.A.H.e.S., L.C.d.S., J.J.d.S., F.J.d.S.L. and E.L.; resources, A.P.d.O. and E.L.; data curation, G.d.A.M.; writing—original draft preparation, G.d.A.M.; writing—review and editing, G.d.A.M. and A.P.d.O.; visualization, G.d.A.M.; supervision, A.P.d.O. and E.L.; project administration, A.P.d.O. and E.L.; funding acquisition, A.P.d.O. and E.L. All authors have read and agreed to the published version of the manuscript.

Funding: This research was sponsored by the Brazilian Research Foundation CAPES (CAPES Financial Code 001), National Council for Scientific and Technological Development (CNPq): 309079/2013-6, 305357/2012-3, 462734/2014-5, 304786/2018-7, 154320/2018-8, 304786/2018-7; São Paulo Research Foundation (FAPESP): 2011/50178-5, 2020/07141-2.

Acknowledgments: We thank Alexandre Sadao Okubo and the SEFAZ technical team for their support throughout the measurement campaign.

Conflicts of Interest: The authors declare no conflict of interest.

References

1. Lange, D.; Tiana-alsina, J.; Saeed, U.; Tomás, S.; Rocadenbosch, F. Using a Kalman Filter and Backscatter Lidar Returns. *IEEE Trans. Geosci. Remote. Sens.* **2014**, *52*, 4717–4728. [CrossRef]
2. Bravo-Aranda, J.A.; de Arruda Moreira, G.; Navas-Guzmán, F.; Granados-Muñoz, M.J.; Guerrero-Rascado, J.L.; Pozo-Vázquez, D.; Arbizu-Barrena, C.; Olmo-Reyes, F.J.; Mallet, M.; Alados-Arboledas, L. A new methodology for PBL height estimations based on lidar depolarization measurements: Analysis and comparison against MWR and WRF model-based results. *Atmos. Chem. Phys.* **2017**, *17*, 6839–6851. [CrossRef]
3. Moreira, G.A.; Guerrero-Rascado, J.L.; Benavent-Oltra, J.A.; Ortiz-Amezcua, P.; Román, R.; Bedoya-Velásquez, A.E.; Bravo-Aranda, J.A.; Olmo Reyes, F.J.; Landulfo, E.; Alados-Arboledas, L. Analyzing the turbulent planetary boundary layer by remote sensing systems: The Doppler wind lidar, aerosol elastic lidar and microwave radiometer. *Atmos. Chem. Phys.* **2019**, *19*, 1263–1280. [CrossRef]
4. Liu, B.; Ma, Y.; Gong, W.; Zhang, M.; Yang, J. Improved two-wavelength Lidar algorithm for retrieving atmospheric boundary layer height. *J. Quant. Spectrosc. Radiat. Transf.* **2019**, *224*, 55–61. [CrossRef]
5. Lopes, F.J.S.; Moreira, G.A.; Rodrigues, P.F.; Guerrero-Rascado, J.L.; Andrade, M.F.; Landulfo, E. Lidar measurements of tropospheric aerosol and water vapor profiles during the winter season campaigns over the metropolitan area of Sao Paulo, Brazil. In Proceedings of the Lidar Technologies, Techniques, and Measurements for Atmospheric Remote Sensing X. International Society for Optics and Photonics, San Diego, CA, USA, 20 October 2014; p. 92460H. [CrossRef]
6. Ma, X.; Wang, C.; Han, G.; Ma, Y.; Li, S.; Gong, W.; Chen, J. Regional Atmospheric Aerosol Pollution Detection Based on LiDAR Remote Sensing. *Rem. Sens.* **2019**, *11*, 2339. [CrossRef]
7. Moreira, G.A.; da Silva Lopes, F.J.; Guerrero-Rascado, J.L.; da Silva, J.J.; Arleques Gomes, A.; Landulfo, E.; Alados-Arboledas, L. Analyzing the atmospheric boundary layer using high-order moments obtained from multiwavelength lidar data: Impact of wavelength choice. *Atmos. Meas. Tech.* **2019**, *12*, 4261–4276. [CrossRef]
8. Hu, Y.; Lu, X.; Zhai, P.; Hostetler, C.A.; Hair, J.W.; Cairns, B.; Sun, W.; Stamnes, S.; Omar, A.; Baize, R.; et al. Liquid phase cloud microphysical property estimates from CALIPSO measurements. *Front. Remote. Sens.* **2021**, *2*, 724615. [CrossRef]
9. Moreira, G.A.; Guerrero-Rascado, J.L.; Bravo-Aranda, J.A.; Foyo-Moreno, I.; Cazorla, A.; Alados, I.; Lyamani, H.; Landulfo, E.; Alados-Arboledas, L. Study of the planetary boundary layer height in an urban environment using a combination of microwave radiometer and ceilometer. *Atmos. Res.* **2020**, *240*, 104932. [CrossRef]
10. Mariano, G.L.; Lopes, F.J.S.; Jorge, M.P.O.M.; Landulfo, E. Assessment of biomass burnings activity with synergy of sunphotometric and LIDAR measurements in São Paulo, Brazil. *Atmos. Res.* **2010**, *98*, 486–499. [CrossRef]

11. Lopez, D.H.; Rabbani, M.R.; Crosbie, E.; Raman, A.; Arellano, A.F.; Sorooshian, A. Frequency and Character of Extreme Aerosol Events in the Southwestern United States: A Case Study Analysis in Arizona. *Atmosphere* **2016**, *7*, 1. [CrossRef]
12. Chan, K.L. Biomass burning source and their contributions to the local air quality in Hong Kong. *Scien. Tot. Environ.* **2017**, *596*, 212–221. [CrossRef]
13. Kovalev, A.V.; Eichinger, E.W. *Elastic Lidar: Theory, Practice and Analysis Methods*; Willey Interscience: Hoboken, NJ, USA, 2004.
14. Baars, H.; Ansmann, A.; Engelmann, R.; Althausen, D. Continuous monitoring of the boundary-layer top with lidar. *Atmos. Chem. Phys.* **2008**, *8*, 10749–10790. [CrossRef]
15. Emeis, S. *Surface-Based Remote Sensing of the Atmospheric Boundary Layer*; Springer: Berlin, Germany, 2011.
16. Ribeiro, F.N.D.; Soares, J.; Oliveira, A.P. The Co-Influence of the sea breeze and the coastal upwelling at Cabo Frio: A numerical investigation using coupled models. *Braz. J. Oceanogr.* **2011**, *59*, 131–144. [CrossRef]
17. Finnigan, J.; Ayotte, K.; Harman, I.; Katul, G.; Oldroyd, H.; Patton, E.; Poggi, D.; Ross, A.; Taylor, P. Boundary-Layer Flow over Complex Topography. *Bound.-Layer Meteorol.* **2020**, *177*, 247–313. [CrossRef]
18. Barlow, J.F.; Halios, C.H.; Lane, S.E.; Wood, C.R. Observations of urban boundary layer structure during a strong urban heat island event. *Env. Fluid Mech.* **2015**, *15*, 373–398. [CrossRef]
19. De Wekker, S.F.J.; Kossmann, M. Convective Boundary Layer Heights over Mountainous Terrain—A Review of Concepts. *Front. Earth Sci.* **2015**, *3*, 77. [CrossRef]
20. Fernández, A.J.; Sicard, M.; Costa, M.J.; Guerrero-Rascado, J.L.; Gómez-Amo, J.L.; Molero, F.; Barragán, R.; Basart, S.; Bortoli, D.; Bedoya-Velásquez, A.E.; et al. Extreme, wintertime Saharan dust intrusion in the Iberian Peninsula: Lidar monitoring and evaluation of dust forecast models during the February 2017 event. *Atmos. Res.* **2019**, *228*, 223–241. [CrossRef]
21. Gonzalez, M.E.; Garfield, J.G.; Corral, A.F.; Edwards, E.-L.; Zeider, K.; Sorooshian, A. Extreme Aerosol Events at Mesa Verde, Colorado: Implications for Air Quality Management. *Atmosphere* **2021**, *12*, 1140. [CrossRef]
22. Marques, J.F.; Alves, M.B.; Silveira, C.F.; Silva, A.A.; Silva, T.A.; Santos, V.J.; Calijuri, M.L. Fires dynamics in the Pantanal: Impacts of anthropogenic activities and climate change. *J. Environ. Manag.* **2021**, *299*, 113586. [CrossRef]
23. da Silva, S.S.; Oliveira, I.; Morello, T.F.; Anderson, L.O.; Karlokoski, A.; Brando, P.M.; de Melo, A.W.F.; da Costa, J.G.; de Souza, F.S.C.; da Silva, I.S.; et al. Burning in southwestern Brazilian Amazonia, 2016–2019. *J. Environ. Manag.* **2021**, *286*, 112189. [CrossRef]
24. Landulfo, E.; Lopes, F.J.S.; Mariano, G.L.; Torres, A.S.; Jesus, W.C.; Nakaema, W.M.; Jorge, M.P.P.M.; Mariani, R. Study of the Properties of Aerosols and the Air Quality Index Using a Backscatter Lidar System and Aeronet Sunphotometer in the City of São Paulo, Brazil. *J. Air Waste Manag. Assoc.* **2010**, *60*, 386–392. [CrossRef] [PubMed]
25. Landulfo, E.; Lopes, F.; Landulfo, E.; Mariano, E.; Martins, M.P. Impacts of Biomass burning in the atmosphere of the southeastern region of Brazil using remote sensing systems. In *Atmospheric Aerosol—Regional Characteristics—Chemistry and Physics*; IntechOpen: London, UK, 2003.
26. Oliveira, P.L.; de Figueiredo, B.R.; Cardoso, A.A. Atmospheric pollutants in São Paulo state, Brazil and effects on human health—A review. *Geochim. Bras.* **2011**, 17–24.
27. Squizzato, R.; Nogueira, T.; Martins, L.D.; Astolfo, R.; Machado, C.B.; Andrade, M.F.; Freitas, E.D. Beyond megacities: Tracking air pollution from urban areas and biomass burning in Brazil. *npj Clim. Atmos. Sci.* **2021**, *4*, 17. [CrossRef]
28. Mapbiomas Brasil. Project. Available online: https://mapbiomas.org/en/project (accessed on 12 December 2021).
29. Secretaria Municipal do Verde e do Meio Ambiente. Available online: https://www.prefeitura.sp.gov.br/cidade/secretarias/meio_ambiente/ (accessed on 12 December 2021).
30. Umezaki, A.S.; Ribeiro, F.N.D.; Oliveira, A.P.; Soares, J.; Miranda, R.M. Numerical characterization of spatial and temporal evolution of summer urban heat island intensity in São Paulo, Brazil. *Urban Clim.* **2020**, *32*, 100615. [CrossRef]
31. Oliveira, A.P.; Bornstein, R.; Soares, J. Annual and diurnal wind patterns in the city of Sao Paulo. *Water Air Soil Pollut.* **2003**, *3*, 3–15. [CrossRef]
32. Ribeiro, F.N.D.; Oliveira, A.P.; Soares, J.; Miranda, R.M.; Barlage, M.; Chen, F. Effect of sea breeze propagation on the urban boundary layer of the metropolitan region of Sao Paulo, Brazil. *Atmos. Res.* **2018**, *214*, 174–188. [CrossRef]
33. Instituto Brasileiro de Geografia e Estatística. Available online: http://ibge.gov.br (accessed on 1 December 2021).
34. Oliveira, A.P.; Marques Filho, E.P.; Ferreira, M.J.; Codato, G.; Ribeiro, F.N.D.; Landulfo, E.; Moreira, G.A.; Pereira, M.M.R.; Mlakar, P.; Boznar, M.Z.; et al. Assessing urban effects on the climate of metropolitan regions of Brazil—Preliminary results of the MCITY BRAZIL project. *Explor. Environ. Sci. Res.* **2020**, *1*, 38–77. [CrossRef]
35. Joint Polar Satellite System: Mission and Instruments. Available online: https://www.jpss.noaa.gov/mission_and_instruments.html (accessed on 30 October 2021).
36. Torres, O.; Bhartia, P.K.; Herman, J.R.; Ahmad, Z.; Gleason, J. Derivation of aerosol properties from satellite measurements of backscattered ultraviolet radiation: Theoretical basis. *J. Geophys. Res.* **1998**, *103*, 17,099–17,110. [CrossRef]
37. Holben, B.N.; Eck, T.F.; Slutsker, I.; Tanré, D.; Buis, J.P.; Setzer, A.; Vermote, E.; Reagan, J.A.; Kaufman, Y.J.; Nakajima, T.; et al. AERONET—A Federated Instrument Network and Data Archive for Aerosol Characterization. *Remote Sens. Environ.* **1998**, *66*, 1–16. [CrossRef]
38. Dubovik, O.; Smirnov, A.; Holben, B.N.; King, M.D.; Kaufman, Y.J.; Eck, T.F.; Slutsker, I. Accuracy assessments of aerosol optical properties retrieved from Aerosol Robotic Network (AERONET) Sun and sky radiance measurements. *J. Geophys. Res.* **2000**, *105*, 9791–9806. [CrossRef]

39. Dubovik, O.; King, M.D. A flexible inversion algorithm for retrieval of aerosol optical properties from Sun and sky radiance measurements. *J. Geophys. Res.* **2000**, *105*, 20673–20696. [CrossRef]
40. D'Amico, G.; Amodeo, A.; Mattis, I.; Freudenthaler, V.; Pappalardo, G. EARLINET Single Calculus Chain—Technical—Part 1: Pre-processing of raw lidar data. *Atmos. Meas. Tech.* **2016**, *9*, 491–507. [CrossRef]
41. Moreira, G.A.; Lopes, F.J.S.; Guerrero-Rascado, J.L.; Granados-Muñoz, M.J.; Bourayou, R.; Landulfo, E. Comparison between two algorithms based on different wavelets to obtain the Planetary Boundary Layer height. In Proceedings of the Lidar Technologies, Techniques, and Measurements for Atmospheric Remote Sensing X. International Society for Optics and Photonics, Amsterdam, The Netherlands, 20 October 2014; 2014; p. 92460D. [CrossRef]
42. Stull, R.B. A theory for mixed-layer-top levelness over irregular topography. In Proceedings of the 10th AMS Symposium on Turbulence and Diffusion (Portland), Portland, OR, USA, 29 September–2 October 1992.
43. Shi, S.; Cheng, T.; Gu, X.; Guo, H.; Wu, Y.; Wang, Y. Biomass burning aerosol characteristics for different vegetation types in different aging periods. *Environ. Int.* **2019**, *126*, 504–511. [CrossRef]
44. Liu, L.; Cheng, Y.; Wang, S.; Wei, C.; Pöhlker, M.L.; Pöhlker, C.; Artaxo, P.; Shrivastava, M.; Andreae, M.O.; Pöschl, U.; et al. Impact of biomass burning aerosols on radiation, clouds, and precipitation over the Amazon: Relative importance of aerosol–cloud and aerosol–radiation interactions. *Atmos. Chem. Phys.* **2020**, *20*, 13283–13301. [CrossRef]
45. Cazorla, A.; Bahadur, R.; Suski, K.J.; Cahill, J.F.; Chand, D.; Schmid, B.; Ramanathan, V.; Prather, K.A. Relating aerosol absorption due to soot, organic carbon, and dust to emission sources determined from in-situ chemical measurements. *Atmos. Chem. Phys.* **2013**, *13*, 9337–9350. [CrossRef]
46. Moosmüller, H.; Chakrabarty, R.K. Technical Note: Simple analytical relationships between Ångström coefficients of aerosol extinction, scattering, absorption, and single scattering albedo. *Atmos. Chem. Phys.* **2011**, *11*, 10677–10680. [CrossRef]
47. Cappa, C.D.; Kolesar, K.R.; Zhang, X.; Atkinson, D.B.; Pekour, M.S.; Zaveri, R.A.; Zelenyuk, A.; Zhang, Q. Understanding the optical properties of ambient sub- and supermicron particulate matter: Results from the CARES 2010 field study in northern California. *Atmos. Chem. Phys.* **2016**, *16*, 6511–6535. [CrossRef]
48. Pereira, G.M.; Caumo, S.E.S.; Grandis, A.; Nascimento, E.Q.M.; Correia, A.L.; Barbosa, H.M.J.; Marcondes, M.A.; Buckeridge, M.S.; Vasconcellos, P.C. Physical and chemical characterization of the 2019 "black rain" event in the Metropolitan Area of São Paulo, Brazil. *Atmos. Environ.* **2021**, *248*, 118229. [CrossRef]
49. Stein, A.F.; Draxler, R.R.; Rolph, G.D.; Stunder, B.J.B.; Cohen, M.D.; Ngan, F. NOAA's HYSPLIT atmospheric transport and dispersion modeling system. *Bull. Amer. Meteor. Soc.* **2015**, *96*, 2059–2077. [CrossRef]

atmosphere

MDPI

Article

Effects of Periodic Tidal Elevations on the Air-Sea Momentum and Turbulent Heat Fluxes in the East China Sea

Yuting Han [1,2,†], Yuxin Liu [1,3,*,†], Xingwei Jiang [3,4], Mingsen Lin [3,4], Yangang Li [5], Bo Yang [5], Changsan Xu [1,5], Lingling Yuan [6], Jingxin Luo [7], Kexiu Liu [7], Xingrong Chen [2], Fujiang Yu [2] and Xiangzhou Song [1]

[1] Key Laboratory of Marine Hazards Forecasting, Ministry of Natural Resources, Hohai University, Nanjing 210024, China; hyting@nmefc.cn (Y.H.); xucs@ecs.mnr.gov.cn (C.X.); xzsong@hhu.edu.cn (X.S.)
[2] National Marine Environment Forecasting Center, Ministry of Natural Resources, Beijing 100081, China; luckychen@nmefc.cn (X.C.); yufj@nmefc.cn (F.Y.)
[3] National Satellite Ocean Application Service, Ministry of Natural Resources, Beijing 100081, China; xwjiang@mail.nsoas.org.cn (X.J.); mslin@mail.nsoas.org.cn (M.L.)
[4] Key Laboratory of Space Ocean Remote Sensing and Application, Ministry of Natural Resources, Beijing 100081, China
[5] East China Sea Bureau of Ministry of Natural Resources, Shanghai 200137, China; liyangang@ecs.mnr.gov.cn (Y.L.); yangb@ecs.mnr.gov.cn (B.Y.)
[6] National Center of Ocean Standards and Metrology, Ministry of Natural Resources, Tianjin 300112, China; yuanoo81@163.com
[7] National Marine Data and Information Service, Ministry of Natural Resources, Tianjin 300171, China; luo_jingxin@163.com (J.L.); kxliu@163.com (K.L.)
* Correspondence: liuyuxin@mail.nsoas.org.cn
† These authors contributed equally to this work.

Abstract: Using bulk formulas, two-year platform (fastened to the seabed) hourly observations from 2016 to 2017 in the East China Sea (121.6° E, 32.4° N) are used to investigate the role of the tide-induced surface elevation in changing the fixed observational height and modifying the momentum and air-sea turbulent heat fluxes. The semidiurnal tide-dominated elevation anomalies ranging from −3.6 to 3.9 m change the fixed platform observational height. This change causes hourly differences in the wind stress and latent and sensible heat fluxes between estimates with and without considering surface elevation, with values ranging from -1.5×10^{-3} Nm^{-2}, −10.2 Wm^{-2}, and −3.6 Wm^{-2} to 2.2×10^{-3} Nm^{-2}, 8.4 Wm^{-2}, and 4.6 Wm^{-2}, respectively. More significant differences occur during spring tides. The differences show weak dependence on the temperature, indicating weak seasonal variations. The mean (maximum) difference percentage relative to the mean magnitude is approximately 3.5% (7%), 1.5% (3%), and 1.5% (3%) for the wind stress and latent and sensible heat fluxes, respectively. The boundary layer stability (BLS) can convert from near-neutral conditions to stable and unstable states in response to tide-induced changes in the observational height, with a probability of occurrence of 2%. Wind anomalies play dominant roles in determining the hourly anomalies of the latent heat flux, regardless of the state of the BLS. Extreme cases, including the cold air outbreak in 2016, tropical cyclones Meranti in 2016, and Ampil in 2018, are also examined. This study will facilitate future observation-reanalysis comparisons in the studied coastal region where ocean–atmosphere-land interactive processes are significant.

Keywords: air-sea turbulent heat fluxes; tidal processes; wind stress; boundary layer stability; platform observations

Citation: Han, Y.; Liu, Y.; Jiang, X.; Lin, M.; Li, Y.; Yang, B.; Xu, C.; Yuan, L.; Luo, J.; Liu, K.; et al. Effects of Periodic Tidal Elevations on the Air-Sea Momentum and Turbulent Heat Fluxes in the East China Sea. *Atmosphere* **2022**, *13*, 90. https://doi.org/10.3390/atmos13010090

Academic Editors: Yubin Li and Jie Tang

Received: 24 November 2021
Accepted: 16 December 2021
Published: 6 January 2022

1. Introduction

The turbulent fluxes of momentum (wind stress), latent heat flux (Q_{LH}), and sensible heat flux (Q_{SH}) at the air-sea interface are fundamental in virtually every atmosphere–ocean feedback process and undoubtedly important for understanding the air-sea interaction and climate change. The sea surface wind stress (τ) drives the local and basin-scale surface Ekman current, transport [1,2], and pumping, which determine the general ocean circulation [3,4] in terms of the wind stress curl. The air-sea turbulent heat fluxes, including evaporative Q_{LH} and conductive fluxes, balance the incoming solar radiation to obtain a net surface heat flux [5,6], which contributes to the seasonal cycle of sea surface temperature [7,8]. Thus, constructing accurate air-sea momentum and turbulent heat fluxes is vital for understanding upper ocean dynamics.

However, the momentum and heat flux estimates at global to regional scales suffer from significant uncertainties [9–14] arising from both the observation of physical variables in the air-sea boundary layers and the empirical estimates of parameters [15]. Errors can be generated by approximations in bulk formulas (described in Section 2). For example, the Q_{LH} obtained from eddy covariance measurements must be corrected by the mean vertical heat flux as a result of the requirement of the zero balanced net dry mass flux. This is called the Webb correction [16] and has been found to be approximately 2–3% of Q_{LH} [17]. The other climatological impacts of approximations and assumptions in the parameterization of bulk formulas on the momentum flux and turbulent heat fluxes have been discussed in a recent study by Brodeau et al. [18]. Khanna and Brasseur found that the effect of the boundary layer height on the dimensionless wind gradient under unstable conditions indirectly affects the exchange coefficient [19]. Johansson [20] and Sahlée's study [21] demonstrated the importance of the boundary layer height in the bulk algorithm for calculating the global mean air-sea flux. Gryning proposed that wind profile deviations based on the surface-layer theory and Monin–Obukhov scaling increasingly occur at 50–80 m above the surface, and these deviations are at least related to the boundary layer height under stable conditions [22]. Wind profile correction based on boundary layer height has been applied in many subsequent studies [23,24]. Song et al. introduced wave-induced components into the marine atmospheric boundary layer and calculated the wind profiles under two conditions: monochromatic wave and fully developed wind-generated sea. They studied the influence of surface waves on stable near-surface wind profiles [25]. Approximately 10–15% uncertainty is found for the momentum flux and turbulent heat fluxes related to the choice of the algorithm, which accounts for the significant heat flux discrepancies among different products [15]. These studies indicate that parameterized physics can cause potential bias or uncertainty in estimating wind and heat fluxes, which impedes our understanding of the surface energy balance from regional to global scales.

Errors can be found and identified by direct comparisons between point-to-point observations and flux products [10,12,26,27], which helps illuminate the uncertainties in products and increase confidence in their use. In a recent study by Song and Yu [28], it was found that, compared with low-resolution models, high-resolution models can resolve the complex coastline and provide more physically reasonable surface heat flux estimates. It is suggested that the air-sea heat fluxes along a coast with complex geography and air-sea interaction processes should have a significant effect on the regional and global energy balance. The near-coast platform observations fastened to the seabed in the East China Sea (ECS, Figure 1a) provide reliable air-sea variables that can be used to diagnose current products of wind stress and turbulent air-sea heat fluxes. Yet, such work has seldom been done in this region. The main objectives of the platform location selection in this study are as follows: First, the platform provides the opportunity to diagnose the wind stress and turbulent heat flux estimates in atmospheric reanalysis and objectively analyzed products in the coastal region, where the land–sea mask boundary is located. This will help assess the model's ability to resolve land–sea interactions at a high frequency. Second, this platform provides the basis for the comparisons between models and observations under lower air-sea turbulent heat fluxes (Figure 1a, coloured background) on the continental shelf, where the sea surface temperature (SST) is relatively lower than that in the open ocean or even in the Kuroshio Current in the ECS.

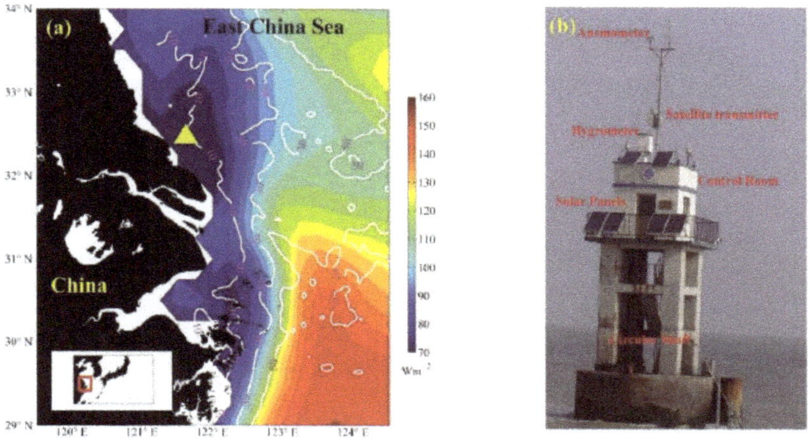

Figure 1. (**a**): The topography (white contours) of the East China Sea (ECS) with data from the General Bathymetric Chart of the Oceans (GEBCO, http://www.gebco.net accessed on 15 December 2021). The yellow triangle represents the location of the Huo Xing Sha platform. The coloured background is the annual mean Q_{LH} in 2016 and 2017 based on the ECMWF ERA-5 reanalysis. The incorporated frame (red) represents the study area of the ECS. (**b**): The in situ architecture of the platform with the structure labelled with red text.

However, one of the most important ocean dynamic processes on the continental shelf in the ECS is tidal currents and the associated changes in elevation at semidiurnal and diurnal frequencies [29–31]. The dominant semidiurnal rotating tidal currents can change the air-sea relative wind speed instead of the absolute wind speed during the estimates of the turbulent heat fluxes, which affects the diurnal cycles of these air-sea fluxes, which range from 0 to 24 Wm^{-2} [27] based on bulk formulas. Similarly, tidal elevation changes along the coast may change the heights of platform-based observations. The heights of the air-sea variables measured by the platform are fixed relative to the Earth (fastened to the seabed) but change relative to the tide-dominated sea surface elevation (SSE). Thus, the effective momentum and turbulent heat fluxes can be affected by a changing boundary height (z) induced by tidal elevation changes, and this is because the variables and parameters in bulk formulas are strongly dependent on the observational height z and boundary layer stability (BLS). This paper aims to solve a problem in estimating surface wind stress and turbulent heat fluxes by including tidal elevation based on platform observations. This study investigates how much uncertainty in the wind stress and turbulent heat fluxes is introduced by tidal elevation changes and how the BLS changes in response to periodic sea surface changes.

In this study, the role of tidal SSE (η) in modifying the platform-based air-sea momentum and turbulent heat fluxes is quantitatively identified along the coast of the ECS (Figure 1a), where tidal movements are significant. The goal of this paper is to evaluate how much uncertainty the tide-influenced SSE introduces into the estimates of τ and $Q_{LH,SH}$ in terms of the Coupled Ocean-Atmospheric Response Experiment (COARE) bulk flux algorithm version 3.0 (hereafter COARE 3.0) [32,33]. The remainder of this paper is organized as follows: Section 2 introduces the data and the bulk algorithms for estimating the air-sea momentum and turbulent heat flux. The observed air-sea variables and tidal SSE at the platform are analyzed in Section 3. In Section 4, the general role of the tidal SSE in modifying the high-resolution wind stress and turbulent heat fluxes is identified, and the BLS changes in terms of the tidal SSE are specifically analyzed. Section 5 presents the effects of tidal SSE on the modification of extreme air-sea momentum and turbulent heat fluxes under synoptic weather scale processes of cold-air outbreaks and tropical cyclones. Section 6 shows the wind and thermal effect on the high-resolution anomalies of Q_{LH} and

Q_{SH} under different air-sea boundary layer states. Finally, a summary and discussion are given in Section 7.

2. Data and Method

2.1. Description of Platform Observations: Construction and Instruments

Figure 1 shows the location (121.6° E, 32.4° N) and field situation of the platform that is used for operational air-sea observations. The platform has been built to help mitigate marine disasters and study climate change. Since tropical cyclones (TCs), simply called typhoons, land in China frequently every summer season, such platforms are designed to have a high level of wind resistance. The Huo Xing Sha platform was constructed in 2010 on the coast of the ECS and was maintained by the State Oceanic Administration (SOA), Ministry of Natural Resources (MNR), China. The instruments on the platform are marked in Figure 1b and listed in Table 1. The system was supplied with power from solar panels, and all the data were transmitted by the satellite. The two-year observations in this paper span from 1 January 2016 to 31 December 2017, with an hourly resolution. In April 2016 and August to September 2017, observations were missing due to platform maintenance. Thus, although there should be 17,544 hourly samples over a two-year period, only 14,358 samples are available for the analyses.

Table 1. Information on the observed variables and associated instruments on the platform.

	Variables	Manufacturer	Type	Range	Accuracy & Resolution	Height
1	SST	National Ocean Technology Center (NOTC)	YZY4 T-S Sensor	(−5~+50) °C	0.05 °C	0 m
2	SSS	NOTC	YZY4 T-S Sensor	8~36	0.1	0 m
3	SSE	Switzerland DIMETIX and NOTC	DLS-A30 SCA11	(0~1000) cm	0.1 cm	0 m
4	Wind Speed	NOTC	XFY3 Wind Sensor	(0~75) m/s	0.1 m/s	22 m
5	Wind Direction	NOTC	XFY3 Wind Sensor	(0~360)°	1°	22 m
6	SAT	Finland, VAISALA	HMP45A T-RH Sensor	(−40~+60) °C	0.1 °C	17 m
7	RH	Finland, VAISALA	HMP45A T-RH Sensor	(0~100)%	1%	17 m
8	SLP	Setra, USA	278 Barometric Sensor	(850~1050) hPa	0.1 hPa	13 m

Figure 2 shows the plan-view architecture of the fastened platform. The platform is supported by six prestressed high-intensity concrete (PHC) pipe piles with a length of 46 m and two auxiliary piles with a length of 36 m. The diameters of the PHC pipe piles and auxiliary piles are 1.2 m and 1 m, respectively. The diameter of the base is 6.5 m. The height of the platform base is 7 m, which is slightly higher than the local water depth of 5.3 m (analyses in Section 2). The observed physical variables and parameters, including the observed accuracy and heights, are listed in Table 1. The SST and sea surface salinity (SSS) were observed by a YZY T-S sensor manufactured by the National Ocean Technology Center (NOTC, Tianjing, China) of the MNR. The SSE was measured by a DLS-A30 SCA11 from the NOTC (Tianjing, China). The heights of the SST and SSS measurements were 0 m due to the float-type T-S sensor in the circular shaft. The wind speed (WS) and direction were obtained by an SFY4 wind sensor on the top of the control room, with a height of

22 m above the mean sea surface. The HMP45A T-RH sensor for surface air temperature (SAT) and relative humidity (RH) was approximately 17 m above the mean sea surface and thus lower than the wind sensor. The sea level pressure (SLP) was recorded by a 278 barometric sensor (Setra, Boxborough, MA, USA) at the height of 13 m in the control room. The temporal resolutions of SST, SAT, SLP, RH, WS, and SSE were recorded hourly. All the instruments and sensors were calibrated by the National Center of Ocean Standards and Metrology, MNR. Data quality control was strictly performed by the National Marine Data and Information Service, MNR.

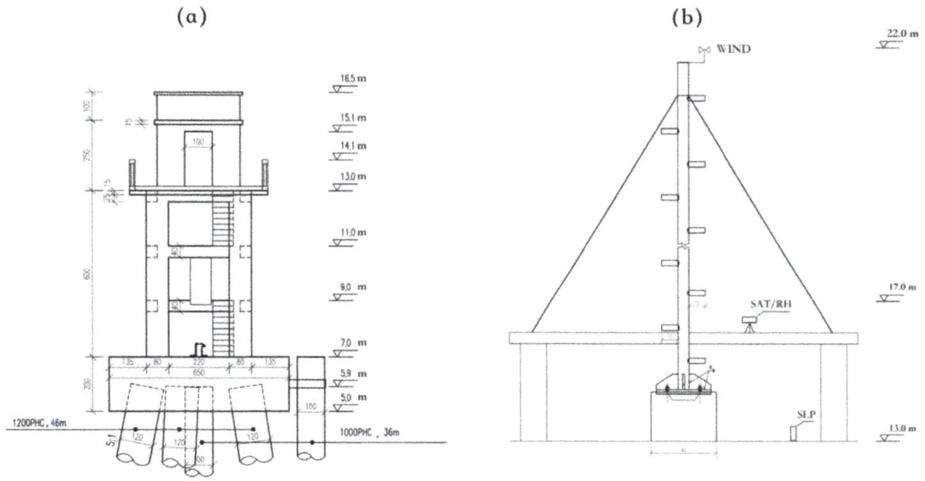

Figure 2. The plan-view architecture of the platform (**a**) and the observational air-sea parameter sensors (**b**). Note that the numbers followed by units are in meters, but those without units are in centimetres. Detailed information on the observational heights of the sensors is listed in Table 1.

2.2. Method of Air-Sea Flux Calculation: Bulk Formulas

Following the Monin–Obukhov similarity theory (MOST) [34], momentum and turbulent heat fluxes, namely, τ, Q_{LH} and Q_{SH}, are conventionally estimated by bulk Equations [32,33,35–37]:

$$\tau = \rho c_D \mathbf{u}_z |\mathbf{u}_z| \tag{1}$$

$$Q_{LH} = \rho L_e c_E |\mathbf{u}_z| (q_s - q_a) \tag{2}$$

$$Q_{SH} = \rho c_p c_H |\mathbf{u}_z| (T_s - \theta_a) \tag{3}$$

where ρ is the density of air, L_e is the Q_{LH} of evaporation, c_p is the specific heat capacity of air, and $\mathbf{u}_z$ is the wind vector at height z. The drag coefficient for wind stress τ is denoted by c_D. The turbulent exchange coefficients for the Q_{LH} and Q_{SH} are denoted by c_E and c_H, respectively. $\Delta q = q_s - q_a$ and $\Delta T = T_s - \theta_a$ represent the sea–air humidity and temperature difference, respectively. The surface and near-surface atmospheric specific humidity are denoted by q_s and q_a, respectively. T_s is the SST. q_s is the saturation-specific humidity at T_s and includes a 2% reduction to account for the reduction in vapour pressure caused by a typical salinity of 34 psu: $q_s \cong 0.98 q_{sat}(T_s)$. θ_a includes a correction for the adiabatic lapse rate ϑ, $\theta_a = T_z + \vartheta z$, and T_z is the air temperature at height z. The parameters in Equations (1)–(3), such as the WS, RH, and air temperature, are closely related to the observation height z in the vertical profile. Furthermore, the drag coefficient c_D, moisture transfer coefficient c_E, and temperature transfer coefficient c_H are also related to the surface roughness lengths.

According to the MOST, the wind speed, humidity, and temperature profiles can be made dimensionless, and their dimensionless profile functions can be expressed as:

$$\varphi_m(\xi) = \frac{kz}{u_*}\frac{\partial u}{\partial z} \tag{4}$$

$$\varphi_q(\xi) = \frac{kz}{q_*}\frac{\partial q}{\partial z} \tag{5}$$

$$\varphi_h(\xi) = \frac{kz}{\theta_*}\frac{\partial \theta}{\partial z} \tag{6}$$

where m, q, and h are momentum, humidity, and sensible heat, respectively. u_* is the friction velocity, q_* is the humidity scaling parameter, θ_* is the temperature scaling parameter, $\xi = z/L$ is the Monin–Obukhov stability parameter. The estimation of the momentum and turbulent heat fluxes are closely associated with the air-sea BLS, which is determined by the Monin–Obukhov stability parameter, where z is the height of the turbulent transfer coefficient and L is the Monin–Obukhov length scale:

$$L = \frac{u_*^2}{\kappa \frac{g}{\bar{\theta}} \theta_*} \tag{7}$$

In Equation (7), $\kappa \approx 0.4$ is the von Karman constant, g is the gravitational acceleration, $\bar{\theta}$ is the mean temperature in the boundary layer, and $\theta_* = -\frac{\overline{w'\theta'}}{u_*}$. $\overline{w'\theta'}$ is the Reynolds stress term, where θ and w represent the temperature and vertical motion, respectively, and the prime symbol denotes a fluctuation. The Monin–Obukhov length scale (L) represents the ratio of the work performed by the Reynolds stress to that performed by buoyancy forces. The unstable, near-neutral, and stable boundary conditions are determined by $\xi < -0.4$, $-0.4 < \xi < 0.1$ and $\xi > 0.1$, respectively. The Monin–Obukhov length (L) and other scaling parameters have been reviewed by previous studies [32,33,38–41]. This paper focuses on the modification of the observational height z by SSE, η, and thus, the boundary layer parameter (z/L) may vary as a result of changes in tidal elevation. An analysis of how the changes in SSE affect the BLS and air-sea momentum and turbulent heat fluxes is presented in Section 4.

The exchange coefficients in Equations (1)–(3) can be obtained by using Equations (4)–(6):

$$c_D = \left(\frac{\kappa}{ln(z/z_0) - \psi_m(\xi)}\right)^2 \tag{8}$$

$$c_E = \left(\frac{\kappa}{ln(z/z_0) - \psi_m(\xi)}\right)\left(\frac{\kappa}{ln(z/z_{0q}) - \psi_q(\xi)}\right) \tag{9}$$

$$c_H = \left(\frac{\kappa}{ln(z/z_0) - \psi_m(\xi)}\right)\left(\frac{\kappa}{ln(z/z_{0T}) - \psi_h(\xi)}\right) \tag{10}$$

where z_0 is the aerodynamic roughness lengths, z_{0T} and z_{0q} are the roughness lengths for temperature and humidity, respectively. ψ_x is the integrated dimensionless profile function:

$$\psi_x == \int_0^{z/L} \frac{(1 - \varphi_x(\xi))}{\xi} d(\xi) \tag{11}$$

where $x = m, q, h$, respectively.

In the estimation of turbulent fluxes, the profiles of WS, RH, temperature, SLP, and even the coefficients in bulk formulas are dependent on the observation height (z). In this paper, the COARE bulk flux algorithm version 3.0 (Φ) [32,33,37] was used as in the Objectively Analyzed Air-sea Flux project (OAFlux) [42], which includes synthesized observations and estimates from various sources, including satellite observations and reanalysis. By convention, the wind stress $\tau(z)$ and turbulent heat fluxes $Q_{LH,SH}$ are estimated by the COARE algorithm in terms of WS $\mathbf{u}(z_w)$ at height z_w, RH γ at z_γ, SAT θ_a at z_θ, and SLP p and SST T_s at z_s:

$$\begin{aligned} \tau(z+\eta) &= \Phi_\tau[\mathbf{u}(z_w+\eta), \gamma(z_\gamma+\eta),\ \theta_a(z_\theta+\eta),\ p(z_s),\ T_s(z_s)] \\ Q_{LH,SH}(z+\eta) &= \Phi_{LH,SH}[\mathbf{u}(z_w+\eta), \gamma(z_\gamma+\eta),\ \theta_a(z_\theta+\eta),\ p(z_s),\ T_s(z_s)] \end{aligned} \tag{12}$$

where η is the SSE associated with tidal motions and $\eta = 0$ when tides are not considered.

The differences in τ, Q_{LH}, and Q_{SH} between the estimates with and without SSE η are estimated by the following Equations:

$$[\tau, Q_{LH,SH}]_{DIFF} = [\Phi_\tau(z_w+\eta), \Phi_{LH,SH}(z+\eta)] - [\Phi_\tau(z_w), \Phi_{LH,SH}(z)] \tag{13}$$

3. Observations of Air-Sea Variables and SSE

3.1. Direct Observations of Air-Sea Variables

Figure 3 shows the two-year time series of air-sea variables of SST (T), SAT (θ_a), RH (γ), SLP (p), and WS ($|\mathbf{u}_z|$), which can be used to estimate the air-sea momentum fluxes. The two-year mean, maximum, and minimum T (Figure 3a) were 16.2 °C, 32.4 °C, and 1.1 °C, while the two-year mean, maximum, and minimum θ_a were 15.0 °C, 33.8 °C, and −10.4 °C, respectively. The mean θ_a (Figure 3b) was less than the mean T, indicating that heat loss occurred moving from the ocean to the atmosphere by heat conduction, namely, Q_{SH}. However, due to synoptic weather processes, the maximum θ_a was higher than the maximum T. Both T and θ_a showed significant seasonal cycles from 2016 to 2017. The SST demonstrated a clear diurnal cycle, while θ_a, which exhibited peaks, was mostly affected by synoptic-scale weather processes. The mean RH (γ) was 74%. The SLP (p) was lower in the warm season from May to October than in the cold season. Unlike the other variables, WS (Figure 3e) did not show any particular cycles over the two-year duration, although peaks were occasionally associated with synoptic-scale processes. These five key air-sea parameters provide the basis for calculating the traditional momentum and turbulent heat fluxes.

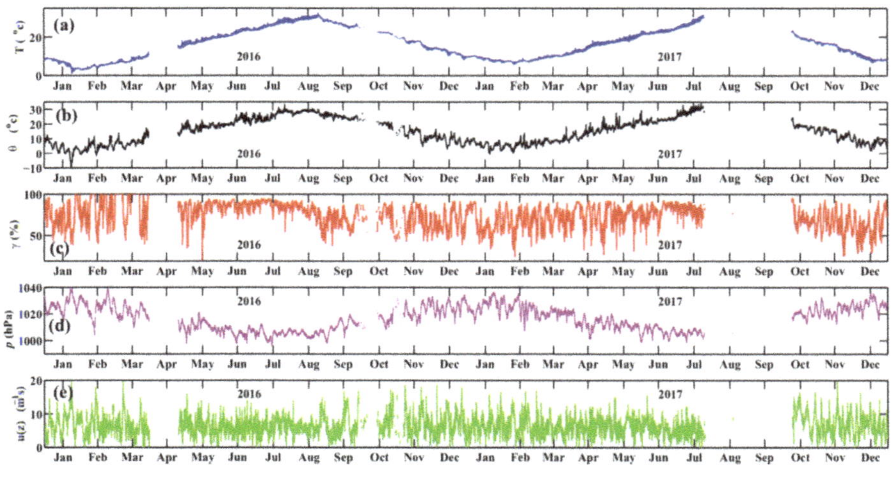

Figure 3. Time series of the air-sea variables observed at the platform from 2016 to 2017. The hourly SST (dark blue), SAT (black), RH (red), SLP (purple) and WS (green) are shown in panels (**a**–**e**), respectively.

3.2. Observations of SSE and Harmonic Analysis

Together with the air-sea variables at the platform, the SSE was also obtained with a significant periodic signal (Figure 4a). As the tidal signal was dominantly periodic, the plots only showed the results of June 2016. From 2016 to 2017, the mean water depth (AWD) was 5.3 m, with a maximum of 9.2 m and a minimum of 1.7 m, which indicated a maximum difference in SSE at 7.5 m. Using harmonic analysis, the main tidal constituents in this location were semidiurnal and diurnal components (Table 2). Six main components were obtained, including M2, S2, N2, K2, K1, and O1. The M2 component was dominant over the other components, namely, S2, N2, K2, K1, and O1. The amplitude of M2 was 1.9 m, which was larger than the sum of the other five. The sum of the amplitudes of the six main components was 3.5 m.

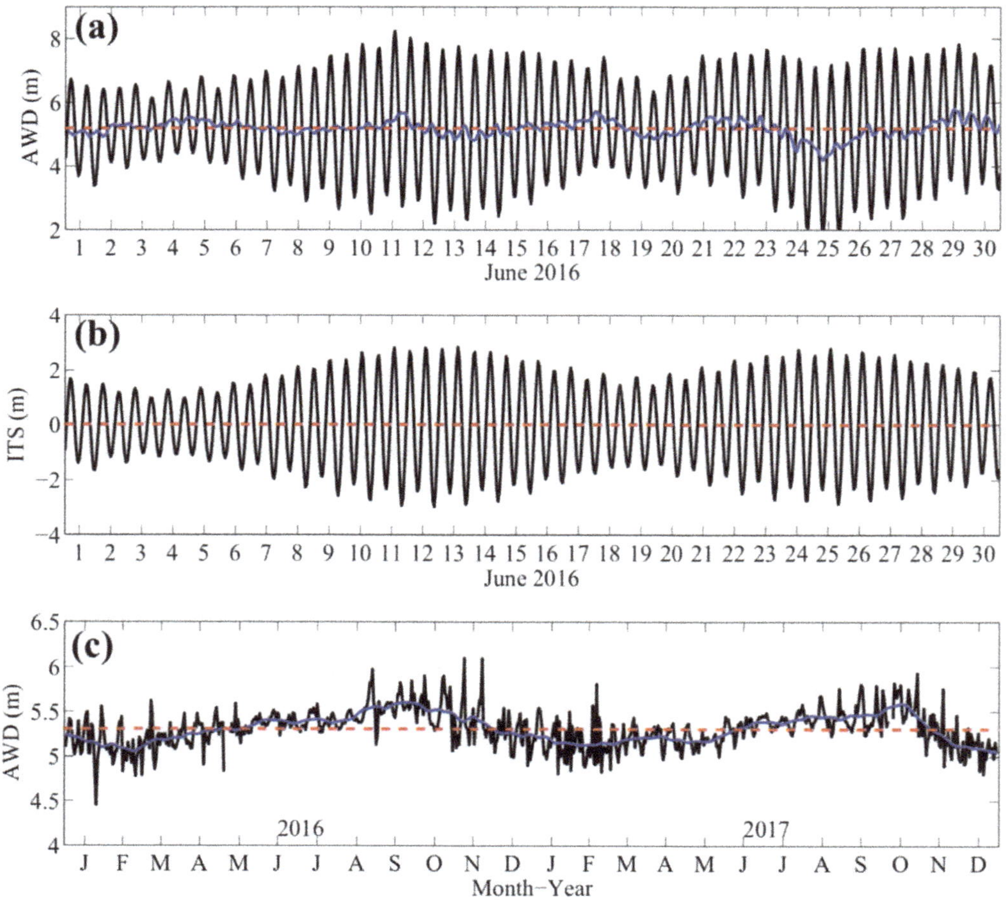

Figure 4. (**a**): The hourly water depth (WD, unit: m) at the Huo Xing Sha platform (black solid) and its mean value (red dash) in June 2016. (**b**): The integrated tidal signal (ITS, solid black, unit: m) by harmonic analysis in June 2016. The blue solid line in (**a**) represents the residual signal between the AWD and ITS. (**c**): The daily mean AWD (unit: m) from 2016 to 2017 (red curve) and the monthly mean results (dark blue). Detailed information on tidal constituents is listed in Table 2.

Table 2. Characteristics of the six tide constituents by harmonic analysis with a signal-noise ratio greater than 100. M2, S2, N2 and K2 are semidiurnal tidal components, while K1 and O1 are diurnal components.

Tide Constituent	Frequency ($hour^{-1}$)	Amplitude (m)	Amplitude Error (m)	Phase (°)	Phase Error (°)	Signal Noise Ratio
M2	8.1×10^{-2}	1.9	2.0×10^{-2}	239.0	0.5	1.5×10^{4}
S2	8.3×10^{-2}	0.8	2.0×10^{-2}	28.9	1.2	2.8×10^{3}
N2	7.9×10^{-2}	0.3	1.6×10^{-2}	32.9	2.8	4.6×10^{2}
K1	4.2×10^{-2}	0.2	1.0×10^{-2}	277.0	2.6	6.4×10^{2}
K2	8.4×10^{-2}	0.2	2.0×10^{-2}	317.1	5.2	1.4×10^{2}
O1	3.9×10^{-2}	0.1	1.0×10^{-2}	165.6	4.7	1.9×10^{2}

Figure 4b shows the integrated tidal signal (ITS) of the above six components in June. The neap tides for this regard occurred around 4 and 18 June, and the spring tides occurred around 12 and 25 June. The maximum ITS amplitude was 3 m during spring tide, indicating a maximum 6 m SSE difference over the whole day. Note that the maximum amplitude of the ITS was slightly smaller than that of the sum of amplitudes of six tidal components due to the interactions among different components. The minimum ITS amplitude was only 1 m, with a 2 m SSE difference. The residual signal of the difference between the SSE and ITS indicates the effect of other ocean dynamics beyond tidal processes, such as mesoscale eddies and coastal waves (blue in Figure 4a). However, the magnitude of the residual signal was no more than 0.2 m, which was much weaker than that of the ITS. In this paper, the original SSE observations were incorporated for calculating momentum fluxes. Figure 4c shows the daily mean AWD and the monthly mean results from 2016 to 2017. Significant seasonal variations can be found: a low AWD of 5 m was observed in February, and a high AWD of approximately 5.5 m was observed in September. However, there is no clear, dynamic explanation of these seasonal variations in terms of the coastal sea level in the ECS, and this is beyond the scope of this paper.

3.3. Estimations of Air-Sea Turbulent Heat Fluxes Based on Bulk Formulas

Using the two-year observations of air-sea variables at the Huo Xing Sha platform, Figure 5 shows the platform-based daily mean Q_{LH} and Q_{SH} compared with the OAFlux product and the scatter plots of daily mean $Q_{LH,SH}$ between the platform observations and OAFlux. The mean Q_{LH} from two-year observations without considering SSE was 78 Wm^{-2} (positive values indicate upward heat flux), with a daily standard deviation (STD) of 70 Wm^{-2}, while the mean Q_{LH} of OAFlux was 63 Wm^{-2}, with a daily STD of 60 Wm^{-2}. Furthermore, the two-year mean platform-based Q_{SH} was 14 Wm^{-2} (STD = 25 Wm^{-2}), while the mean Q_{SH} of OAFlux was approximately 12 Wm^{-2}, with an STD of 31 Wm^{-2}. Although the same algorithm was used for the flux calculations, the estimates of turbulent heat fluxes based on the platform observations were slightly higher than those of OAFlux. The reason for this discrepancy might be attributed to the uncertainty in the air-sea variables between the OAFlux and platform observations in July and August of 2016. The air-sea momentum and turbulent heat fluxes and associated air-sea variables will be compared among the platform measurements and other products in the following study. The correlation coefficients (root mean square and STD values) of Q_{LH} and Q_{SH} between the platform estimates and OAFlux were 0.83 (42 and 39 Wm^{-2}) and 0.90 (14 and 14 Wm^{-2}), respectively.

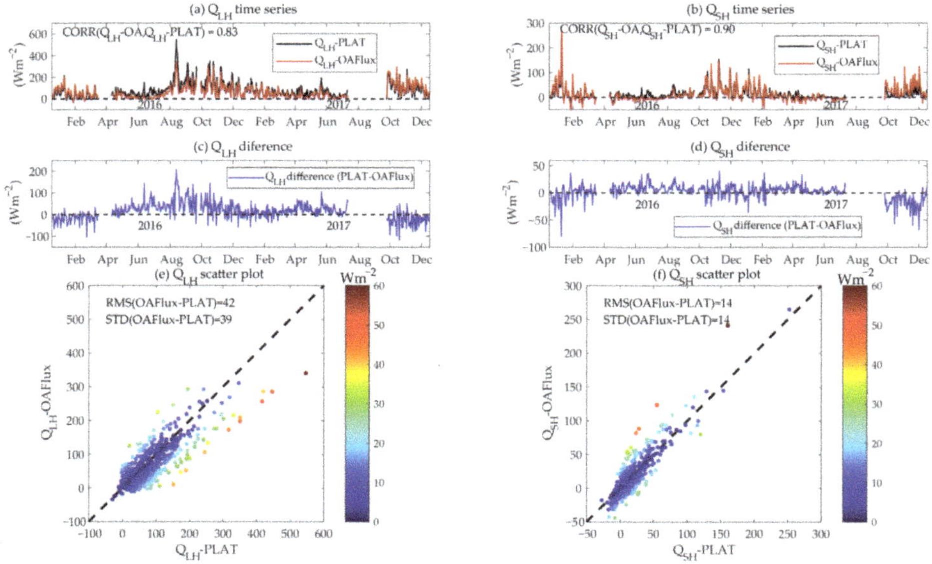

Figure 5. Estimated daily mean Q_{LH} and Q_{SH} ((**a**,**b**), unit: Wm^{-2}) from 2016 to 2017 based on platform observations in ECS using COARE 3.0 bulk formulas (black) and OAFlux (red) data. (**c**,**d**): Estimated daily mean Q_{LH} and Q_{SH} differences between OAFlux and platform. (**e**,**f**): Scatter plots for the comparisons with the colours of the absolute difference between coastal platform observations and OAFlux.

Figure 5 also shows the significant seasonal variations in Q_{LH} and Q_{SH}. During boreal summer, when the thermal states of the air (θ_a and q_s) are closer to those of the near-surface Q_{LH} and Q_{SH} are relatively low. However, in boreal winter, when the air-sea temperature and humidity contrasts are more significant, and the winds are stronger, the Q_{LH} and Q_{SH} are high. The peaks of Q_{SH} in the platform were closely associated with the cold-air outbreak in winter (25 January 2016), which was consistent with previous studies [43–45]. The extreme Q_{LH} might be associated with the TC Meranti on 9–16 September 2016. The maximum Q_{LH} and Q_{SH} were 664 Wm^{-2} and 345 Wm^{-2}, respectively, based on platform observations from 2016 to 2017. The extreme momentum and turbulent heat fluxes under extreme weather conditions (cold-air outbreaks in winter and TCs in summer) and the role of the tidal SSE in modifying them are analyzed in Section 5.

4. Effect of Tidal SSE on the Air-Sea Momentum and Turbulent Heat Fluxes

4.1. Mean Result

In terms of Equation (13), the differences in wind stress, Q_{LH}, and Q_{SH} between the estimates with and without SSE ranged from -1.5×10^{-3} Nm^{-2}, -10.2 Wm^{-2}, and -3.6 Wm^{-2} to 2.2×10^{-3} Nm^{-2}, 8.4 Wm^{-2}, and 4.6 Wm^{-2}, respectively, indicating the significant role of tidal SSE in modifying the high-frequency (hourly) air-sea momentum and turbulent heat fluxes. However, in the mean state, the differences were much more minor than the mean values due to the elimination of the tidal SSE during the two-year observations. Figure 6 shows the interpolated differences in wind stress, Q_{LH}, and Q_{SH} with anomalies in tidal SSE and SST from 2016 to 2017. It is evident that the tidal SSE plays an important role in contributing to the calculation of the wind stress, Q_{LH}, and Q_{SH}. With tidal SSE anomalies from -3.5 to 3.5 m, the wind stress, Q_{LH}, and Q_{SH} differences between the estimates with and without SSE varied from -0.015 Nm^{-2}, -8 Wm^{-2}, and -3 Wm^{-2} to 0.02 Nm^{-2}, 8 Wm^{-2}, and 3 Wm^{-2}, respectively. Taking the wind stress as an example, when the SSE rises, the effective height of the observed wind stress acting on the sea surface decreases. The resultant sea surface wind stress is enhanced. On the other hand, the effective wind is weakened by an increased observational height when the SSE drops. The Q_{LH} (Figure 6b) exhibits a pattern similar to that of the wind stress (Figure 6a); however, there are a number of exceptional points for Q_{SH} (Figure 6c). Q_{SH} can be high under low $z + \eta$ conditions and can be low under high $z + \eta$ conditions. This may be determined by the changes in BLS associated with the surface elevations η and the changes in air temperature $\theta_a(z) = T + \vartheta z$ associated with the lapse rate, which is analyzed explicitly in Section 4.2.

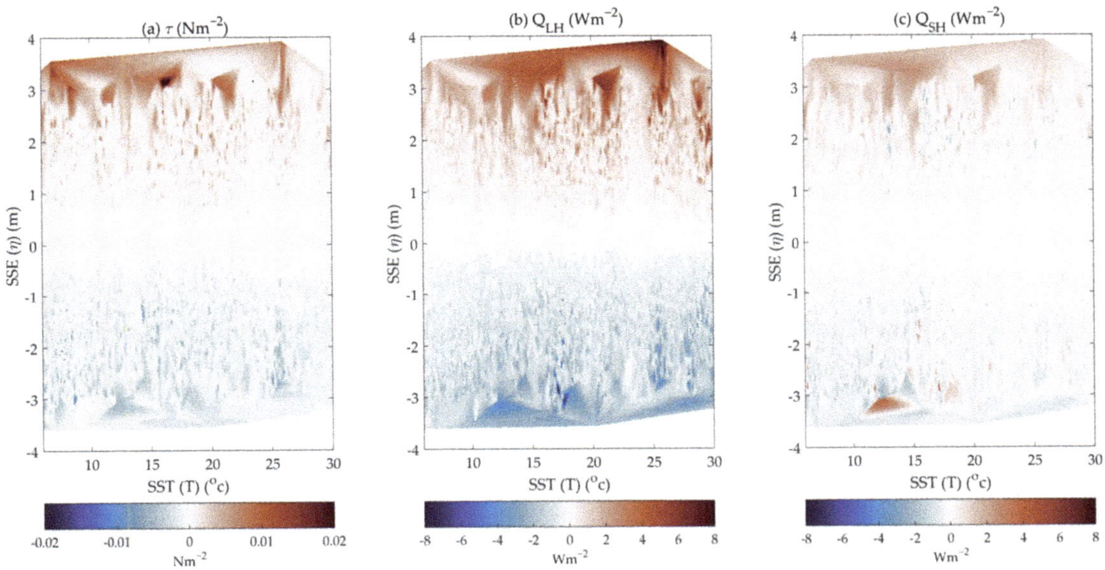

Figure 6. (**a**) Wind stress differences between magnitudes with and without the inclusion of SSE with variations in magnitudes of SSE (y-axis) and SST (x-axis). The colours represent the magnitudes of differences. (**b**,**c**) are the same as (**a**) but for latent heat flux (unit: Wm^{-2}) and sensible heat flux (unit: Wm^{-2}), respectively. Note that the results in (**a**–**c**) are interpolated based on the platform observations in the ECS with some missing ranges.

However, the differences in wind stress, Q_{LH}, and Q_{SH} between the estimates with and without SSE show weak dependence on SST (uncorrelated), indicating weak seasonal variations. Thus, the regression of the above differences onto the tidal SSE anomaly can be individually performed regardless of the seasonal cycle. Figure 7 shows the mean differences in wind stress, Q_{LH}, and Q_{SH} between the estimates with and without SSE from 2016 to 2017. On average, with the tidal SSE ranging from −3 to 3 m, the differences in wind stress, Q_{LH}, and Q_{SH} ranged from -1.4×10^{-3} Nm^{-2}, -1.5 Wm^{-2}, and -0.3 Wm^{-2} to 3.5×10^{-3} Nm^{-2}, 2.2 Wm^{-2}, and 0.7 Wm^{-2}, respectively. It should be noted that the slight asymmetries in the differences can be attributed to the nonlinear relationships between parameters in the bulk formulas (Equation (7)). For example, BLS changes (Section 4.2) and, in particular, seasonal variations in SSE, with a maximum of approximately 0.5 m. In addition, the interpolation of the observational samples might also introduce asymmetric uncertainties.

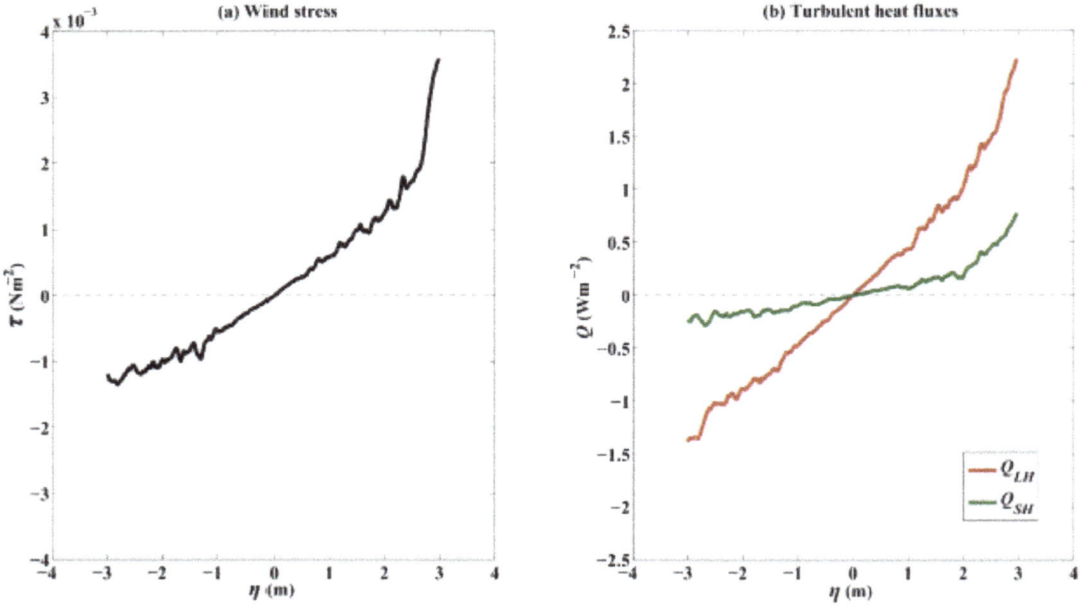

Figure 7. (**a**) Wind stress differences with variations in the magnitude of SSE (*x-axis*) based on two-year observations. (**b**) is the same as (**a**) but for Q_{LH} (unit: Wm^{-2}) and Q_{SH} (unit: Wm^{-2}), respectively. Note that the results in (**a**) and (**c**) are the temperature mean results (x-axis mean) of Figure 6.

To evaluate the contributions of the wind stress and heat flux differences to the mean values, uncertainty percentages were used to show the relative importance. Figure 8 shows the uncertainty percentages for the wind stress and turbulent heat fluxes. The two-year results demonstrate that the percentages are nearly symmetric for rising and sinking SSEs. The percentage for Q_{SH} (R_{SH}) was approximately twice as large as those for wind stress (R_{τ}) and Q_{LH} (R_{LH}) due to the relatively small mean magnitude of Q_{SH}. R_{LH} was equivalent to but slightly higher than that of R_{τ}. The mean (maximum) values of R_{τ}, R_{LH}, and R_{SH} with SSEs ranging from −3 to 3 m were approximately 1.5% (3%), 1.5% (3%), and 3.5% (7%), respectively. The linear trends of R_{τ}, R_{LH}, and R_{SH} with SSE changes were 1% per meter (pm), 1% pm, and 2.5% pm, respectively.

4.2. *Effect of Tidal SSE under BLS Changes*

The tidal SSE effectively affects the BLS associated with the Monin–Obukhov stability parameter $\xi = z/L$, determined by the nonlinear effects of frictional velocity–temperature scales, mean temperature and Monin–Obukhov length. The observations indicate that 254 hourly samples with BLS changes were collected during the two-year observation period, with 14,358 samples in total (Table 3, Figures 9 and 10). The probability of occurrence of BLS changes was approximately 2%, in other words, once every two days at an hourly resolution. However, only near-neutral boundary conditions can be converted into stable and unstable boundary conditions, while no conversion occurs between stable and unstable BLS conditions. Conversions between near-neutral and unstable BLS occur more easily, with a threefold higher probability of occurrence than conversion between near-neutral and stable BLS (Table 1).

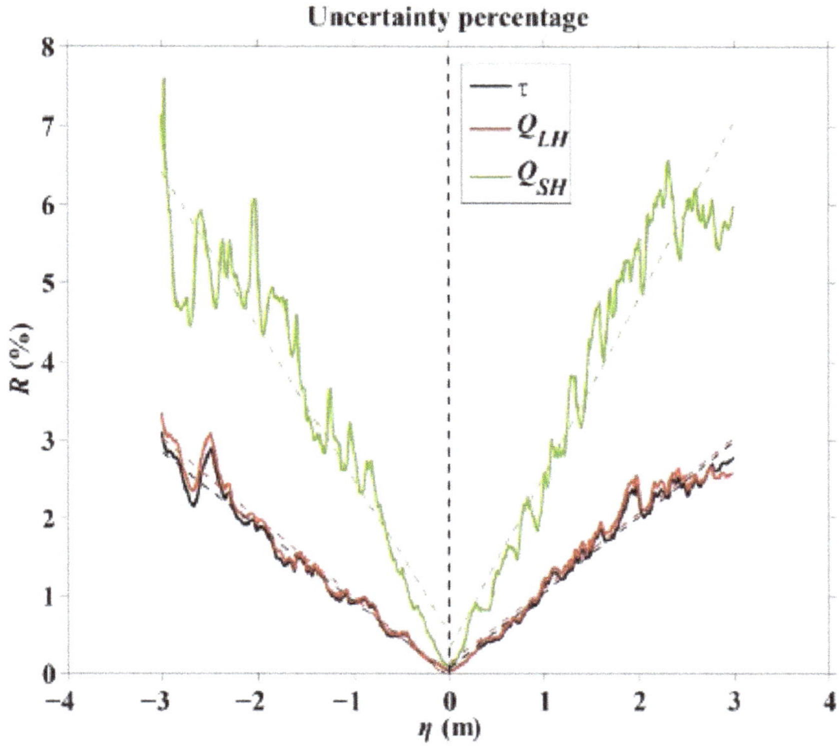

Figure 8. The uncertainty percentages (R) of wind stress (solid green), latent heat flux (solid red), and sensible heat flux (solid black) with variations in SSE (η) ranging from −3 to 3 m. The linear trends are incorporated with dashed lines.

Figure 9 shows the hourly differences in wind stress, Q_{LH} and Q_{SH} between the estimates with and without SSE under different BLS changes. When the SSE drops, the observational height increases. The BLS can change from near-neutral to stable or unstable conditions. If the Monin–Obukhov length L (Equation (7)) decreases (positive ξ), thus dominating the increase in the Monin–Obukhov stability parameter ξ, the BLS changes from near-neutral to stable conditions (green triangles) with a lower frictional velocity u_* scale and a higher temperature T scale, resulting in relatively higher mean temperatures (Figure 10, Table 3). On average, the wind stress and Q_{LH} flux decreased by 1.3×10^{-3} Nm^{-2} and 1 Wm^{-2}, respectively. However, the Q_{SH} had a weak mean increase, with a magnitude of 4×10^{-2} Wm^{-2}, and Q_{SH} increased approximately half the time. This result can explain the noise in the differences in Q_{SH} between the estimates with and without SSE in Figure 6c. If the Monin–Obukhov length L is indeterminate (negative ξ), the reduced frictional velocity (u_*) and mean temperature ($\overline{T}$) result in a smaller ξ. The BLS then shifts from near-neutral to unstable conditions. In this situation, the wind stress and turbulent heat fluxes generally decrease (Table 3).

When the SSE increases, the observational height decreases. The BLS changes from stable or unstable conditions to a near-neutral state (Figure 10). In this situation, a smaller Monin–Obukhov stability parameter ξ (positive ξ) causes the BLS to change from a stable to a near-neutral state (magenta circles in Figures 9 and 10). The Monin–Obukhov length L significantly increases at a magnitude of 20 m, and a smaller Monin–Obukhov stability parameter ξ is obtained with a reduced observational height. Otherwise, when ξ is negative, the unstable state changes into near-neutral conditions. The wind stress and Q_{LH} increased at magnitudes of 1×10^{-3} Nm^{-2} and 1 Wm^{-2}, respectively, regardless of whether the conditions were stable or unstable. The Q_{SH} increased by approximately 0.4 Wm^{-2} when the boundary condition shifted from an unstable to a near-neutral state but decreased slightly by approximately 2×10^{-2} Wm^{-2} (approximately half the time, Figure 9) when the BLS changed from stable to near-neutral conditions.

Table 3. The boundary layer parameter (Equation (7)) changes during shifts in the boundary layer stability (BLS, $\Delta(\frac{z}{L})$) in terms of tidal elevation. The number of samples of the BLS state conversions is incorporated in the last column of the table.

$\Delta(\frac{z}{L})$	$\Delta\tau$ (Nm^{-2})	$\Delta(Q_{LH})$ (Wm^{-2})	$\Delta(Q_{SH})$ (Wm^{-2})	u_* (ms^{-1})	T_* (C)	$\overline{T}$(C)	$\Delta(L)$ (m)	N
$N \rightarrow S$	-1.3×10^{-3}	-1	4×10^{-2}	-2.5×10^{-3}	2.6×10^{-4}	17	-19	35
$N \rightarrow U$	-8.5×10^{-4}	-1	-4×10^{-1}	-1.2×10^{-3}	7.7×10^{-4}	13	–	93
$S \rightarrow N$	1.4×10^{-3}	1	-2×10^{-2}	2.8×10^{-3}	-2.9×10^{-4}	19	20	36
$U \rightarrow N$	9.1×10^{-4}	1	4×10^{-1}	1.3×10^{-3}	-8.8×10^{-4}	15	–	90

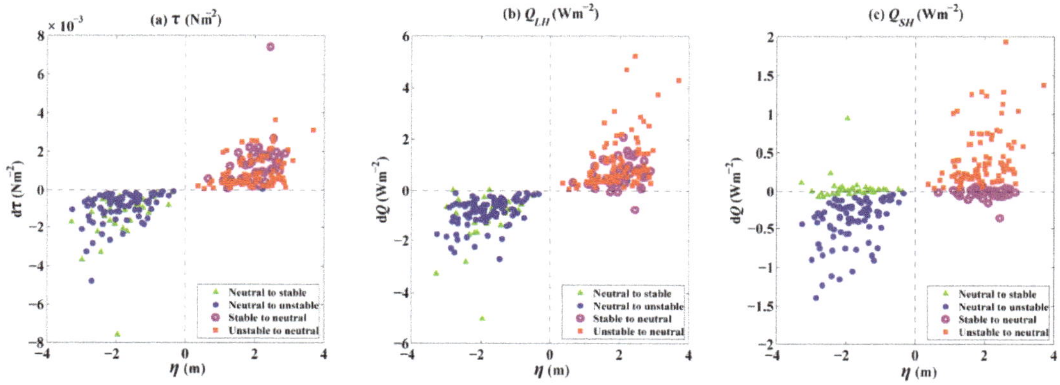

Figure 9. Scatter plots of hourly differences in wind stress (Nm^{-2}, (**a**)), Q_{LH} (Wm^{-2}, (**b**)), and Q_{SH} (Wm^{-2}, (**c**)) between magnitudes with and without the inclusion of SSE (η, x-axis) along with boundary layer stability (BLS) shifts. Changes in the state of the BLS are marked by different colours and symbols. Green triangles and blue dots represent shifts from near-neutral conditions ($-0.4 < \xi < 0.1$) to stable ($\xi > 0.1$) and unstable ($\xi < -0.4$) conditions, respectively, while magenta circles and red squares represent shifts from the stable and unstable boundary conditions to near-neutral conditions, respectively.

Figure 11 shows the schematic of the summarized processes of the BLS changes associated with the tidal SSE. Regardless of whether the SSE rises or drops, the changes in BLS between near-neutral and stable conditions are determined by the changes in both observational height z and Monin–Obukhov length L. In this situation, the changed variables of frictional velocity and temperature scales, which are dependent on the observational height, contribute to a changed L and play a positive role (feedback) in determining the BLS changes. However, the changes in BLS between near-neutral and unstable conditions are determined mainly by the tide-induced changes in observational height z. In this situation, the Monin–Obukhov length L is indeterminate (also seen in Figure 10d).

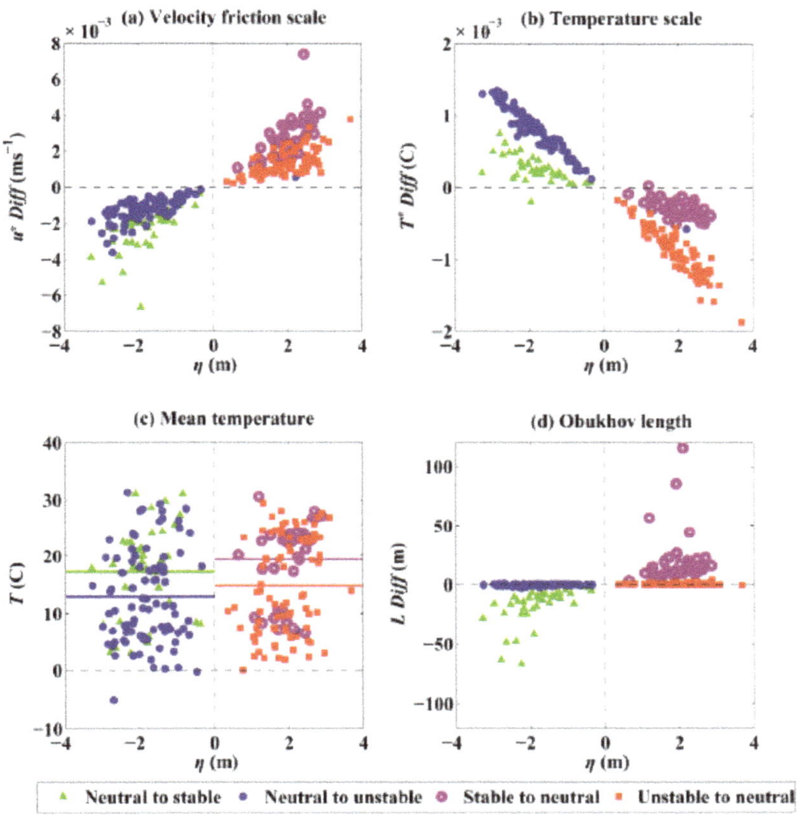

Figure 10. Scatter plots of hourly differences in frictional velocity (u^*, ms^{-1}, (**a**)) and temperature scale (T^*, C, (**b**)), mean temperature (T, C, (**c**)), and Obukhov length scale (L, m, (**d**)) between magnitudes with and without the inclusion of SSE (η, x-axis). The ensemble mean temperature in (**c**) is plotted by solid lines with the same colours as the scatters. Changes in the state of the BLS are marked by different colours and symbols. Green triangles and blue dots represent shifts from near-neutral conditions ($-0.4 < \xi < 0.1$) to stable ($\xi > 0.1$) and unstable ($\xi < -0.4$) conditions, respectively, while magenta circles and red squares represent shifts from the stable and unstable boundary conditions to near-neutral conditions, respectively.

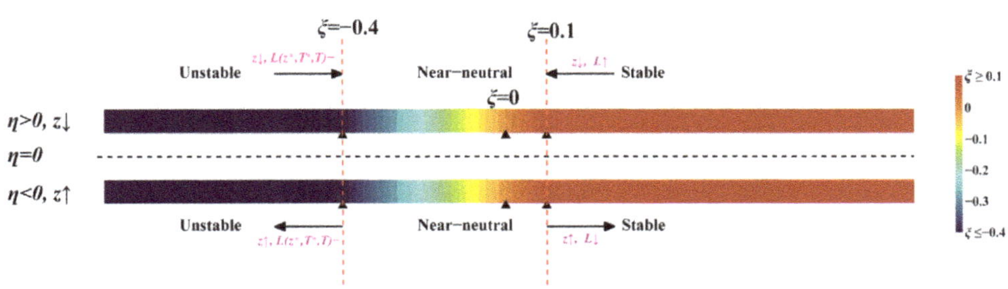

Figure 11. Schematic of BLS switches in terms of tidal elevations based on platform observations. η, z, ξ, L, z^*, T^* and T represent the surface elevation, observational height, Monin–Obukhov stability parameter, Monin–Obukhov length, frictional velocity, temperature scale and mean temperature, respectively.

5. Effect of Tidal SSE on Wind Stress and Turbulent Heat Fluxes under Extreme Weather Conditions

5.1. Effect of Tidal SSE on τ and $Q_{LH,SH}$ during Cold-Air Outbreaks

A cold-air outbreak occurred on 25 January 2016. The air temperature (θ_a) was approximately −10 °C, and the SLP (p) reached the highest value during the two-year observations at 1040 hPa (Figure 3). The WS was 20 ms^{-1} with maximum wind stress τ of approximately 0.8 Nm^{-2}. The highest Q_{SH} was 350 Wm^{-2}, approximately 100 Wm^{-2} higher than that of the Q_{LH} (Figure 12). From 24 to 26 January 2016, the tidal SSE ranged from −3 to 3 m and was dominated by the semidiurnal frequency. More significant differences in the wind stress and turbulent heat fluxes between the estimates with and without SSE were observed when the SSE and magnitudes of τ and $Q_{LH,SH}$ were higher. When the cold-air outbreak weakened on 26 January 2016, the mean values and differences in τ and $Q_{LH,SH}$ became much lower than those on January 25 because the differences were not only dependent on the observational heights but also the magnitudes of the turbulent fluxes. The differences in τ, Q_{LH}, and Q_{SH} ranged from −0.015 Nm^{-2}, −4 Wm^{-2}, and −2 Wm^{-2} to 0.01 Nm^{-2}, 3 Wm^{-2}, and 2 Wm^{-2}, respectively.

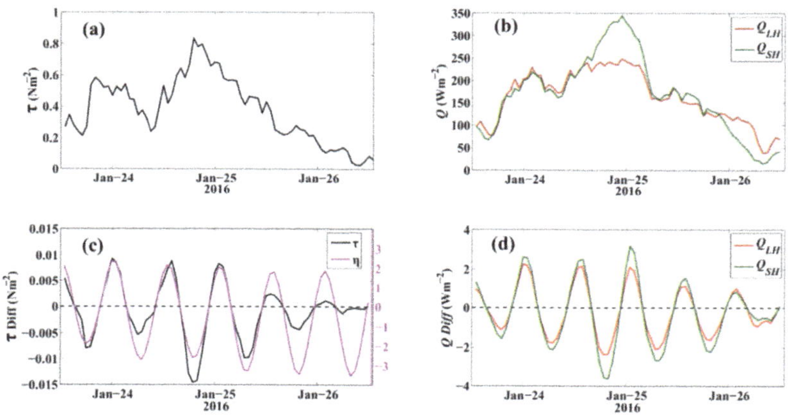

Figure 12. The wind stress (Nm^{-2}, (**a**)) and turbulent heat fluxes (Wm^{-2}, (**b**)) during the cold-air outbreak at the end of January 2016. (**c**): Wind stress differences (left y-axis) with variations in the magnitude of SSE (magenta, right y-axis) during cold-air outbreaks. (**d**) is the same as (**c**) but for Q_{LH} (red, unit: Wm^{-2}) and Q_{SH} (green, unit: Wm^{-2}).

5.2. Effect of Tidal SSE on τ and $Q_{LH,SH}$ during Tropical Cyclones

No TC directly passed the Huo Xing Sha platform during the two-year observation period in this study. However, TC Meranti, which originated in the western Pacific on 9 September 2016, transitioned from a super typhoon to a tropical depression quickly after making landfall in Fujian Province (Figure 13a). The tropical depression passed the platform as it travelled from land to sea on 16 September 2016. The maximum WS was approximately 16 ms^{-1}, with a maximum RH of γ = 90% and a minimum SLP of p = 1005 hPa (Figure 3). On 16 September 2016, the maximum wind stress was approximately 0.4 Nm^{-2}, which was smaller than that during the cold-air outbreak in winter (Figure 14). The Q_{SH} was quite weak, with a mean magnitude of 10 Wm^{-2} from September 15 to 17, when the air-sea temperature difference was slight. However, the Q_{LH} (evaporation) during the tropical depression was enhanced due to a high RH and high WS. The maximum Q_{LH} was 326 Wm^{-2} on September 16 but decreased sharply to approximately 80 Wm^{-2} as a result of a sharp decrease in WS. The semidiurnal tidal SSE ranged from −2.5 to approximately 3.5 m, which caused maximum differences in wind stress, Q_{LH}, and Q_{SH} between the estimates with and without SSE of 0.01 Nm^{-2}, 5 Wm^{-2}, and 1 Wm^{-2}, respectively.

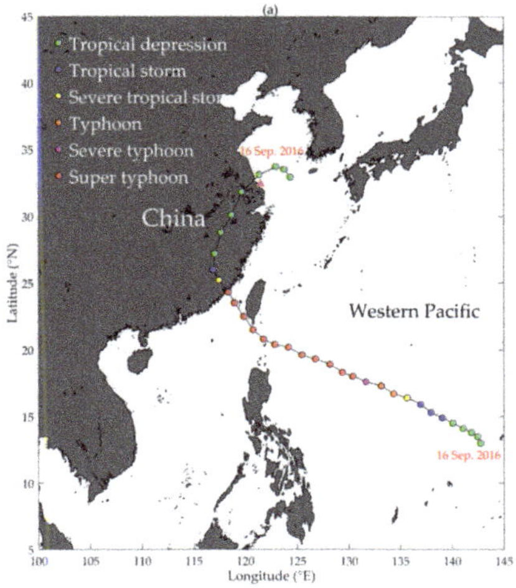

Figure 13. The trajectory of tropical cyclone (TC) Meranti (**a**) during 9~16 September 2016 and TC Ampil; (**b**) during 18~22 July 2018. The TC categories are labelled by circles in different colours. The Huo Xing Sha platform (magenta triangle) is near the trajectory of Meranti on 16 September 2016, when the TC category changed to a tropical depression and that of Ampil on 22 July 2018, when the TC category changed from severe tropical storm to tropical storm. The TC trajectory data were obtained from the China Meteorological Administration (CMA).

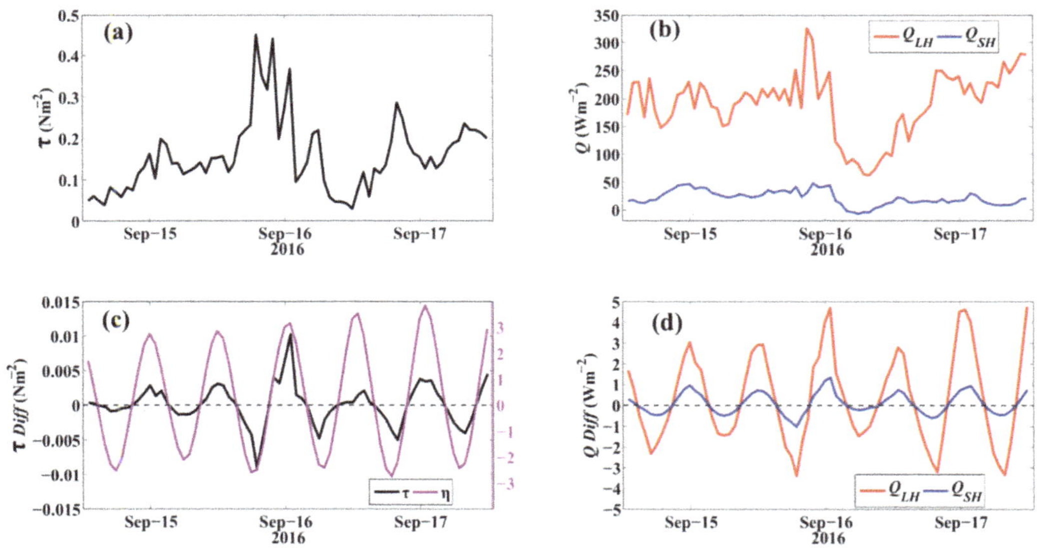

Figure 14. The wind stress (Nm^{-2}, (**a**)) and turbulent heat fluxes (Wm^{-2}, (**b**)) during TC Meranti from 15 to 17 September 2016. (**c**): Wind stress differences (left y-axis) with variations in the magnitude of SSE (magenta, right y-axis) during TC Meranti. (**d**) is the same as (**c**) but for Q_{LH}(red, unit: Wm^{-2}) and Q_{SH} (blue, unit: Wm^{-2}).

Using the additional observations from 22 to 23 July 2018, the wind stress and turbulent heat fluxes with and without tidal SSE were estimated during TC Ampil (Figure 15). TC Ampil landed in Jiangsu Province on 22 July 2018, and the TC category changed immediately from severe tropical storms to tropical storms (Figure 13b). Figure 15 shows that the wind stress approached 1 Nm^{-2} on the afternoon of 22 July 2018, when the WS reached 23 ms^{-1}, and the maximum Q_{LH} and Q_{SH} were 80 and 20 Wm^{-2}, respectively. However, the magnitudes of the turbulent heat fluxes were relatively low in July due to the small air-sea thermal/humidity difference. The two-year observations from 2016 to 2017 indicate that the SAT (θ_a) is highest in July (Figure 3) and higher than SST, which causes stable boundary conditions and negative Q_{SH} ($T - \theta_a < 0$). When TC Ampil passed the Huo Xing Sha platform on 22 July 2018, an ebb tide was observed, with an SSE anomaly ranging from −1.2 to 2 m compared with the two-year mean AWD of 5.3 m. Although the WS was highest at 14:00 on 22 July 2018, the differences in the wind stress and turbulent heat fluxes between the estimates with and without tidal SSE were minor due to low SSE η. The maximum difference in wind stress, Q_{LH} and Q_{SH} occurred when the η anomaly was approximately 2 m, with values of 0.012 Nm^{-2}, 0.6 Wm^{-2}, and 0.5 Wm^{-2}, respectively. The tidal SSE-induced uncertainties in the turbulent heat fluxes were small during TC Ampil due to the weak tidal elevation magnitudes and low mean values in July.

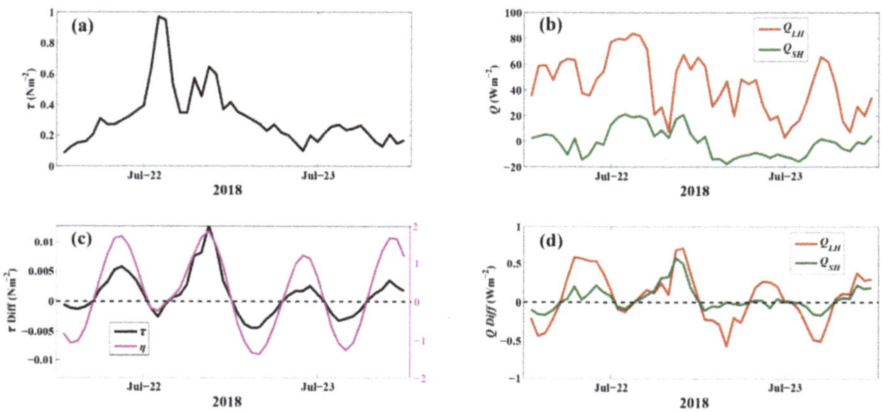

Figure 15. The wind stress (Nm^{-2}, (**a**)) and turbulent heat fluxes (Wm^{-2}, (**b**)) during TC Ampil from 22 to 23 July 2018. (**c**): Wind stress differences (left y-axis) with variations in the magnitude of SSE (magenta, right y-axis) during TC Ampil. (**d**) is the same as (**c**) but for Q_{LH} (red, unit: Wm^{-2}) and Q_{SH} (green, unit: Wm^{-2}).

6. Wind and Thermal Effects on the Hourly Anomalies of Q_{LH} and Q_{SH} for Unstable and Stable BL States

Figure 16 shows the hourly air-sea temperature difference (ΔT) and the estimated boundary layer stability (BLS, $\xi = z/L$). The BLS is closely associated with the air-sea temperature difference. Normally, the SST is higher than the SAT for unstable BLSs. However, in the summer season, the SAT related to synoptic weather processes is frequently higher than the SST, which helps generate frequent stable air-sea boundary layers ($\xi > 0.1$). In this section, the individual contributions of the wind and thermal effects (ΔT and Δq) to the anomalies of the Q_{LH} and Q_{SH}, respectively, on the time scale of hours are discussed. Ten-day hourly observations of Q_{LH}, Q_{SH}, and associated air-sea physical variables were used to examine the roles of the wind effect and thermal effects in determining the hourly variations of the Q_{LH} and Q_{SH} under different states of BLS. Assuming the groups of coefficients in Equations (2) and (3) are all constants, namely, $C_1 = \rho L_e c_E$ for Q_{LH} and $C_2 = \rho c_p c_H$ for Q_{SH}, Equations (2) and (3) for hourly anomalies of Q_{LH} and Q_{SH} are obtained:

$$\underbrace{Q'_{LH}}_{TermA} = C_1 \left(\underbrace{|\mathbf{u_z}|' \cdot \overline{(\Delta q)}}_{TermB} + \underbrace{\overline{|\mathbf{u_z}|} \cdot (\Delta q)'}_{TermC} + \underbrace{|\mathbf{u_z}|' \cdot (\Delta q)'}_{TermD} \right) \tag{14}$$

$$\underbrace{Q'_{SH}}_{TermA} = C_2 \left(\underbrace{|\mathbf{u_z}|' \cdot \overline{(\Delta T)}}_{TermB} + \underbrace{\overline{|\mathbf{u_z}|} \cdot (\Delta T)'}_{TermC} + \underbrace{|\mathbf{u_z}|' \cdot (\Delta T)'}_{TermD} \right) \quad (15)$$

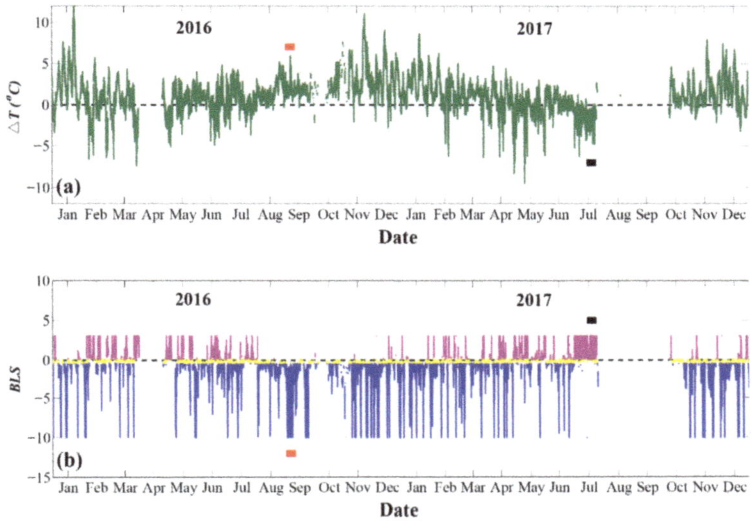

Figure 16. (**a**). The hourly air-sea temperature differences (ΔT) from January 2016 to December 2017. (**b**). The hourly boundary layer stability (BLS) associated with the Monin–Obukhov stability parameter $\xi = z/L$. Note that the stable boundary conditions ($\xi > 0.1$) are plotted in magenta, while the unstable ($\xi < -0.4$) and near-neutral ($-0.4 < \xi < 0.1$) boundary conditions are plotted in blue and yellow, respectively. The ξ smaller than -10 are set to -10 for visual examination. The hourly ten-day unstable BLSs (blue) and positive air-sea temperature differences are marked by red bars from September 2 to 11 September in 2016, while the stable BLCs (magenta) and negative air-sea temperature differences are marked by black bars from 16 July to 25 July in 2017.

In Equations (14) and (15), terms A, B, C, and D represent the hourly anomalies of turbulent heat fluxes, the contributions of wind anomalies, the contributions of thermal difference anomalies, and the nonlinearity between B and C, respectively. The primes in the above two equations denote the hourly anomalies with respect to the ten-day mean values under unstable (2 to 11 September in 2016) or stable (16 to 25 July in 2017) BLSs. The ten-day period was chosen based on the continuous states of the BLS (Figure 16b) and the air-sea temperature difference (Figure 16a).

A visual examination of the four diagrams in Figure 17 reveals that the variations of Q_{LH} and Q_{SH} are dominated by the wind anomalies regardless of the states of BLS based on the analysis of the different terms in Equations (14) and (15). However, the contributions of the thermal effects (term C) and the nonlinear terms (term D) are secondary compared to the wind anomalies (term B). Under unstable (stable) BLSs, the correlation coefficient (r) between wind anomaly and Q_{LH} anomaly was 92% (95%) at a confidence level of 95%, while the coefficients between the thermal/nonlinear effect and Q_{LH} were approximately equivalent, with r = 25% indicating much weaker correlations compared with the wind effect. When the BL is stabilized, the wind turbulence helps overcome the buoyancy state and contributes to the anomaly of the evaporation and Q_{LH}. Thus, the correlation coefficients between wind anomalies and Q_{LH} anomalies during stable BL conditions are slightly higher than those during unstable conditions.

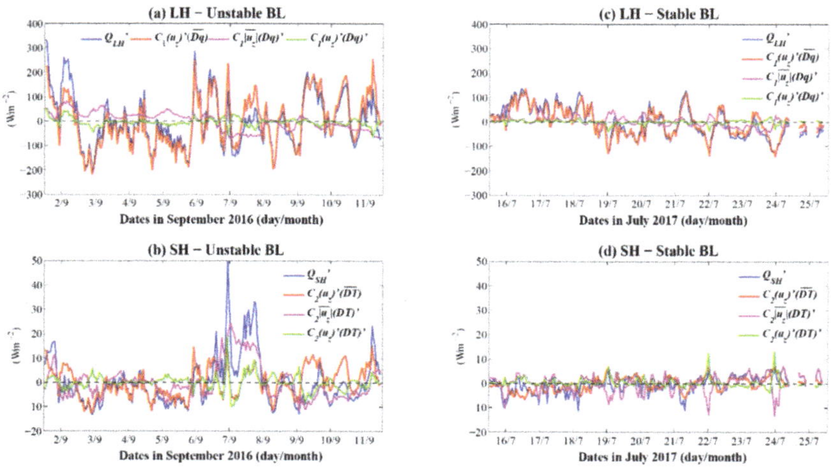

Figure 17. The hourly variations of the terms in Equations (14) and (15). (**a**): the time series of the anomalies of the Q_{LH} (term A in Equation (14), dark blue), the anomalies of Q_{LH} induced by wind anomalies (term B in Equation (14), red) and air-sea humidity difference anomalies (term C in Equation (14), magenta) and the nonlinear term (term D in Equation (14), green) under unstable BLSs in September of 2016. (**b**): the same as in (**a**) but for anomalies of the Q_{SH} and associated terms. (**c**) and (**d**) are the same as in (**a**,**b**), but for anomalies under stable BLSs in July of 2017.

The results for the Q_{SH} anomalies were different from those of the Q_{LH} anomalies. The hourly air-sea temperature difference anomalies play equivalent roles in determining the Q_{SH} anomalies. The contributions of wind anomalies to the Q_{SH} anomalies (r = 61%) were also higher than those of the anomalies of ΔT (r = 46%) under stable BL conditions. A mechanism similar to that used for Q_{LH} can be applied. However, under unstable BL conditions, the temperature difference anomalies dominated the Q_{SH} anomalies with r = 76%, followed by wind anomalies (r = 57%). In the four cases, as shown in Figure 17, the roles of the nonlinear effects ($|\mathbf{u_z}|' \cdot (\Delta q)'$ and $|\mathbf{u_z}|' \cdot (\Delta T)'$) are nonnegligible in contributing to the turbulent heat fluxes with r ranging from 16% to 33%.

These findings are also different from the results of the monthly anomalies. Over the global oceans, the decadal variability of Q_{LH} and Q_{SH} is determined primarily by the variations in the air-sea thermal difference, namely, the ΔT and Δq. Physically, the winds contribute to the anomalies of Q_{LH} and Q_{SH} under the background air-sea thermal differences. However, the direct role of the wind anomalies is secondary compared to the thermal effects [46,47], especially for the Q_{SH}. The wind indirectly drives the basin-scale ocean dynamics and determines the SST, which helps change the properties of the air-sea Q_{LH} and Q_{SH}. The results in this paper indicate that the high-resolution observations of wind speed play a dominant role in determining the turbulent heat fluxes, especially the Q_{LH}.

7. Summary and Discussion

Based on the platform observations of air-sea boundary variables in the ECS, the bulk formulas of COARE 3.0 were used to estimate the two-year air-sea momentum fluxes, including wind stress, Q_{LH}, and Q_{SH}. The effect of tidal SSE, which was dominated by the semidiurnal frequency, on the wind stress and turbulent heat fluxes was studied. The main findings in this paper are as follows:

First, the periodic tidal SSE affects the hourly estimates of wind stress and turbulent heat fluxes based on the Huo Xing Sha platform observations. Larger tidal magnitudes result in more significant differences. The maximum (minimum) wind stress difference between the estimates with and without tidal SSE was 2.2×10^{-3} (-1.5×10^{-3}) Nm^{-2}, and the maximum (minimum) Q_{LH} and Q_{SH} differences were 8.4 and 4.6 Wm^{-2} (−10.2 and −3.6 Wm^{-2}), respectively. With respect to the two-year annual mean state, the magnitudes of the wind stress, Q_{LH}, and Q_{SH} differences were approximately 5×10^{-3} Nm^{-2}, 4 Wm^{-2}, and 1 Wm^{-2}, respectively, with a tidal SSE amplitude of approximately 6 m. In response to SSE changes, the uncertainty percentages for these variables were 1% pm, 1% pm, and 2.5% pm.

Second, the BLS can shift from a state of near-neutral conditions to stable or unstable conditions. The probability of the occurrence was approximately 2% during the two-year observations from 2016 to 2017. The BLS cannot shift between stable and unstable conditions due to changes in the tidal SSE. When the SSE increases, the reduced observational height z dominates the shift in the BLS conditions in terms of the Monin–Obukhov stability parameter $\xi = z/L$. Under positive ξ, the Monin–Obukhov length L increases, which helps the BLS change from stable to near-neutral conditions. In this situation, the stable and unstable BLS conditions change into the near-neutral state with increasing wind stress and Q_{LH} (summarized in Figure 18). However, the changes in the Q_{SH} strongly depend on the boundary layer conditions. The opposite results are obtained when the SSE decreases. The mechanism of the hourly anomalies of the Q_{LH} and Q_{SH} was analyzed in terms of the different states of the BLS. It is found that the hourly anomalies of the Q_{LH} are dominated primarily by the wind anomalies over the air-sea thermal effects regardless of the states of the BLS; however, the anomalies of Q_{SH} are determined by the combined wind and thermal effects. The role of the nonlinear effect is nonnegligible.

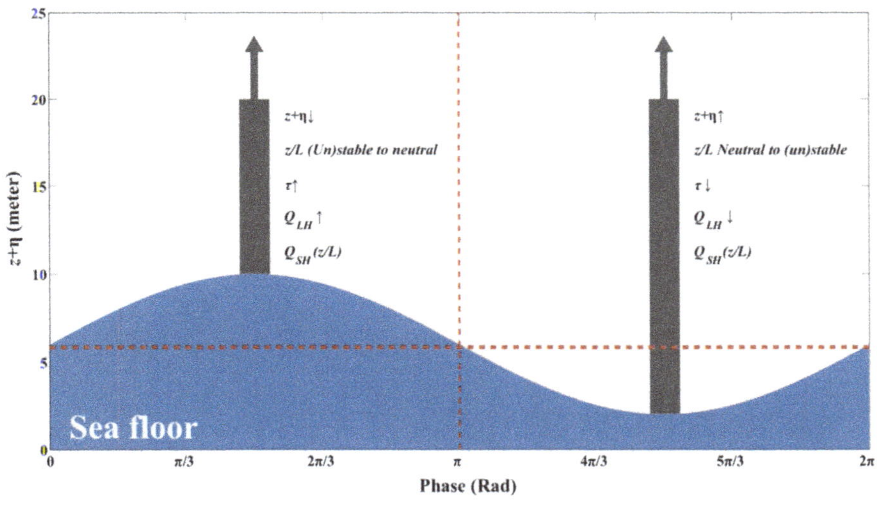

Figure 18. Schematic of the effect of tidal elevations on platform-based air-sea turbulent fluxes. The x-axis represents the tidal phase, while the y-axis represents the SSE. The turbulent fluxes and the boundary condition shifts are marked in the figure. The horizontal red dotted line represents the mean tidal elevations, and the vertical red dotted line represents the dividing line between SSE rising and falling.

Third, using platform observations, stronger Q_{SH} and Q_{LH} fluxes can be found during extreme weather processes, including a cold-air outbreak and two TCs (Meranti in 2016 and Ampil in 2018), as indicated in previous studies. Significant uncertainties in the wind stress and turbulent heat fluxes can also be caused by the tidal SSE during synoptic-scale weather processes. The steady platform observations provide an important benchmark for comparisons of atmospheric reanalysis and model simulations, especially for the study of ocean–land–atmosphere interactions. It is expected that this study will help to obtain more accurate air-sea momentum fluxes for a better understanding of the air-sea interactions in the ECS. The corrected air-sea momentum and turbulent heat fluxes and associated air-sea variables will be used for the observation and reanalysis comparisons in the following study.

The results in this paper show that the high-frequency (semidiurnal) tidal SSE primarily affects the diurnal variations in the wind stress and turbulent heat fluxes over a diurnal period. However, if the WS is relatively steady and the air-sea variables remain unchanged over a diurnal period, the daily mean contribution of the tidal SSE to the momentum and turbulent heat fluxes diminishes. The idealized extreme contributions of tidal SSE to the mean magnitudes of wind stress and turbulent heat fluxes can be obtained by the hourly random WS during a diurnal period. This platform is near the coast of the ECS, where small-scale ocean–land–atmosphere processes are active. The measured air-sea variables and turbulent heat fluxes can be used for future observation–reanalysis comparisons.

This paper investigated the role of the tide-induced SSE in modifying the platform-based estimates of wind stress and turbulent heat fluxes in terms of changed observational height z in the vertical direction. In another publication (Song 2020), the authors investigated the detailed contributions of horizontal tidal currents to the estimates of the wind stress and turbulent heat flux by modifying the relative WS between the absolute WS relative to the Earth and the sea surface current. With different mechanisms, these twin studies demonstrate that although the mean contributions of tidal currents and SSEs are relatively weak over the course of a diurnal period, they can affect the diurnal variations in the wind stress and turbulent heat fluxes, thereby helping to provide accurate estimates of the detailed upper ocean dynamics that occur on the continental shelf of the ECS.

Author Contributions: Conceptualization, X.S., Y.H., and Y.L. (Yuxin Liu); methodology, Y.H. and Y.L. (Yuxin Liu); software, Y.L. (Yuxin Liu) and C.X.; validation, L.Y., X.C., and B.Y.; formal analysis, Y.L. (Yuxin Liu); investigation, X.C., and B.Y.; resources, Y.L. (Yangang Li), J.L., K.L., X.C., and B.Y.; data curation, Y.H. and Y.L. (Yuxin Liu); writing—original draft preparation, Y.H. and Y.L. (Yuxin Liu); writing—review and editing, Y.H. and Y.L. (Yuxin Liu); visualization, X.S. and C.X.; supervision, X.J., M.L., F.Y., and X.S.; project administration, Y.L. (Yuxin Liu); funding acquisition, X.S. All authors have read and agreed to the published version of the manuscript.

Funding: This study is funded by the National Key Research and Development Program of China (2018YFB0505000), the National Natural Science Foundation of China (42076016) and the Fundamental Research Funds for the Central Universities (2019B02814).

Institutional Review Board Statement: Not applicable.

Informed Consent Statement: Not applicable.

Data Availability Statement: The data presented in this study are available on request from the corresponding author.

Acknowledgments: The authors extend their thanks to the crew of the ECS branch of the SOA, MNR. The OAFlux, ERA5 and GEBCO data were downloaded from http://www.oaflux.whoi.edu/ accessed on 15 December 2021, https://www.ecmwf.int/ accessed on 15 December 2021, and http://www.gebco.net/ accessed on 15 December 2021, respectively. The authors appreciate the constructive comments and suggestions from the anonymous reviewers.

Conflicts of Interest: The authors declare no conflict of interest.

References

1. Ekman, V.W. On the influence of the Earth's rotation on ocean-currents. *Ark. Mat. Astron. Fys.* **1905**, *2*, 1–52.
2. Price, J.F.; Weller, R.A.; Schudlich, R.R. Wind-driven ocean currents and Ekman transport. *Science* **1987**, *238*, 1534–1538. [CrossRef]
3. Pedlosky, J. *Geophysical Fluid Dynamics*, 2nd ed.; Springer: New York, NY, USA, 1987; 710p.
4. Huang, R.X. *Ocean Circulation: Wind-Driven and Thermohaline Processes*; Cambridge University Press: Cambridge, UK, 2011; 828p.
5. Cayan, D.R. Latent and sensible flux over the north oceans: The connection to monthly atmosphere circulation. *J. Clim.* **1992**, *5*, 354–369. [CrossRef]
6. Carton, J.; Zhou, Z. Annual cycle of sea surface temperature in the tropical Atlantic Ocean. *J. Geophys. Res. Space Phys.* **1997**, *102*, 27813–27824. [CrossRef]

7. Moisan, J.R.; Niiler, P.P. The seasonal heat budget of the North Pacific: Net heat flux and heat storage rates (1950–1990). *J. Phys. Oceanogr.* **1998**, *28*, 401–421. [CrossRef]
8. Yu, L.; Jin, X.; Weller, R.A. Role of Net Surface Heat Flux in Seasonal Variations of Sea Surface Temperature in the Tropical Atlantic Ocean. *J. Clim.* **2006**, *19*, 6153–6169. [CrossRef]
9. Weare, B.C. Uncertainties in estimates of surface heat fluxes derived from marine reports over the tropical and subtropical oceans. *Tellus* **1989**, *41*, 357–370. [CrossRef]
10. Gleckler, P.J.; Weare, B.C. Uncertainties in global ocean surface heat flux climatologies derived from ship observations. *J. Clim.* **1997**, *10*, 2764–2781. [CrossRef]
11. Grist, J.P.; Josey, S.A. Inverse Analysis Adjustment of the SOC Air-sea Flux Climatology Using Ocean Heat Transport Constraints. *J. Clim.* **2003**, *16*, 3274–3295. [CrossRef]
12. Brunke, M.A.; Wang, Z.; Zeng, X.; Bosilovich, M.G.; Shie, C.-L. An Assessment of the Uncertainties in Ocean Surface Turbulent Fluxes in 11 Reanalysis, Satellite-Derived, and Combined Global Datasets. *J. Clim.* **2011**, *24*, 5469–5493. [CrossRef]
13. Song, X.; Yu, L. How much net surface heat flux should go into the Western Pacific Warm Pool? *J. Geophys. Res. Oceans* **2013**, *118*, 3569–3585. [CrossRef]
14. Weller, R.A.; Farrar, J.; Buckley, J.; Mathew, S.; Venkatesan, R.; Lekha, J.S.; Chaudhuri, D.; Kumar, N.S.; Kuman, B.P. Air-Sea Interaction in the Bay of Bengal. *Oceanography* **2016**, *29*, 28–37. [CrossRef]
15. Yu, L. Global Air-sea Fluxes of Heat, Fresh Water, and Momentum: Energy Budget Closure and Unanswered Questions. *Annu. Rev. Mar. Sci.* **2019**, *11*, 227–248. [CrossRef] [PubMed]
16. Webb, E.K.; Pearman, G.I.; Leuning, R. Correction of the flux measurements for density effects due to heat and water vapor transfer, Quart. *J. R. Meteorol. Soc.* **1980**, *106*, 85–100. [CrossRef]
17. Liebethal, C.; Foken, T. On the Significance of the Webb Correction to Fluxes. *Bound. Layer Meteorol.* **2003**, *109*, 99–106. [CrossRef]
18. Brodeau, L.; Barnier, B.; Gulev, S.K.; Woods, C. Climatologically Significant Effects of Some Approximations in the Bulk Parameterizations of Turbulent Air-sea Fluxes. *J. Phys. Oceanogr.* **2016**, *47*, 5–28. [CrossRef]
19. Khanna, S.; Brasseur, J.G. Analysis of Monin–Obukhov similarity from large-eddy simulation. *J. Fluid Mech.* **1997**, *345*, 251–286. [CrossRef]
20. Johansson, C.; Smedman, A.; Högström, U.; Brasseur, J.G.; Khanna, S. Critical test of the validity of Monin-Obukhov similarity during convective conditions. *J. Atmos. Sci.* **2001**, *58*, 1549–1566. [CrossRef]
21. Sahlée, E.; Smedman, A.S.; Högström, U. Influence of the boundary layer height on the global air-sea surface fluxes. *Clim. Dyn.* **2009**, *33*, 33–44. [CrossRef]
22. Gryning, S.E.; Batchvarova, E.; Brümmer, B.; Jørgensen, H.E.; Larsen, S.E. On the extension of the wind profile over homogeneous terrain beyond the surface boundary layer. *Bound. Layer Meteorol.* **2007**, *124*, 251–268. [CrossRef]
23. Peña, A.; Gryning, S.E.; Hasager, C.B. Comparing mixing-length models of the diabatic wind profile over homogeneous terrain. *Theor. Appl. Climatol.* **2010**, *100*, 325–335. [CrossRef]
24. Optis, M.; Monahan, A.; Bosveld, F.C. Moving Beyond Monin–Obukhov Similarity Theory in Modelling Wind-Speed Profiles in the Lower Atmospheric Boundary Layer under Stable Stratification. *Bound. Layer Meteorol.* **2014**, *153*, 497–514. [CrossRef]
25. Song, J.; Fan, W.; Li, S.; Zhou, M. Impact of Surface Waves on the Steady Near-Surface Wind Profiles over the Ocean. *Bound. Layer Meteorol.* **2015**, *155*, 111–127. [CrossRef]
26. Josey, S.A. A Comparison of ECMWF, NCEP–NCAR, and SOC Surface Heat Fluxes with Moored Buoy Measurements in the Subduction Region of the Northeast Atlantic. *J. Clim.* **2001**, *14*, 1780–1789. [CrossRef]
27. Song, X. The Importance of Relative Wind Speed in Estimating Air-sea Turbulent Heat Fluxes in Bulk Formulas: Examples in the Bohai Sea. *J. Atmos. Ocean. Technol.* **2020**, *37*, 589–603. [CrossRef]
28. Song, X.; Yu, L. Air-sea heat flux climatologies in the Mediterranean Sea: Surface energy balance and its consistency with ocean heat storage. *J. Geophys. Res. Oceans* **2017**, *122*, 4068–4087. [CrossRef]
29. Guo, X.; Yanagi, T. Three-dimensional structure of tidal current in the East China Sea and the Yellow Sea. *J. Oceanogr.* **1998**, *54*, 651–668. [CrossRef]
30. Niwa, Y.; Hibiya, T. Three-dimensional numerical simulation of M_2 internal tides in the East China Sea. *J. Geophys. Res.* **2004**, *109*, C4027. [CrossRef]
31. Song, X.; Wu, D.; Xie, X. Tides and Turbulent Mixing in the North of Taiwan Island. *Adv. Atmos. Sci.* **2019**, *36*, 313–325. [CrossRef]
32. Fairall, C.W.; Bradley, E.F.; Rogers, D.P.; Edson, J.B.; Young, G.S. Bulk parameterization of air-sea fluxes for Tropical Ocean-Global Atmosphere Coupled-Ocean Atmosphere Response Experiment. *J. Geophys. Res.* **1996**, *101*, 3747–3764. [CrossRef]
33. Fairall, C.W.; Bradley, E.F.; Hare, J.E.; Grachev, A.A.; Edson, J.B. Bulk Parameterization of Air-Sea Fluxes: Updates and Verification for the COARE Algorithm. *J. Clim.* **2003**, *16*, 571–591. [CrossRef]
34. Monin, A.S.; Obukhov, A.M. The main features of turbulent mixing in the surface atmospheric layer. *Tr. Inst. Geophys. Acad. Sci. USSR* **1954**, *24*, 163–187.
35. Liu, W.T.; KKatsaros, B.; Businger, J.A. Bulk parameterization of the air-sea exchange of heat and water vapor including the molecular constraints at the interface. *J. Atmos. Sci.* **1979**, *36*, 2052–2062. [CrossRef]
36. Large, W.G.; Pond, S. Open ocean momentum flux measurements in moderate to strong winds. *J. Phys. Oceanogr.* **1981**, *11*, 324–336. [CrossRef]

37. Edson, J.B.; Jampana, V.; Weller, R.A.; Bigorre, S.P.; Plueddemann, A.J.; Fairall, C.W. On the exchange of momentum over the open ocean. *J. Phys. Oceanogr.* **2013**, *43*, 1589–1610. [CrossRef]
38. Garratt, J.R. Review of Drag Coefficients over Oceans and Continents. *Mon. Weather Rev.* **1977**, *105*, 915–929. [CrossRef]
39. Panofsky, H.A.; Dutton, J.A. *Atmospheric Turbulence*; Wiley-Interscience: New York, NY, USA, 1984; p. 397.
40. Smith, S.D. Coefficients for sea surface wind stress, heat flux, and wind profiles as a function of wind speed and temperature. *J. Geophys. Res.* **1988**, *93*, 15467–15472. [CrossRef]
41. Weller, R.A.; Bradley, F.; Lukas, R. The Interface or Air-sea Flux Component of the TOGA Coupled Ocean–Atmosphere Response Experiment and Its Impact on Subsequent Air-sea Interaction Studies. *J. Atmos. Ocean. Technol.* **2004**, *21*, 223–257. [CrossRef]
42. Yu, L.; Weller, R.A. Objectively Analyzed air-sea heat Fluxes for the global ice-free oceans (1981–2005). *Bull. Am. Meteor. Soc.* **2007**, *88*, 527–539. [CrossRef]
43. Grossman, R.L.; Betts, A.K. Air-sea Interaction during an Extreme Cold Air Outbreak from the Eastern Coast of the United States. *Mon. Weather Rev.* **1990**, *118*, 324–342. [CrossRef]
44. Xue, H.; Bane, J.M.; Goodman, L.M. Modification of the Gulf Stream through Strong Air-sea Interactions in Winter: Observations and Numerical Simulations. *J. Phys. Oceanogr.* **1995**, *25*, 533–557. [CrossRef]
45. Renfrew, I.A.; Moore, G.W.K. An extreme cold-air outbreak over the Labrador Sea: Roll vortices and air-sea interaction. *Mon. Weather Rev.* **1999**, *127*, 2379–2394. [CrossRef]
46. Yu, L. Global Variations in Oceanic Evaporation (1958–2005): The Role of the Changing Wind Speed. *J. Clim.* **2007**, *20*, 5376–5390. [CrossRef]
47. Song, X.; Yu, L. High-Latitude Contribution to Global Variability of Air-sea Sensible Heat Flux. *J. Clim.* **2012**, *25*, 3515–3531. [CrossRef]

atmosphere

Article

Influences of MJO on the Diurnal Variation and Associated Offshore Propagation of Rainfall near Western Coast of Sumatra

Bojun Zhu [1], Yu Du [2,3,4,*] and Zhiqiu Gao [1,5]

1 Climate and Weather Disasters Collaborative Innovation Center, School of Applied Meteorology, Nanjing University of Information Science & Technology, Nanjing 210044, China; bojun.zhu@nuist.edu.cn (B.Z.); zgao@mail.iap.ac.cn (Z.G.)
2 School of Atmospheric Sciences, and Guangdong Province Key Laboratory for Climate Change and Natural Disaster Studies, Sun Yat-sen University, Zhuhai 519082, China
3 Southern Marine Science and Engineering Guangdong Laboratory (Zhuhai), Zhuhai 519082, China
4 Key Laboratory of Tropical Atmosphere-Ocean System (Sun Yat-sen University), Ministry of Education, Zhuhai 519082, China
5 State Key Laboratory of Atmospheric Boundary Layer Physics and Atmospheric Chemistry, Institute of Atmospheric Physics, Chinese Academy of Sciences, Beijing 100029, China
* Correspondence: duyu7@mail.sysu.edu.cn

Abstract: Madden-Julian Oscillation (MJO) plays an important role in modulating precipitation at Maritime Continent (MC) not only on a larger scale, but also in the diurnal cycle. Diurnal rainfall offshore propagation is one of the most evident features near coasts. This study investigates the impacts of MJO on diurnal rainfall and its offshore propagation at the western coast of Sumatra during boreal winters using ERA5 reanalysis. The real-time multivariate MJO (RMM) index was applied to locate the active MJO convection through eight different phases, in the western hemisphere and Africa in P8–P1, at the Indian Ocean in P2–P3, at MC in P4–P5, and the western Pacific Ocean in P6–P7. The rainfall characteristics, including the daily rate, the absolute and normalized diurnal variation amplitudes, and the strengths of diurnal offshore propagation, not only depend on active/inactive MJO stages but also vary under different MJO phases, through the combined modulations of large-scale backgrounds and local-scale land–sea circulations. The offshore rainfall propagation is associated with meso-large-scale gravity waves generated from land–sea thermal contrast and thus is affected by the radiation effect of cloud under different MJO phases. The stronger wave signals in P8–P1 and P6–P7 enhance the diurnal rainfall variation amplitudes away from the coast, while the strong coupling of moist convection with gravity waves contributes greatly to the diurnal rainfall cycle in P2–P3.

Keywords: MJO; rainfall; diurnal variation; land–sea circulation; gravity wave

Citation: Zhu, B.; Du, Y.; Gao, Z. Influences of MJO on the Diurnal Variation and Associated Offshore Propagation of Rainfall near Western Coast of Sumatra. *Atmosphere* **2022**, *13*, 330. https://doi.org/10.3390/atmos13020330

Academic Editor: Stefan Liess

Received: 13 January 2022
Accepted: 14 February 2022
Published: 16 February 2022

1. Introduction

Madden–Julian Oscillation (MJO) [1,2] is a dominant intra-seasonal tropical oscillation, characterized by an eastward propagation of about 5 m s^{-1} and a zonal shift of 12,000–20,000 km within 30–100 days [3]. A number of previous studies have documented noticeable influences of MJO on global precipitation by the modulation of background-wind convergence and vapor transportation [4–8]. The most significant oscillations and influences usually occurred during boreal winters [9–13]. Those findings improved our understanding of the formation and development of precipitation controlled by the large-scale forces during an MJO event.

Maritime Continent (MC) is the general designation of tropical seas and islands between Indian Ocean and Pacific Ocean, with a zonal span of around 5000 km [14] and variations in topography, shape and area [15]. As one of the regions with most active convection and great influences by factors such as MJO or convectively coupled equatorial waves (CCEWs), MC generates frequent precipitation all over the years, and the associated latent heat from rainfall becomes one of the main energy sources driving global circulation [16], which plays nonnegligible roles in global weather and climate [17–20]. The diurnal variation of rainfall is a common phenomenon in coastal regions, including MC; more specifically, it is often characterized by offshore propagation near the coasts, whose mechanisms are closely related to gravity waves [21–24], background-wind modulations [21,23,25], land–sea breezes [26–29], and cold pool outflows [30–32].

Previous studies revealed the MJO's influences on the diurnal variation of coastal rainfall at MC via sea surface temperature (SST), background winds and convection using reanalysis, satellite or radar data [33–37], which are as important as the roles of other factors, for example, different types of CCEWs that modulate sub-seasonal or intra-seasonal convective activities uniquely, across synoptic or longer timescales [38–42]. However, a controversial question has remained regarding which MJO phase undertakes the most significant diurnal precipitation variation, as some studies suggested the arrival of the MJO's active convection at MC [34,35], while others recognized the preconditioning stage narrowly prior to that arrival [43,44]. Additionally, Rauniyar and Walsh [45] concluded that the MJO convective disturbance was favorable for coastal precipitation but unfavorable for landside precipitation. Although these studies on various MJO events might draw distinguished conclusions, very limited analysis has been conducted for detailed explanation. The remarks above reflect the difficulty to regard MC as a broad whole since the impacts of each MJO phase could vary among different MC islands.

The Sumatra Island investigated in this paper lies to the west of Indonesia, as one of the giant islands at MC that exhibits the earliest responses to the MJO as it moves eastward, and has attracted increasing attention for studying the MJO's impacts on diurnal rainfall variations [24,25,29,36,37,46–48] and rainfall extremes [49,50]. The study from Zhu et al. [51] also reveals the difficulty of improving numerical rainfall simulations through local-scale data assimilation in this area under the MJO active phase, due to the large-scale convective envelope as a dominating convection driver. The inconsistency between coastal and surrounding features of diurnal precipitation variation, together with multi-scale synoptic modulations, were exhibited in some studies [39,41], but their variations among different MJO phases still lack enough discussion. In addition, most of the studies focus on how the MJO affects the diurnal variation at Sumatra, but the effects on the diurnal offshore propagation are generally unclear. According to the study from Vincent and Lane [36], the diurnal propagation at western Sumatra is much stronger than that at other MC islands such as Java and New Guinea, as one nonnegligible characteristic of coastal Sumatra weather. They further concluded that the offshore migration of the diurnal cycle peaks was distinct (indistinct) when the MJO was over the Indian Ocean (had passed Sumatra), similar to the findings from Kamimera et al. [47], indicating that the modulations from the Indian Ocean might play critical roles. These proposed phenomena require detailed analysis, including a comparison of dominating factors among different MJO stages. Besides, the gravity wave signal, as one mechanism of the offshore rainfall migration, was found evident even after the MJO active stage when there was little precipitation, as reflected in the study in New Guinea from Vincent and Lane [23].

Unlike most of the prior studies, the present study is to compare the multi-scale influences of MJO on diurnal rainfall variation and associated offshore propagation among different MJO phases, near the western coast of Sumatra during the boreal winter seasons (December, January, and February) of the latest 10 years (2011–2020). To achieve this goal, the diurnal rainfall variations among different MJO phases are compared from perspectives of large-scale backgrounds and local-scale land–sea circulations associated with gravity waves. The rest of the paper is organized as follows. Section 2 introduces the data and methodology used in this study. Section 3.1 describes the spatial and temporal variations of rainfall during different MJO phases. The effects of large-scale circulations and local-scale land–sea circulations are analyzed in Sections 3.2 and 3.3, respectively. Section 3.4 explores the mechanisms of offshore diurnal propagation and its relationship with MJO. Finally, Section 4 presents a discussion and Section 5 summarizes the conclusions.

2. Data and Methodology

The coastal area around western Sumatra was selected as the analysis region in the present paper (Figure 1a). The Barisan Mountains lie along the west coast of Sumatra, with the highest elevation of around 1.5 km above mean sea level (AMSL). Figure 1b presents the vertical cross section of topography perpendicular to the Barisan Mountains averaged over the red box in Figure 1a.

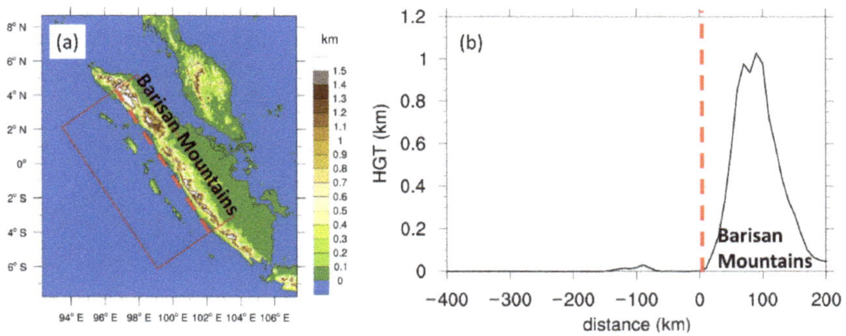

Figure 1. (**a**) The spatial distribution and (**b**) the vertical cross section (perpendicular to the coast) of topography (km) near western Sumatra coast. The red solid box is used for all the Hovmöller diagrams in this paper, with the distance ranging from −400 km offshore to 200 km onshore. The red dashed line marks the location of western coast with the distance of 0 km in (**b**).

The 0.1° × 0.1° Integrated Multi-satellitE Retrievals for GPM (IMERG) V06B final run precipitation data [52,53] was used to present the rainfall distribution and its diurnal variation. A comparison among different products (early run, late run and final run) indicated very limited differences between one another, and the final run data was utilized subsequently, with the consideration that it underwent additional adjustments with gauge data and climatology, compared with the early run or late run products. The half-hourly IMERG data was combined each hour to estimate the hourly precipitation rate. The 0.25° × 0.25° ECMWF Reanalysis 5th Generation (ERA5) data [54] during winter seasons (month DJF) of 2011–2020 was adopted for the background circulation analysis. The hourly ERA5 data began at 1000 hPa with the largest vertical interval of 50 hPa in the upper half of the troposphere.

The daily real-time multivariate MJO (RMM) index provided by the Australian Government Bureau of Meteorology is utilized to identify MJO phases, was calculated with outgoing long-wave radiation and 850-hPa/200-hPa zonal wind within 15° N–15° S, based on the method from Wheeler and Hendon [55]. The MJO amplitude was estimated using the RMM1 and RMM2 series generated by the empirical orthogonal functions (EOF) analysis. An MJO amplitude exceeding 1.0 indicates a strong MJO event. The MJO phases ranging from P1 to P8 correspond to varying locations of the active tropical convective disturbance: western hemisphere and Africa in P8–P1P8–P1, Indian Ocean in P2–P3, MC in P4–P5, and western Pacific Ocean in P6–P7. The days during strong MJO events in DJF 2011–2020 were classified into these four groups.

3. Results

3.1. Rainfall Variations Modulated by MJO

3.1.1. Spatial Distribution

Figure 2 shows the spatial distribution of the precipitation rate averaged in DJF 2011–2020 and the rainfall anomalies in different MJO phases. The islands of Sumatra, Borneo and Java, together with adjacent seas, had the highest precipitation rate of above 0.4 mm h^{-1} due to the windward side of MC (Figure 2a). The locations of positive rainfall anomalies are consistent with the MJO convective disturbances among different MJO phases found in previous studies [43,46]. The rainfall anomalies in our analysis region (green box) and surrounding environments are consistently positive in P2–P3 (Figure 2c) and negative in P6–P7 (Figure 2e), which corresponds to the active phase and the suppressed phase, respectively, that are significantly affected by the large-scale background circulation associated with the MJO (which will be discussed in Section 3.2). In contrast, the rainfall anomalies in P8–P1P8–P1 are positive near the coastal region of Sumatra but are negative in the surrounding areas; P4–P5 exhibits opposite characteristics of the drier coast and the wetter environment. Both P8–P1 and P4–P5 are transitional phases between the active phase (P2–P3) and the suppressed phase (P6–P7), suggesting the lower impacts of MJO convective disturbances and indicating other possible factors that might take the lead (which will be discussed in Section 3.3).

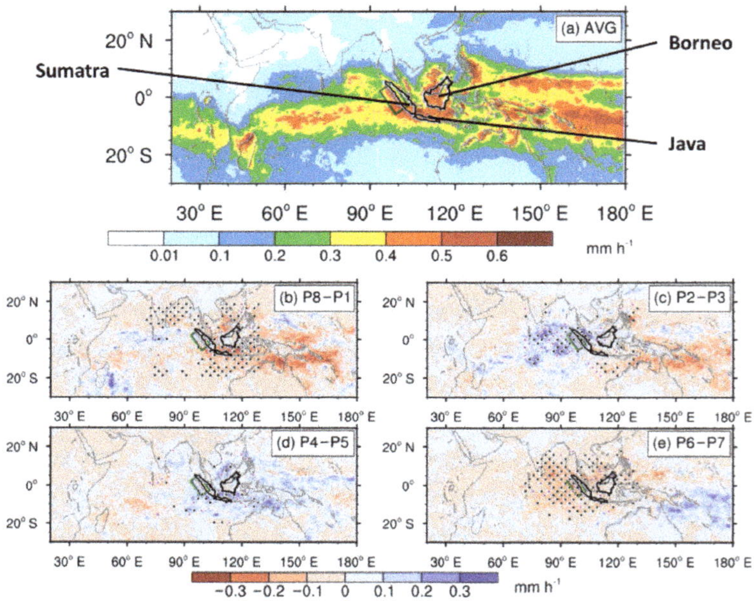

Figure 2. The precipitation rate (mm h^{-1}) (**a**) averaged in DJF 2011–2020 and its anomalies in (**b**) P8–P1, (**c**) P2–P3, (**d**) P4–P5 and (**e**) P6–P7. The black outlines mark the islands of Sumatra, Borneo and Java. The green box at western Sumatra coast marks the analysis region in this paper. The purple and black dots indicate the 90% and 95% confidence at 20° N–20° S, 70° E–130° E, according to the Student's *t*-test.

3.1.2. Diurnal Variation

Figure 3 contains the Hovmöller diagrams of precipitation rates among different MJO phases. In general, the diurnal variation of coastal rainfall stands out with a daytime landside peak followed by a nighttime seaside peak. The diurnal amplitude varies among different MJO phases, as strongest in P2–P3 (Figure 3b) and weakest in P6–P7 (Figure 3d), similar to the findings from Fujita et al. [46]. The diurnal variation of rainfall is characterized by evident offshore propagation, relatively obscured in P4–P5 (Figure 3c). The landside rainfall begins around 1300 LST around 50 km onshore, and migrates offshore with a phase speed of 3–4 m s^{-1}, until approximately ~30 km offshore around 0100 LST. Subsequently, the offshore propagation with a higher phase speed of 8–9 m s^{-1} occurs from ~30 km offshore to ~300 km offshore until ~1000 LST in the following day. In P4–P5, a diurnal phase-locking pattern with a maximum at ~0600 LST is found around −370–−160 km offshore.

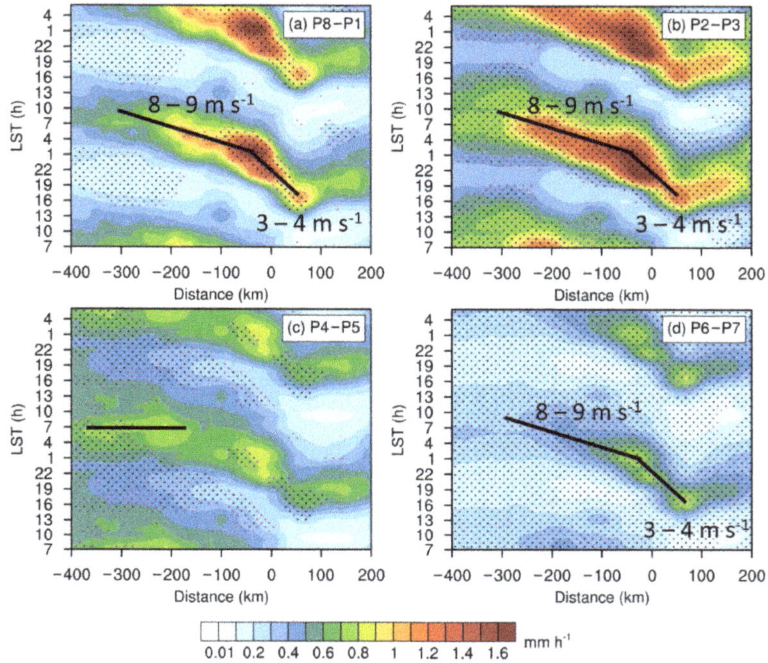

Figure 3. The time-distance Hovmöller diagrams of hourly precipitation rate (mm h^{-1}) in (**a**) P8–P1, (**b**) P2–P3, (**c**) P4–P5 and (**d**) P6–P7 averaged in DJF 2011–2020. The purple and black dots indicate the 90% and 95% confidence according to the Student's *t*-test.

In addition to the absolute diurnal amplitudes described above, we compare the relative diurnal amplitudes. Figure 4 includes the Hovmöller diagrams of normalized hourly precipitation deviation, which is calculated as the ratio of diurnal deviation to daily mean in each MJO stage. Although the absolute diurnal amplitude in P2–P3 is slightly higher than that in P8–P1 (Figure 3), the relative diurnal amplitude in P2–P3 is much lower than that in P8–P1 (Figure 4). Similar features exist in the comparison of P4–P5 and P6–P7. The sinusoidal regression averaged within −100–0 km from the coast quantitatively achieves the offshore absolute diurnal amplitudes (A_o) and normalized diurnal amplitudes (A_n), as displayed in Table 1. The A_o in P2–P3 is much higher than that in P6–P7 due to the significant large-scale contrast of the wettest and driest stages in MJO, but A_n becomes similar in the two stages due to the strengthened diurnal signal in P6–P7, which is also available in P8–P1, in which A_n is much higher than that in P4–P5. The ratio R_a, calculated as A_o divided by A_n is also exhibited, representing hourly precipitation rates (mm h^{-1}) averaged within −100–0 km offshore. Furthermore, R_a indicates larger difference between P8–P1 and P2–P3 as well as between P4–P5 and P6–P7 than A_o does, because A_o is the product of R_a and A_n for which the features contrast each other. For example, R_a is higher in P2–P3 (active stage) than in P8–P1 (transitional stage), while A_n is in the opposite possibly due to local effects, but their difference in A_o is relatively minor. The comparison of R_a better reflects the influences from the large-scale background circulation associated with MJO, than that of A_o.

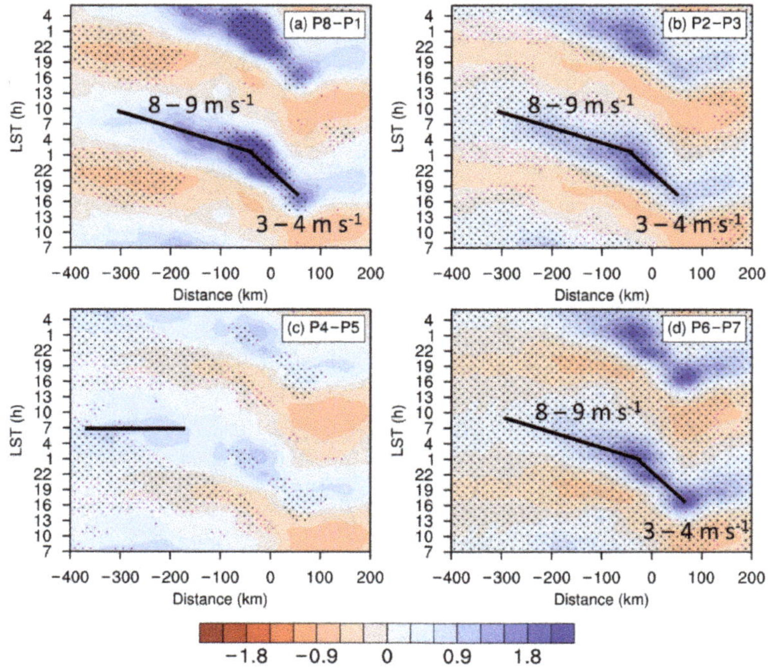

Figure 4. As in Figure 3 but for normalized hourly precipitation deviation in (**a**) P8–P1, (**b**) P2–P3, (**c**) P4–P5 and (**d**) P6–P7 averaged in DJF 2011–2020.

Table 1. Diurnal amplitudes from precipitation (A_o, mm h^{-1}) and from normalized precipitation (A_n) averaged within −100–0 km offshore via sinusoidal regressions, together with R_a (mm h^{-1}) calculated as A_o divided by A_n.

	P8–P1	P2–P3	P4–P5	P6–P7
$\mathbf{A_o}$	0.599	0.648	0.248	0.266
$\mathbf{A_n}$	0.896	0.799	0.491	0.762
$\mathbf{R_a}$	0.668	0.812	0.505	0.349

3.2. Effect of Large-Scale Forcing

Based on the characteristics in Section 3.1, the large-scale background circulations among different MJO phases are compared. Figure 5a shows the seasonal mean of zonal winds and vertical motions at 850-hPa, whereby the upward motions are marked with diagonal lines. The westerly winds from northern Indian Ocean, and the northeasterly winds from the north of Borneo, bring abundant oceanic moisture and converge at the western Sumatra coast. The upward motions dominate at and around the coast. Both the thermodynamic and dynamic conditions are favorable for coastal rainfall, consistent with the high rainfall rate indicated in Figure 2a.

Figure 5b–e present the anomalies of zonal winds and vertical motions at 850 hPa under different MJO phases, where the positive (upward) anomalies are marked with diagonal lines. In P2–P3 (Figure 5c), the strengthened westerly winds to the west and the strengthened easterly winds to the northeast result in the stronger convergence near the Sumatra Island, while the wind anomalies in P6–P7 exhibit an opposite scenario and weakened convergence. In addition, the vertical motions near the Sumatra Island in P2–P3 are much stronger than that in P6–P7, while the anomalies of 850-hPa water vapor mixing ratio in P2–P3 (P6–P7) are positive (negative) near the Sumatra Island (Figure 6b,d), further influencing the strength of moist advection and convection (Figure 2b–e). The results above could explain why P2–P3 and P6–P7 are the wettest and driest MJO stages at Sumatra.

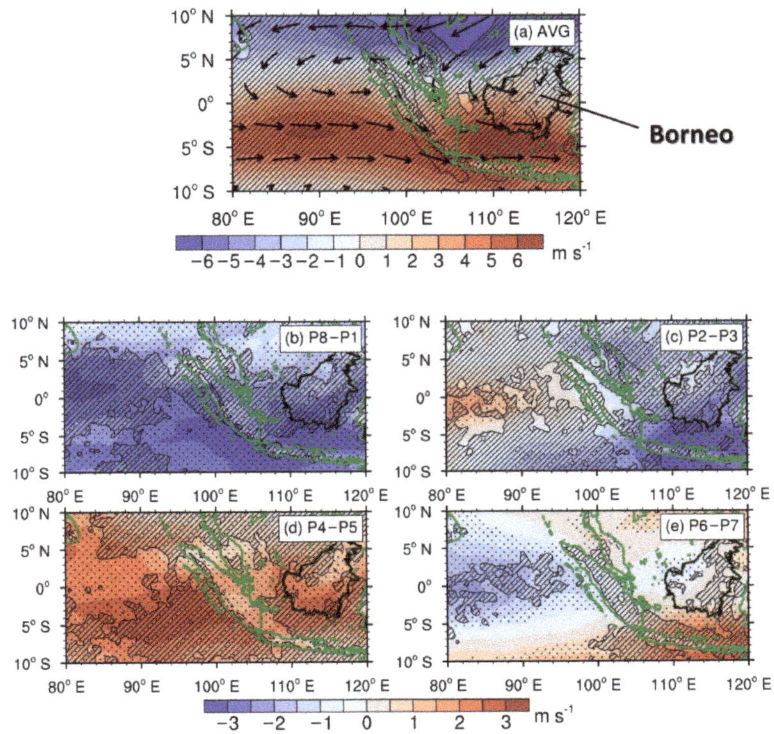

Figure 5. The horizontal distributions of zonal wind (shaded, m s^{-1}) and vertical motions (contour, 0.5 cm s^{-1}) at 850 hPa (**a**) averaged in DJF 2011–2020 as well as its anomaly in (**b**) P8–P1, (**c**) P2–P3, (**d**) P4–P5 and (**e**) P6–P7. The black outline marks the island of Borneo. The diagonal lines indicate the positive (upward) averages or phase anomalies for vertical motions. The black dots indicate the 95% confidence according to the Student's *t*-test.

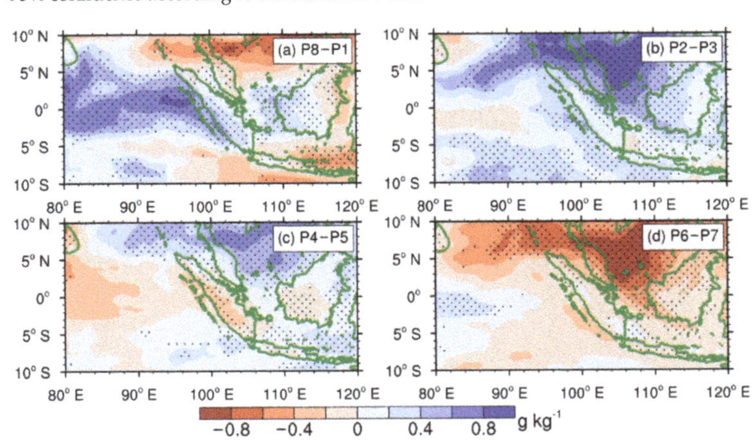

Figure 6. The anomalies of water vapor mixing ratio at 850 hPa (g kg^{-1}) in (**a**) P8–P1, (**b**) P2–P3, (**c**) P4–P5 and (**d**) P6–P7. The black dots indicate the 95% confidence according to the Student's *t*-test.

In contrast, P8–P1 and P4–P5 have both conducive and unconducive conditions affecting the convection. As shown in Figure 5b,d, the wind anomalies to the west and northeast of Sumatra are consistent, and result in relatively weaker convergence anomalies. The differences in the precipitation rate (higher in P8–P1 than in P4–P5) arise mostly from the upstream vapor anomalies at the northern Indian Ocean (Figure 6a,c).

Therefore, the large-scale conditions for precipitation are favorable in the active stage P2–P3, unfavorable in the suppressed stage P6–P7, and relatively neutral in the transitional stages P8–P1 and P4–P5, influenced largely by local factors.

3.3. *Effect of Local-Scale Land-Sea Circulation*

Figure 7 presents the anomalies of total cloud cover under different MJO phases, which is in agreement with the precipitation anomalies (Figure 2). The cloud cover anomalies may lead to a variation in the normalized diurnal precipitation amplitude (A_n) among different MJO phases by modulating the land–sea thermal contrast and the local-scale land–sea circulation via the solar radiative heating: less cloud leads to higher thermal contrast and stronger local circulation via enhanced radiative heating. The absolute diurnal precipitation amplitude (A_o) and the coastal precipitation rate (R_a) in P6–P7 are much lower than those in P2–P3, but the normalized diurnal precipitation variation amplitudes (A_n) become similar in the two stages (Table 1) due to the lower cloud cover in P6–P7 (Figure 7). Similarly, A_o and R_a are lower in P8–P1 than in P2–P3, but A_n is higher in P8–P1 with less cloud. Therefore, the diurnal rainfall signals are influenced not only by large-scale background conditions but also by local-scale radiative heating or cooling, where the cloud cover plays a nonnegligible role, as indicated by previous studies [34,45,56–60].

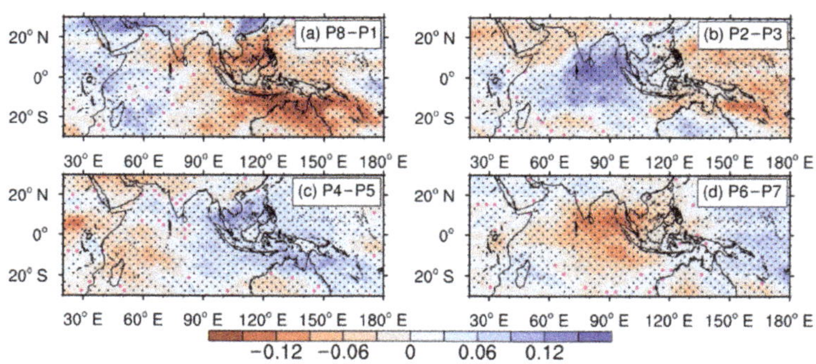

Figure 7. As in Figure 6 but for the anomalies of total cloud cover in (**a**) P8–P1, (**b**) P2–P3, (**c**) P4–P5 and (**d**) P6–P7. The purple and black dots indicate the 90% and 95% confidence according to the Student's *t*-test.

Figure 8 compares the diurnal variations (Figure 8a) and diurnal amplitudes (Figure 8b) of land–sea temperature contrast at 850 hPa, as the driving factor of land–sea circulation, among different MJO phases. The daytime positive temperature contrast (warmer on land) and the nighttime negative temperature contrast (warmer over ocean), as presented in Figure 8a, derives from the radiative heating and cooling influenced by relatively lower landside heat capacity. The time series in P8–P1 are very close to the total average, which is a possible explanation for the insignificance through the *t* test. The diurnal variation is strongest in P6–P7 and weakest in P2–P3 (Figure 8b). As is known, low-level temperature is modulated largely by cloud cover and precipitation. In P6–P7 the cloud cover is lowest (Figure 7d), leading to most of the apparent daytime radiative heating and nighttime cooling (Figure 8a), meanwhile P6–P7 is also the driest stage due to its divergence anomalies (Figure 5e) and negative moisture anomalies (Figure 6d) so that the daytime precipitation cooling over the land is weakest, further enlarging the diurnal amplitude (Figure 8b). In contrast, the higher cloud cover in P2–P3 (Figure 7b) lowers the radiative heating and increases the daytime landside precipitation cooling, both resulting in the weakened diurnal temperature variation (Figure 8b). In other words, the radiative modulation and precipitation cooling are both conducive (unconducive) for enlarging the diurnal temperature amplitude in P6–P7 (P2–P3), so the distinction between these two stages is most evident. Similar features are reflected via the comparison between P8–P1 and P2–P3 as well as between P4–P5 and P6–P7. Note that the difference between P8–P1 and P4–P5 is minor where the conducive and unconducive conditions are mixed: the stronger cloud-affected radiative modulation is more favorable for the diurnal temperature variations in P8–P1, but the weaker landside precipitation cooling is more favorable in P4–P5.

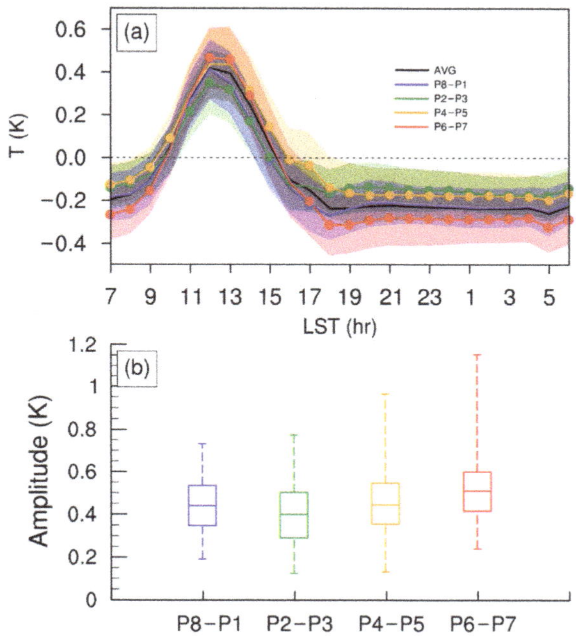

Figure 8. (**a**) The diurnal variations of 850-hPa land–sea temperature contrast (line, K) and its standard deviation (shaded, K) for total average and phase averages, calculated as the landside mean subtracted by seaside mean within 100 km from western Sumatra coast; (**b**) the box figures of diurnal amplitudes (K). The dots in Figure 8a indicate the 95% confidence according to the Student's *t*-test.

Figure 9 compares the diurnal variations of low-level land–sea circulations by presenting the differences between the vertical circulation at 0100 LST and at 1300 LST. The seasonal mean land–sea nocturnal circulation (Figure 9a) reveals upward motions over the ocean and downward motions over the land, with a distinct boundary close to the western Sumatra coast. The cloud cover had a noticeable impact below 1.5 km AMSL, and the P8–P1 and P6–P7 stages with lower cloud cover had enhanced land–sea circulations (Figure 9b,e), whereas the higher cloud cover in P2–P3 and P4–P5 compressed the land–sea circulations (Figure 9c,d), consistent with the higher (lower) 850-hPa diurnal temperature amplitudes in P8–P1 and P6–P7 (P2–P3 and P4–P5) (Figure 8), revealing the dependency of land–sea circulation on the diurnal land–sea thermal contrast. The boundaries between positive and negative vertical motion anomalies near the coast are clearer in the transitional MJO phases, P8–P1 and P4–P5, reflecting larger impacts of cloud cover on the local-scale circulation and the diurnal variation of precipitation in these phases. The consistently (both seaside and landside) positive (negative) vertical motion anomalies near the coast, of above 1.5 AMSL in P2–P3 (P6–P7) reflect the dominating modulations by the large-scale MJO circulations.

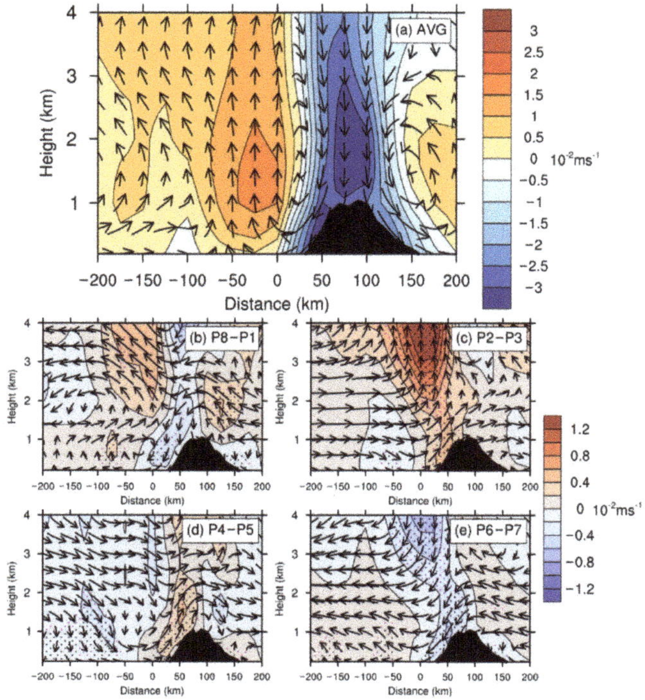

Figure 9. The differences of vertical motion (shaded, 10^{-2} m s^{-1}, positive means upward) and wind vectors (the horizontal wind normal to western Sumatra coast with the vertical motion) at 0100 LST from those at 1300 LST (**a**) averaged in DJF 2011–2020 as well as its anomaly in (**b**) P8–P1, (**c**) P2–P3, (**d**) P4–P5 and (**e**) P6–P7. The black shade marks the topography of coastal mountains. The purple and black dots indicate the 90% and 95% confidence according to the Student's *t*-test.

The positive middle-level offshore vertical motion anomalies in the diurnal circulation enhance the nighttime offshore convection and enlarge the diurnal offshore variation both in P8–P1 and P2–P3 (Figure 9b,c), but their mechanisms might be different. In P8–P1, the stronger offshore convection is driven by the strengthened offshore transportation of upward motions from coastal topography (stronger gravity waves), attributed to a greater land–sea contrast of low-level temperature arising from enhanced cloud-affected radiative heating. In P2–P3, however, such offshore gravity waves are less evident due to relatively weaker radiative heating and stronger daytime precipitation cooling over the land. Nevertheless, the enhanced low-level onshore background wind and its convergence with local land surges contribute to the stronger offshore convection in P2–P3. In other words, the normalized diurnal precipitation variation in P8–P1 takes larger effects from the local-scale circulations (or gravity waves) under the cloud cover modulations, while the large-scale backgrounds play more important roles in P2–P3. Similar characteristics are revealed in the comparison between P4–P5 and P6–P7.

3.4. Diurnal Propagation Associated with Gravity Waves

As shown in Figures 3 and 4, the average phase speed of diurnal offshore precipitation propagation is 3–4 m s^{-1} near the coast, similar with the magnitudes in the land breeze or the cold pool outflow [28,30,32,61]. The phase speed of 8–9 m s^{-1} further offshore is noticeably distinguished from the magnitude of background wind, and indicates other possible factors such as gravity waves, with a similar speed of roughly 8–15 m s^{-1}, as generated by land–sea thermal contrast or convection [21,24,62–65]. The mechanism of this quicker propagation affecting the diurnal rainfall farther offshore will be explored in this section. Figure 10 shows the diurnal signals of offshore propagation in 700-hPa temperature by subtracting the spatial mean from the temporal deviations. The wavelike bands of positive signals (solid ellipses) and negative signals (dashed ellipses) parallel to the western Sumatra coast reflects the influences of the coastline as well as the coastal mountains on the temperature variations, similar with the findings from Ruppert and Zhang [66] as a horizontal view of geography-affected gravity wave activities. The landside positive band develops and is separated from the coast since 0700 LST; afterwards, it moves offshore and gradually dissipates to around 0700 LST the following day. The offshore migration speed from 1900 LST to 0400 LST is estimated to be about 9.7 m s^{-1}, similar to the magnitude of the precipitation migration from 0100 LST to 0700 LST which exhibits a phase lag of approximately 6 h. The landside rainfall is evident since 1600 LST, after the strong thermal convection and the cloud formation around 1300 LST. The offshore migration of rainfall begins at 2200 LST when the thermal disturbance moves offshore, and this migration sustains until 1000 LST in the next day, similar to the beginning and ending of the migration exhibited in Figures 3 and 4.

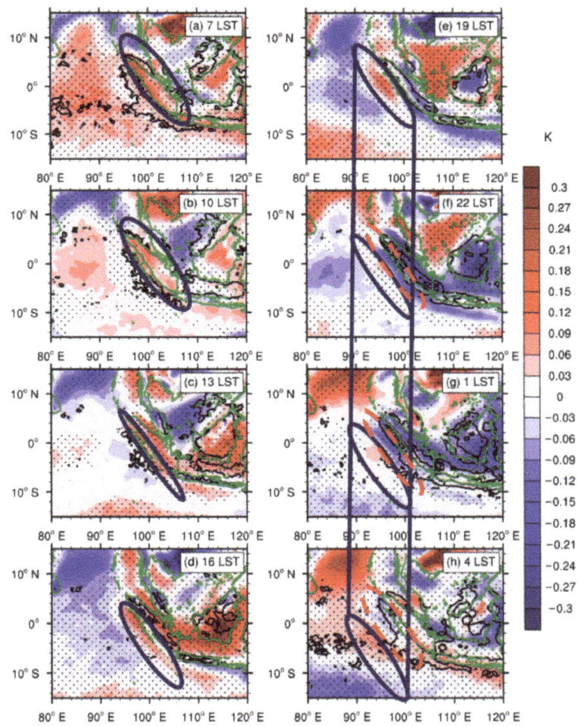

Figure 10. The diurnal variation of 700-hPa temporal temperature disturbance (K) with spatial mean subtracted (shaded) as well as precipitation rate (contour, interval 0.5 mm h^{-1}), as averaged at (**a**) 07 LST, (**b**) 10 LST, (**c**) 13 LST, (**d**) 16 LST, (**e**) 19 LST, (**f**) 22 LST, (**g**) 01 LST, and (**h**) 04 LST, in DJF 2011–2020. The local standard time (LST) is calculated with a Coordinated Universal Time (UTC) offset of +7. The solid (dashed) ellipses mark the positive (negative) signals. The grey and black dots indicate the 90% and 95% confidence according to the Student's *t*-test.

The vertical structures of the wavelike signals are presented in Figure 11. The land–sea contrast of daytime radiative heating induces the low-level landside warm disturbance (Figure 11b–d), of which the vertical structure tilts to the offshore side at 1300 LST (Figure 11c). Then, this vertically tilted phase line (or ray path) propagates offshore and downward until 2200 LST (Figure 11d–f). Meanwhile, the cool tilted disturbance from landside radiative cooling forms at night and repeats the offshore migration (Figure 11e–h). These warm and cool tilted disturbances might come from the gravity waves driven by diurnal land–sea thermal contrast [29,63,64,67]. The offshore low-level warm disturbance and midlevel cool disturbance (Figure 11e–h) helps to destabilize the local atmosphere and creates more conducive environments for the nighttime development of offshore convection. Those patterns are in general agreement with the vertical structures in the study of Bai et al. [65]. The continuous propagation of warm and cool signals until 0700 LST on the next day are considered a main factor of the offshore rainfall migration, with the phase speeds (peak values) estimated as 9.3 m s^{-1} at 900 hPa and 11.6 m s^{-1} at 600 hPa.

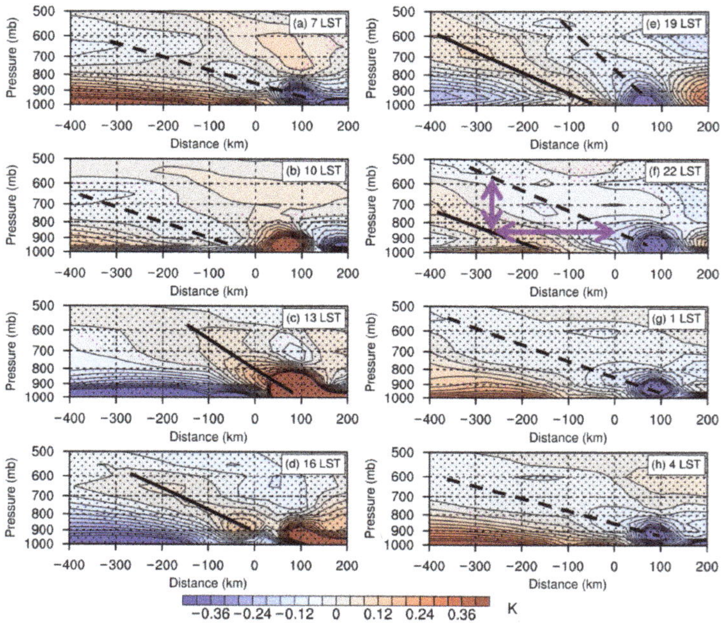

Figure 11. The temporal evolution of the vertical cross section of temperature deviations (K) as averaged at (**a**) 07 LST, (**b**) 10 LST, (**c**) 13 LST, (**d**) 16 LST, (**e**) 19 LST, (**f**) 22 LST, (**g**) 01 LST, and (**h**) 04 LST, in DJF 2011–2020. The purple and black dots indicate the 90% and 95% confidence according to the Student's *t*-test.

The phase speed c of offshore migration (precipitation or temperature disturbance) was estimated as the speed of peak values, as well as with the dispersion relation included in the study from Du and Rotunno [63] as follows:

$$c = \frac{\omega}{k} \tag{1}$$

$$\omega = \sqrt{f^2 + N^2\frac{k^2}{m^2}} \tag{2}$$

where ω is the frequency, k and m is the dominant horizontal and vertical wavenumber, f is the Coriolis parameter, and N is the buoyancy frequency. According to the utilized ERA5 reanalysis data in our study region, $f \approx 5.2 \times 10^{-6}$ s^{-1} and $N \approx 0.012$ s^{-1}. The leaning vertical structure of the gravity waves in Figure 11f exhibits the horizontal and vertical wavelength to be approximately 500 km and 3.3 km (marked with purple vectors), while the corresponding wavenumber is $k \approx 1.26 \times 10^{-5}$ m^{-1} and $m \approx 9.52 \times 10^{-4}$ m^{-1}. Hence, the phase speed (dispersion relation) was estimated as roughly 12.6 m s^{-1}. These estimations are close to those of precipitation far offshore in Figures 3 and 4 (8–9 m s^{-1}), as well as with the estimates in existing literature mentioned at the beginning of this section. The horizontal wavelength of ~500 km suggests the gravity waves in this study to be meso-large-scale.

The differences in the gravity wave signals among various MJO phases and their influences on coastal rainfall propagation are also compared. Figure 12 presents the Hovmöller diagrams of differences between 850-hPa and 600-hPa temperature, as the index of low-level thermal instability. This index also reflects the modulations of the wavelike signals associated with gravity waves, as exhibited in Figure 11. The landside instability increases with accumulated radiative heating and reaches its maximum at around 1300 LST, after which it propagates offshore and destabilizes the seaside atmosphere as the gravity waves propagate offshore. The increased offshore thermal instability contributes to the convection development. The increasing offshore precipitation rate before its peak is accompanied by the positive diurnal deviation of the thermal instability, and their orthogonal relation indicates a phase lag of 6–9 h. The rainfall rate then reaches its peak shortly before the offshore thermal instability (Figure 12a) falls onto the daily mean.

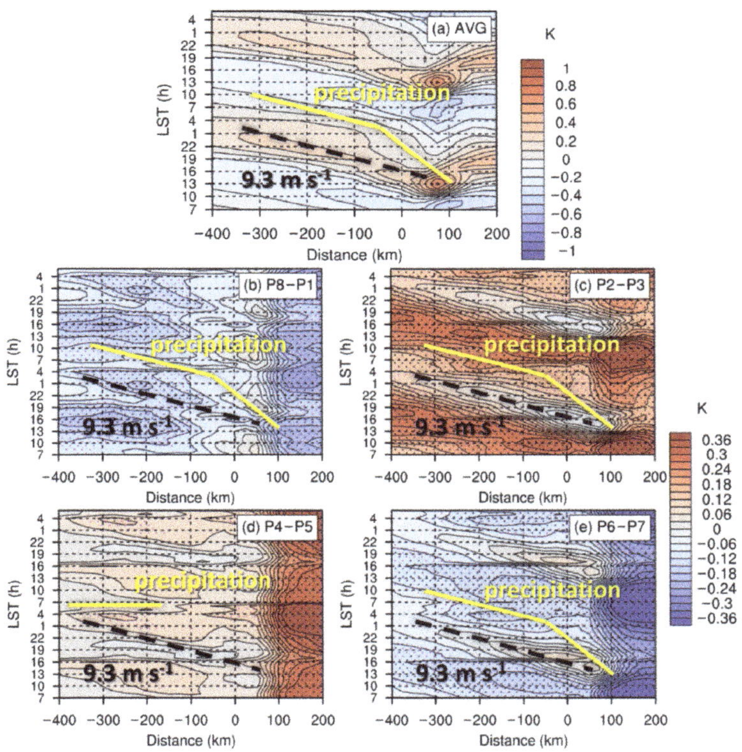

Figure 12. The Hovmöller diagrams of temperature difference (K) between 850 hPa and 600 hPa (**a**) averaged in DJF 2011–2020 with daily mean subtracted, as well as its anomaly in (**b**) P8–P1, (**c**) P2–P3, (**d**) P4–P5 and (**e**) P6–P7. The black dashed lines mark the daily maximum of temperature difference in (**a**). The yellow solid lines mark the diurnal maximum of precipitation as indicated in Figure 3. The purple and black dots indicate the 90% and 95% confidence according to the Student's *t*-test.

The positive anomalies of low-level thermal instability generally dominate in P2–P3 and P4–P5 (Figure 12c,d) with the higher cloud cover (Figure 7). Since the 850-hPa temperature is lowered, owing to the weaker radiative heating during these stages, the higher vertical temperature gradient should result from the larger decrease of 600-hPa temperature. The larger fraction of cloud that is apparently colder than the surrounding 600-hPa air noticeably lowers the average temperature, and vice versa in P8–P1 and P6–P7 (Figure 12b,e) with lower cloud fraction. As revealed in the averaged diurnal cycle (Figure 12a), the offshore migration of warm disturbance as the signals of gravity waves causes the increase in low-level instability along the track of migration. The lowered differences between diurnal peaks and daily averages in P2–P3 and P4–P5 (Figure 12c,d) indicates weakened gravity wave signals, due to the suppressed land–sea thermal contrast under the lowered radiative heating owing to the thicker cloud. In contrast, the thinner cloud in P8–P1 and P6–P7 (Figure 12b,e) is conducive to enhancing the land–sea thermal contrast as well as the associated gravity-wave signals via radiative heating, as indicated by the enhanced differences between diurnal peaks and daily averages, and therefore, has positive effects on the diurnal rainfall variation, consistent with the noticeably high normalized diurnal variation amplitudes A_n exhibited in Table 1.

In addition to the weakened gravity waves, P4–P5 is the only stage with apparently strengthened westerly wind on the upstream side (Figure 5), which also possibly prevents the offshore migration of precipitation during this stage, and thus leads to the phase-locking pattern of diurnal precipitation off the western coast (Figures 3 and 4).

4. Discussion

The noticeable difference between precipitation rates at the western Sumatra coast and its surroundings, as revealed in Figure 2, was also exhibited in the study by Peatman et al. [43]. They mentioned the non-zero or positive precipitation anomalies over Sumatra in MJO P1, defined by Wheeler and Hendon [55] whereby negative anomalies dominated MC, meanwhile, the diurnal amplitudes of precipitation reflected similar features with the absolute diurnal amplitude A_o in our Table 1. Other previous studies indicate prominent diurnal rainfall cycles and offshore migrations at or around P2–P3, with an MJO active convection at the northern Indian Ocean [36,46,47], also in agreement with our findings. Nevertheless, these studies hardly focus on normalized diurnal variation amplitudes (equally A_n in our Table 1), including the higher A_n in P8–P1 than P2–P3, as well as the relatively high A_n in the driest stage P6–P7. The study from Sakaeda et al. [39] exhibits an inconsistency between absolute amplitudes and normalized amplitudes of diurnal coastal precipitation variation, while we further compared this inconsistency during different MJO phases. A comparison and analysis of A_o and A_n among different MJO phases in this study, is expected to further acknowledge the detailed influences of the MJO on the diurnal precipitation cycle.

Various-scale factors for the diurnal rainfall cycle, as indicated within the existing literature introduced in the introduction, usually coexist with one another. This study further focuses on answering the question, that how these factors affect the coastal precipitation during different MJO phases. According to the analysis in Sections 4 and 5, the large-scale impacts of background wind and moisture on the absolute diurnal amplitude dominate during the MJO active (P2–P3) or suppressed (P6–P7) phases at Sumatra, while the modulations of the local-scale land–sea circulation influencing normalized diurnal amplitude are more distinct during the transitional phases (P8–P1 and P4–P5).

In terms of the offshore propagation, several studies [23,29,31] presented distinguished migration near and farther away the coast, whose speeds and ranges are close to those in this study. The wavelike signals (Figure 10) and the leaning vertical structures (Figure 11), generated from local land–sea thermal contrast, resemble the features of gravity waves in existing studies [65,66]. The variation among different MJO phases (Figure 12) reveals the consistency of lower cloud cover (Figure 7) with stronger gravity waves, according to the modulation of radiative heating and cooling. On the other hand, the strengthened waves are not always related with apparent rainfall migration. According to our analysis, the gravity waves were relatively weak in P2–P3, but the environment was dominated by enhanced vertical vapor transportation due to low-level high moisture, as found critical for offshore precipitation in the research from Du and Rotunno [64], and hence is dynamically (convectively) favorable for the coupling of gravity waves and convection [68,69]. That is a possible reason for the strong offshore rainfall propagation in P2–P3, with the noticeable diurnal rainfall cycle (both A_o and A_n in Table 1). In contrast, the environment is thermodynamically (radiatively) favorable for the strength of gravity waves both in P8–P1 and in P6–P7, with higher (similar) diurnal amplitude of normalized precipitation (A_n) in P8–P1 (P6–P7) compared to that in P2–P3, partly similar with the findings from Vincent and Lane [23]. However, the diurnal offshore migration is relatively (noticeably) weaker in P8–P1 (P6–P7), probably owing to weaker vertical vapor transportation and associated coupling with gravity waves compared to P2–P3. Consequently, the coupling (decoupling) of local thermal gravity waves and convection, under the positive (negative) large-scale synoptic modulation, was considered the main factor for the evident (obscured) diurnal offshore rainfall propagation in P2–P3 (P6–P7). In one word, the gravity waves are from local-scale land–sea thermal contrast but its coupling depends greatly on large-scale moisture and convection, both of which influence the coastal precipitation but reveal unique features during each MJO phase. To summarizr, this study not only compares the diurnal variation, especially the associated offshore migration, of winter rainfall among various MJO phases, but also analyzes the large-scale and local-scale impacts which play varying roles during different MJO phases.

5. Conclusions

This paper focused on the influences of winter MJO on the diurnal rainfall variation and its offshore propagation near western coast of Sumatra during 2011–2020 using ERA5 reanalysis. Days under strong MJO background were separated into four stages according to the locations of convective disturbance, respectively, at western hemisphere in P8–P1, at Indian Ocean in P2–P3, at MC in P4–P5, and at western Pacific Ocean in P6–P7. Features that relate to precipitation daily mean rates, diurnal variations, normalized diurnal variations and offshore diurnal propagation were compared among different MJO stages and were explained using an analysis of both large-scale backgrounds and local-scale land–sea circulation. The conclusions are summarized as follows:

More (Less) precipitation occurs near the coastal area in P8–P1 and P2–P3 (P4–P5 and P6–P7); specifically, in P2–P3 (P6–P7) the precipitation is both enhanced (suppressed) at the coastal area and its surroundings, while in P8–P1 the coastal enhancement is accompanied by the surrounding suppression, and vice versa in P4–P5. Precipitation rates and their absolute diurnal variations are affected greatly by large-scale backgrounds. The low-level convergence and moisture are both favorable (unfavorable) in P2–P3 (P6–P7), producing the most (least) rainfall. The moisture in the northern Indian Ocean plays a critical role in affecting precipitation rates during other stages (P8–P1 and P4–P5) where the large-scale low-level convergence is not favorable as in P2–P3.

Normalized precipitation diurnal variations are not consistent with absolute diurnal variations among different MJO phases. The normalized diurnal amplitude is higher in P8–P1 compared to the P4–P5, arising largely from the enhanced diurnal land–sea circulation variations. Although the absolute diurnal amplitude in P2–P3 is much stronger than P6–P7 due to large-scale MJO circulation, the normalized diurnal amplitude in P6–P7 resembles that in P2–P3, suggesting that the strength of the land–sea diurnal circulation plays an important role in the modulation of normalized diurnal amplitude, under the impact of cloud cover.

The offshore diurnal rainfall propagation is associated with gravity waves generated from land–sea thermal contrast by examining its phase speeds, horizontal wavelike signals as well as vertical structures in warm/cool disturbances. The wave signals are more explicit under lower cloud cover in P8–P1 and P6–P7, which is favorable for apparent diurnal propagation. The noticeable diurnal propagation in P2–P3, accompanied by weakened gravity wave signals, reflects the potential large-scale effects of vertical vapor transportation and the strong coupling of convection and gravity waves. The phase-locking diurnal pattern off the coast in P4–P5 possibly occurs due to the local-scale seaside nighttime cooling of cloud top and the strong westerly. In P4–P5, the large-scale rainfall development is suppressed by weaker convergence and lower moisture meanwhile the local-scale land–sea circulation is prohibited by thicker cloud cover.

Since this paper focuses only on seasonal average precipitation during the past decade, in future studies, we plan to investigate specific cases, for the purpose of developing a better understanding of offshore migration and how it is influenced by the MJO. Numerical simulations as well as sensitivity experiments will also be conducted to verify and expanding our conclusions.

Author Contributions: Conceptualization, B.Z., Y.D. and Z.G.; methodology, B.Z. and Y.D.; data curation: Y.D.; software, B.Z. and Y.D.; validation, B.Z., Y.D. and Z.G.; formal analysis, B.Z. and Y.D.; writing—original draft preparation, B.Z.; writing—review and editing, B.Z., Y.D. and Z.G.; visualization, B.Z.; supervision, Y.D. and Z.G.; funding acquisition, Y.D. All authors have read and agreed to the published version of the manuscript.

Funding: The author Y.D. was supported by the Guangdong Major Project of Basic and Applied Basic Research (2020B0301030004), the National Natural Science Foundation of China (Grant Nos. 42075006, 42122033, and 41875055), and Guangzhou Science and Technology Plan Projects (202002030346).

Institutional Review Board Statement: Not applicable.

Informed Consent Statement: Not applicable.

Data Availability Statement: The IMERG precipitation data was from NASA's Global Precipitation Measurement Mission (https://gpm.nasa.gov/data/imerg; accessed on 10 February 2021). The ERA5 reanalysis data was obtained from NCEI (https://www.ncei.noaa.gov; accessed on 29 January 2021). The MJO RMM index data was from the Australian Bureau of Meteorology (http://www.bom.gov.au/climate/mjo/; accessed on 1 February 2021).

Acknowledgments: Computer support from the Center for High-Performance Computing at Nanjing University of Information Science & Technology is gratefully acknowledged. We sincerely thank all the editors and reviewers for the kind comments and valuable suggestions during the revisions.

Conflicts of Interest: The authors declare no conflict of interest.

References

1. Madden, R.A.; Julian, P.R. Detection of a 40–50 day oscillation in the zonal wind in the tropical Pacific. *J. Atmos. Sci.* **1971**, *28*, 702–708. [CrossRef]
2. Madden, R.A.; Julian, P.R. Description of global-scale circulation cells in the tropics with a 40–50 day period. *J. Atmos. Sci.* **1972**, *29*, 1109–1123. [CrossRef]
3. Zhang, C. Madden-julian oscillation. *Rev. Geophys.* **2005**, *43*, RG2003. [CrossRef]
4. Jeong, J.-H.; Kim, B.-M.; Ho, C.-H.; Noh, Y.-H. Systematic variation in wintertime precipitation in East Asia by MJO-induced extratropical vertical motion. *J. Clim.* **2008**, *21*, 788–801. [CrossRef]
5. Barrett, B.S.; Carrasco, J.F.; Testino, A.P. Madden–Julian oscillation (MJO) modulation of atmospheric circulation and Chilean winter precipitation. *J. Clim.* **2012**, *25*, 1678–1688. [CrossRef]
6. Juliá, C.; Rahn, D.A.; Rutllant, J.A. Assessing the influence of the MJO on strong precipitation events in subtropical, semi-arid north-central Chile (30 S). *J. Clim.* **2012**, *25*, 7003–7013. [CrossRef]
7. Zhou, S.; L'Heureux, M.; Weaver, S.; Kumar, A. A composite study of the MJO influence on the surface air temperature and precipitation over the continental United States. *Clim. Dyn.* **2012**, *38*, 1459–1471. [CrossRef]
8. Zhang, C.; Ling, J. Barrier effect of the Indo-Pacific Maritime Continent on the MJO: Perspectives from tracking MJO precipitation. *J. Clim.* **2017**, *30*, 3439–3459. [CrossRef]
9. Wang, B.; Rui, H. Synoptic climatology of transient tropical intraseasonal convection anomalies: 1975–1985. *Meteorol. Atmos. Phys.* **1990**, *44*, 43–61. [CrossRef]
10. Hartmann, D.L.; Michelsen, M.L.; Klein, S.A. Seasonal variations of tropical intraseasonal oscillations: A 20–25-day oscillation in the western Pacific. *J. Atmos. Sci.* **1992**, *49*, 1277–1289. [CrossRef]
11. Salby, M.L.; Hendon, H.H. Intraseasonal behavior of clouds, temperature, and motion in the tropics. *J. Atmos. Sci.* **1994**, *51*, 2207–2224. [CrossRef]
12. Wu, C.-H.; Hsu, H.-H. Topographic influence on the MJO in the Maritime Continent. *J. Clim.* **2009**, *22*, 5433–5448. [CrossRef]
13. Jia, X.; Chen, L.; Ren, F.; Li, C. Impacts of the MJO on winter rainfall and circulation in China. *Adv. Atmos. Sci.* **2011**, *28*, 521–533. [CrossRef]

14. Mori, S.; Hamada, J.-I.; Hattori, M.; Wu, P.-M.; Katsumata, M.; Endo, N.; Ichiyanagi, K.; Hashiguchi, H.; Arbain, A.A.; Sulistyowati, R. Meridional march of diurnal rainfall over Jakarta, Indonesia, observed with a C-band Doppler radar: An overview of the HARIMAU2010 campaign. *Prog. Earth Planet. Sci.* **2018**, *5*, 47. [CrossRef]
15. Katsumata, M.; Mori, S.; Hamada, J.-I.; Hattori, M.; Syamsudin, F.; Yamanaka, M.D. Diurnal cycle over a coastal area of the Maritime Continent as derived by special networked soundings over Jakarta during HARIMAU2010. *Prog. Earth Planet. Sci.* **2018**, *5*, 1–19. [CrossRef]
16. Ramage, C.S. Role of a tropical "maritime continent" in the atmospheric circulation. *Mon. Weather Rev.* **1968**, *96*, 365–370. [CrossRef]
17. Neale, R.; Slingo, J. The Maritime Continent and its role in the global climate: A GCM study. *J. Clim.* **2003**, *16*, 834–848. [CrossRef]
18. Chang, C.; Harr, P.A.; Chen, H.-J. Synoptic disturbances over the equatorial South China Sea and western Maritime Continent during boreal winter. *Mon. Weather Rev.* **2005**, *133*, 489–503. [CrossRef]
19. Lorenz, P.; Jacob, D. Influence of regional scale information on the global circulation: A two-way nesting climate simulation. *Geophys. Res. Lett.* **2005**, *32*, L18706. [CrossRef]
20. Ando, K.; Syamsudin, F.; Ishihara, Y.; Pandoe, W.; Yamanaka, M.D.; Masumoto, Y.; Mizuno, K. Development of new international research laboratory for maritime-continent seas climate research and contributions to global surface moored buoy array. In Proceedings of the "OeanObs' 09: Sustained Ocean Observations and Information for Society" Conference (Annex), Venice, Italy, 21–25 September 2009.
21. Mori, S.; Jun-Ichi, H.; Tauhid, Y.I.; Yamanaka, M.D.; Okamoto, N.; Murata, F.; Sakurai, N.; Hashiguchi, H.; Sribimawati, T. Diurnal land–sea rainfall peak migration over Sumatera Island, Indonesian Maritime Continent, observed by TRMM satellite and intensive rawinsonde soundings. *Mon. Weather Rev.* **2004**, *132*, 2021–2039. [CrossRef]
22. Love, B.S.; Matthews, A.J.; Lister, G.M. The diurnal cycle of precipitation over the Maritime Continent in a high-resolution atmospheric model. *Q. J. R. Meteorol. Soc.* **2011**, *137*, 934–947. [CrossRef]
23. Vincent, C.L.; Lane, T.P. Evolution of the diurnal precipitation cycle with the passage of a Madden–Julian oscillation event through the Maritime Continent. *Mon. Weather Rev.* **2016**, *144*, 1983–2005. [CrossRef]
24. Yokoi, S.; Mori, S.; Katsumata, M.; Geng, B.; Yasunaga, K.; Syamsudin, F.; Nurhayati; Yoneyama, K. Diurnal cycle of precipitation observed in the western coastal area of Sumatra Island: Offshore preconditioning by gravity waves. *Mon. Weather Rev.* **2017**, *145*, 3745–3761. [CrossRef]
25. Yokoi, S.; Mori, S.; Syamsudin, F.; Haryoko, U.; Geng, B. Environmental conditions for nighttime offshore migration of precipitation area as revealed by in situ observation off Sumatra Island. *Mon. Weather Rev.* **2019**, *147*, 3391–3407. [CrossRef]
26. Zhou, L.; Wang, Y. Tropical Rainfall Measuring Mission observation and regional model study of precipitation diurnal cycle in the New Guinean region. *J. Geophys. Res. Atmos.* **2006**, *111*, D17104. [CrossRef]
27. Fujita, M.; Kimura, F.; Yoshizaki, M. Morning precipitation peak over the Strait of Malacca under a calm condition. *Mon. Weather Rev.* **2010**, *138*, 1474–1486. [CrossRef]
28. Dipankar, A.; Webster, S.; Huang, X.-Y. Understanding biases in simulating the diurnal cycle of convection over the western coast of Sumatra: Comparison with pre-YMC observation campaign. *Mon. Weather Rev.* **2019**, *147*, 1615–1631. [CrossRef]
29. Short, E.; Vincent, C.L.; Lane, T.P. Diurnal cycle of surface winds in the Maritime Continent observed through satellite scatterometry. *Mon. Weather Rev.* **2019**, *147*, 2023–2044. [CrossRef]
30. Wu, P.; Hara, M.; Hamada, J.-i.; Yamanaka, M.D.; Kimura, F. Why a large amount of rain falls over the sea in the vicinity of western Sumatra Island during nighttime. *J. Appl. Meteorol. Climatol.* **2009**, *48*, 1345–1361. [CrossRef]
31. Hassim, M.; Lane, T.; Grabowski, W. The diurnal cycle of rainfall over New Guinea in convection-permitting WRF simulations. *Atmos. Chem. Phys.* **2016**, *16*, 161–175. [CrossRef]
32. Yulihastin, E.; Wahyu Hadi, T.; Sari Ningsih, N.; Ridho Syahputra, M. Early morning peaks in the diurnal cycle of precipitation over the northern coast of West Java and possible influencing factors. *Ann. Geophys.* **2020**, *38*, 231–242. [CrossRef]
33. Johnson, R.H.; Rickenbach, T.M.; Rutledge, S.A.; Ciesielski, P.E.; Schubert, W.H. Trimodal characteristics of tropical convection. *J. Clim.* **1999**, *12*, 2397–2418. [CrossRef]
34. Tian, B.; Waliser, D.E.; Fetzer, E.J. Modulation of the diurnal cycle of tropical deep convective clouds by the MJO. *Geophys. Res. Lett.* **2006**, *33*, L20704. [CrossRef]
35. Ichikawa, H.; Yasunari, T. Propagating diurnal disturbances embedded in the Madden-Julian Oscillation. *Geophys. Res. Lett.* **2007**, *34*, L18811. [CrossRef]
36. Vincent, C.L.; Lane, T.P. A 10-year austral summer climatology of observed and modeled intraseasonal, mesoscale, and diurnal variations over the Maritime Continent. *J. Clim.* **2017**, *30*, 3807–3828. [CrossRef]
37. Seiki, A.; Yokoi, S.; Katsumata, M. The impact of diurnal precipitation over Sumatra Island, Indonesia, on synoptic disturbances and its relation to the Madden-Julian Oscillation. *J. Meteorol. Soc. Jpn. Ser. II* **2021**, *99*, 113–137. [CrossRef]
38. Lubis, S.W.; Jacobi, C. The modulating influence of convectively coupled equatorial waves (CCEWs) on the variability of tropical precipitation. *Int. J. Climatol.* **2015**, *35*, 1465–1483. [CrossRef]
39. Sakaeda, N.; Kiladis, G.; Dias, J. The diurnal cycle of rainfall and the convectively coupled equatorial waves over the Maritime Continent. *J. Clim.* **2020**, *33*, 3307–3331. [CrossRef]
40. Ferrett, S.; Yang, G.Y.; Woolnough, S.J.; Methven, J.; Hodges, K.; Holloway, C.E. Linking extreme precipitation in Southeast Asia to equatorial waves. *Q. J. R. Meteorol. Soc.* **2020**, *146*, 665–684. [CrossRef]

41. Peatman, S.C.; Schwendike, J.; Birch, C.E.; Marsham, J.H.; Matthews, A.J.; Yang, G.-Y. A Local-to-Large Scale View of Maritime Continent Rainfall: Control by ENSO, MJO, and Equatorial Waves. *J. Clim.* **2021**, *34*, 8933–8953. [CrossRef]
42. Lubis, S.W.; Respati, M.R. Impacts of convectively coupled equatorial waves on rainfall extremes in Java, Indonesia. *Int. J. Climatol.* **2021**, *41*, 2418–2440. [CrossRef]
43. Peatman, S.C.; Matthews, A.J.; Stevens, D.P. Propagation of the Madden–Julian Oscillation through the Maritime Continent and scale interaction with the diurnal cycle of precipitation. *Q. J. R. Meteorol. Soc.* **2014**, *140*, 814–825. [CrossRef]
44. Birch, C.; Webster, S.; Peatman, S.; Parker, D.; Matthews, A.; Li, Y.; Hassim, M. Scale interactions between the MJO and the western Maritime Continent. *J. Clim.* **2016**, *29*, 2471–2492. [CrossRef]
45. Rauniyar, S.P.; Walsh, K.J. Scale interaction of the diurnal cycle of rainfall over the Maritime Continent and Australia: Influence of the MJO. *J. Clim.* **2011**, *24*, 325–348. [CrossRef]
46. Fujita, M.; Yoneyama, K.; Mori, S.; Nasuno, T.; Satoh, M. Diurnal convection peaks over the eastern Indian Ocean off Sumatra during different MJO phases. *J. Meteorol. Soc. Jpn. Ser. II* **2011**, *89*, 317–330. [CrossRef]
47. Kamimera, H.; Mori, S.; Yamanaka, M.D.; Syamsudin, F. Modulation of diurnal rainfall cycle by the Madden-Julian Oscillation based on one-year continuous observations with a meteorological radar in west Sumatera. *SOLA* **2012**, *8*, 111–114. [CrossRef]
48. Harjupa, W.; Shimomai, T.; Hashiguchi, H.; Fujiyoshi, Y.; Kawashima, M. Differences in Mechanisms of Orographic Rainfall over West Sumatra (Case Study: 10 April and 23 April 2004). *Jurnal Ilmu Fisika | Universitas Andalas* **2021**, *13*, 8–17. [CrossRef]
49. Xavier, P.; Rahmat, R.; Cheong, W.K.; Wallace, E. Influence of Madden-Julian Oscillation on Southeast Asia rainfall extremes: Observations and predictability. *Geophys. Res. Lett.* **2014**, *41*, 4406–4412. [CrossRef]
50. Muhammad, F.R.; Lubis, S.W.; Setiawan, S. Impacts of the Madden–Julian oscillation on precipitation extremes in Indonesia. *Int. J. Climatol.* **2021**, *41*, 1970–1984. [CrossRef]
51. Zhu, B.; Pu, Z.; Putra, A.W.; Gao, Z. Assimilating C-Band Radar Data for High-Resolution Simulations of Precipitation: Case Studies over Western Sumatra. *Remote Sens.* **2022**, *14*, 42. [CrossRef]
52. Huffman, G.J.; Bolvin, D.T.; Braithwaite, D.; Hsu, K.; Joyce, R.; Xie, P.; Yoo, S.-H. NASA global precipitation measurement (GPM) integrated multi-satellite retrievals for GPM (IMERG). *Algorithm Theor. Basis Doc. (ATBD) Version* **2015**, *4*, 26.
53. Tan, J.; Huffman, G.J.; Bolvin, D.T.; Nelkin, E.J. IMERG V06: Changes to the morphing algorithm. *J. Atmos. Ocean. Technol.* **2019**, *36*, 2471–2482. [CrossRef]
54. Hersbach, H.; Bell, B.; Berrisford, P.; Hirahara, S.; Horányi, A.; Muñoz-Sabater, J.; Nicolas, J.; Peubey, C.; Radu, R.; Schepers, D. The ERA5 global reanalysis. *Q. J. R. Meteorol. Soc.* **2020**, *146*, 1999–2049. [CrossRef]
55. Wheeler, M.C.; Hendon, H.H. An all-season real-time multivariate MJO index: Development of an index for monitoring and prediction. *Mon. Weather Rev.* **2004**, *132*, 1917–1932. [CrossRef]
56. Chen, S.S.; Houze, R.A., Jr. Diurnal variation and life-cycle of deep convective systems over the tropical Pacific warm pool. *Q. J. R. Meteorol. Soc.* **1997**, *123*, 357–388. [CrossRef]
57. Sui, C.; Lau, K.; Takayabu, Y.; Short, D. Diurnal variations in tropical oceanic cumulus convection during TOGA COARE. *J. Atmos. Sci.* **1997**, *54*, 639–655. [CrossRef]
58. Oh, J.-H.; Kim, K.-Y.; Lim, G.-H. Impact of MJO on the diurnal cycle of rainfall over the western Maritime Continent in the austral summer. *Clim. Dyn.* **2012**, *38*, 1167–1180. [CrossRef]
59. Sakaeda, N.; Kiladis, G.; Dias, J. The diurnal cycle of tropical cloudiness and rainfall associated with the Madden–Julian oscillation. *J. Clim.* **2017**, *30*, 3999–4020. [CrossRef]
60. Sakaeda, N.; Powell, S.W.; Dias, J.; Kiladis, G.N. The diurnal variability of precipitating cloud populations during DYNAMO. *J. Atmos. Sci.* **2018**, *75*, 1307–1326. [CrossRef]
61. Wapler, K.; Lane, T.P. A case of offshore convective initiation by interacting land breezes near Darwin, Australia. *Meteorol. Atmos. Phys.* **2012**, *115*, 123–137. [CrossRef]
62. Mapes, B.E.; Warner, T.T.; Xu, M. Diurnal patterns of rainfall in northwestern South America. Part III: Diurnal gravity waves and nocturnal convection offshore. *Mon. Weather Rev.* **2003**, *131*, 830–844. [CrossRef]
63. Du, Y.; Rotunno, R. Thermally driven diurnally periodic wind signals off the east coast of China. *J. Atmos. Sci.* **2015**, *72*, 2806–2821. [CrossRef]
64. Du, Y.; Rotunno, R. Diurnal cycle of rainfall and winds near the south coast of China. *J. Atmos. Sci.* **2018**, *75*, 2065–2082. [CrossRef]
65. Bai, H.; Deranadyan, G.; Schumacher, C.; Funk, A.; Epifanio, C.; Ali, A.; Endarwin; Radjab, F.; Adriyanto, R.; Nurhayati, N. Formation of Nocturnal Offshore Rainfall near the West Coast of Sumatra: Land Breeze or Gravity Wave? *Mon. Weather Rev.* **2021**, *149*, 715–731. [CrossRef]
66. Ruppert, J.H.; Zhang, F. Diurnal forcing and phase locking of gravity waves in the Maritime Continent. *J. Atmos. Sci.* **2019**, *76*, 2815–2835. [CrossRef]
67. Du, Y.; Rotunno, R.; Zhang, F. Impact of Vertical Wind Shear on Gravity Wave Propagation in the Land–Sea-Breeze Circulation at the Equator. *J. Atmos. Sci.* **2019**, *76*, 3247–3265. [CrossRef]
68. Du, Y.; Zhang, F. Banded convective activity associated with mesoscale gravity waves over southern China. *J. Geophys. Res. Atmos.* **2019**, *124*, 1912–1930. [CrossRef]
69. Du, Y.; Zhang, F.; Sun, Y.Q.; Wei, J.; Li, X. Practical and Intrinsic Predictability of Wave-Convection Coupled Bands Over Southern China. *J. Geophys. Res. Atmos.* **2021**, *126*, e2021JD034882. [CrossRef]

MDPI
St. Alban-Anlage 66
4052 Basel
Switzerland
Tel. +41 61 683 77 34
Fax +41 61 302 89 18
www.mdpi.com

Atmosphere Editorial Office
E-mail: atmosphere@mdpi.com
www.mdpi.com/journal/atmosphere

www.ingramcontent.com/pod-product-compliance
Lightning Source LLC
Chambersburg PA
CBHW061436210926
51850CB00008B/1530
9783036574301